# Photoacoustic Infrared Spectroscopy

# CHEMICAL ANALYSIS

A SERIES OF MONOGRAPHS ON ANALYTICAL CHEMISTRY
AND ITS APPLICATIONS

*Editor*
**J. D. WINEFORDNER**

**VOLUME 159**

A JOHN WILEY & SONS PUBLICATION

# Photoacoustic Infrared Spectroscopy

KIRK H. MICHAELIAN

A JOHN WILEY & SONS PUBLICATION

Published by John Wiley & Sons, Inc., Hoboken, New Jersey.
Published simultaneously in Canada.



For general information on our other products and services please contact our Customer Care Department within the U.S. at 877-762-2974, outside the U.S. at 317-572-3993 or fax 317-572-4002.

Wiley also publishes its books in a variety of electronic formats. Some content that appears in print, however, may not be available in electronic format.

***Library of Congress Cataloging-in-Publication Data is available.***

ISBN 0-471-13477-5

Printed in the United States of America

10 9 8 7 6 5 4 3 2 1

For Diane

# CONTENTS

# PREFACE

*"Science is concerned with the rational correlation of experience rather than the discovery of fragments of absolute truth."*

SIR ARTHUR EDDINGTON, *The Philosophy of Physical Science*

The continued growth of the primary scientific literature and the increased level of specialization that characterize modern physical science would both appear to require that forthcoming entries in the Chemical Analysis Series be constrained to topics somewhat narrower than those in previous volumes. The subject of photoacoustic (PA) spectroscopy is certainly not an exception to this trend. To be specific, A. Rosencwaig surveyed this then-emergent field in Volume 57, *Photoacoustics and Photoacoustic Spectroscopy*, a little more than two decades ago. The preparation of a similar treatise today would undoubtedly be a daunting task—one that might be expected to yield several texts. This assertion is corroborated by the 1994 publication of *Air Monitoring by Spectroscopic Techniques* (Volume 127, edited by M. W. Sigrist), followed in 1996 by *Photothermal Spectroscopy Methods for Chemical Analysis* (Volume 134, by S. E. Bialkowski). The present text, which is dedicated to the subject of photoacoustic infrared spectroscopy, is the newest volume in this continuing Chemical Analysis Series on photoacoustic and photothermal spectroscopy.

It is never inappropriate to question the need for a book on a particular analytical method. Two factors are especially pertinent with regard to the current submission. First, while several excellent review articles and book chapters on photoacoustic Fourier transform infrared (FTIR) spectroscopy already exist, much of this work is concerned with specific sample classes (e.g., polymers), numerical methods (mostly dealing with the phase of the PA signal), or instrumentation. The present volume adopts a perspective that is alternately historical and contemporary by reviewing both early and recent literature on an array of applications of PA infrared spectroscopy.

Moreover, a slightly different technological viewpoint is adopted here: This text discusses various implementations of PA spectroscopy at near-, mid- and far-infrared wavelengths, and is not restricted to the topic of "PA FTIR" spectroscopy. For example, spectra of gases obtained with CO or $CO_2$ laser radiation at discrete wavelengths, without the use of an interferometer, are included. Similarly, dispersive near- and mid-infrared PA spectra of condensed-phase materials are discussed in several contexts. This approach to the subject was taken so as to minimize the effects of the seemingly arbitrary description of PA infrared spectra that is based on the specific apparatus used to acquire them.

The second justification for the present work arises from the very significant advances in, and expansion of, PA and photothermal science in recent years. Much of the early work in this discipline consisted of research in PA spectroscopy, albeit primarily at wavelengths shorter than those considered here. On the other hand, the entire field of PA and photothermal research has now grown and diversified to the point where it is justifiable—and perhaps necessary—that its major specializations be summarized more or less independently. The ever-increasing extent of the pertinent scientific literature implies that PA infrared spectroscopy certainly merits this treatment, as do several other related topics. It is hoped that this overview and synthesis of the PA infrared literature will prove useful to readers who may require detailed information on this subject.

This book has several specific objectives. Even though PA infrared spectroscopy is well-established, many infrared spectroscopists tend to regard the method as unconventional or marginal and therefore ascribe second-class status to the technique. This situation is partly attributable to the low intensities associated with many early PA spectra, a factor that has been minimized in recent years by numerous advances in instrumentation and data analysis. It also arises from the usual lack of familiarity associated with any developing method. Hence the first objective of the present work is to clearly demonstrate the viability of PA infrared spectroscopy. This is accomplished by showing its widespread acceptance and reviewing its successful utilization in a variety of disciplines.

As the relevant literature was assembled, it became apparent that a significant number of proven experimental and numerical PA infrared techniques exist. It was noted above that some of these methods have already been discussed in one or more review articles or book chapters. However, there does not appear to be a single source that attempts a more comprehensive review of the relevant literature. This goal has now been addressed: The second objective of the current work was to assemble a reference text on PA infrared spectroscopy, which discusses most of its important applications and provides the interested reader with a point of entry into the original literature on the topic.

Because this is a volume in a series on Chemical Analysis, a nonmathematical approach has been adopted where possible. On the other hand when equations are employed, the notation of the original authors has generally been followed; this ensures that the reader will not encounter unnecessary changes when reading both the present work and the primary literature. Because symbol conventions inevitably vary, the reader may notice a few minor examples of alternative usage within the present work. Insofar as mathematical symbols are defined upon their introduction in a particular context, this situation should not cause any significant confusion.

The organization of this text is based on the straightforward division of the published literature into several categories. To begin, Chapters 1 and 2 provide an introduction and historical review of PA infrared spectroscopy, with particular emphasis on early work. After this perspective is established, Chapter 3 discusses a series of seven different experimental PA methods that have been employed by researchers during the last two decades.

One of the most important capabilities of PA infrared spectroscopy—depth profiling—is described in Chapter 4. In this discussion the principles that underlie several relevant experimental techniques are explained and typical results are presented. Chapter 5 then outlines three important numerical methods that pertain to PA spectroscopy. The phase of the PA signal (which is distinct from the instrumental phase that affects all FTIR spectra) plays a key role in several of these manipulations.

A thorough reading of the appropriate scientific literature confirms that the disciplines in which PA infrared spectroscopy has been applied are particularly diverse. Many analysts may be familiar with one or more specific uses of the technique; in all likelihood, very few will be acquainted with the entire array of research topics in which it has been successfully utilized. Chapter 6, which describes a total of 15 different applications of PA infrared spectroscopy, was written to address this situation. Each section is meant to convey a sense of the research carried out in a specific area and to guide the reader who may wish to pursue the subject in more detail. Typical spectra (reproduced from the published literature or measured in the author's laboratory) are presented in each section of this chapter.

As a spectroscopic technique matures, its use for quantitation generally becomes more viable. The status of PA infrared spectroscopy with regard to quantitative analysis is reviewed in Chapter 7. Indeed, while the majority of the literature on PA spectra of solids and liquids describes qualitative or semi-quantitative studies, this discussion shows that many successful experiments in quantitative analysis have also been carried out.

Finally, Chapter 8 discusses two emerging techniques: PA infrared microspectroscopy and synchrotron infrared PA spectroscopy. Very recent experiments thus demonstrate that the field of PA infrared spectroscopy continues to advance after several decades of active research by a particularly wide community of spectroscopists.

The main part of the book is followed by two appendices. Appendix 1 provides definitions of a number of relevant terms used in the PA literature. Appendix 2 is an alphabetical listing of publications (including journal articles, book chapters, and conference proceedings) on PA infrared spectra of solids and liquids; the particular application(s) discussed in each article are indicated using a classification scheme similar to that in Chapter 6. This information has been collected as an aid to the reader who wishes to consult the original literature on PA infrared spectroscopy of condensed-phase samples.

It is usual to conclude preliminary remarks such as these with a series of acknowledgements. In this regard, the support of Natural Resources Canada during this research and the preparation of this manuscript is greatly appreciated. Similarly, I have benefited considerably from continued cooperation and collaboration with Bruker Optics in the implementation of PA infrared spectroscopy at CANMET during the last two decades. My entry into this field came about as the result of a prescient suggestion approximately 20 years ago. The use of rather tentatively worded acknowledgements is often effected stylistically or deferentially; in this case, however, it is literally necessary. Specifically, I would like to thank Dr. J. C. Donini for ensuring my involvement in PA spectroscopy, but cannot; his sudden passing five years ago has made this impossible.

KIRK H. MICHAELIAN

*Edmonton, Alberta, Canada*
*November 2002*

CHAPTER

1

# INTRODUCTION

*"This [photoacoustic] technique is applicable to a wide range of materials that defy analysis by conventional optical spectroscopy."*

P.-E. Nordal and S. O. Kanstad, *Int. J. Quantum Chem.* **12** (Suppl. 2): 115–121 (1977)

This book describes the role of photoacoustic (PA) infrared spectroscopy in chemical analysis. The PA infrared field has enjoyed more or less continual development since its emergence about 30 years ago: A substantial body of literature—consisting of more than 600 publications—on PA infrared spectroscopy exists today. The present work attempts to review and synthesize this literature; one of its principal objectives is to summarize the current status of PA infrared spectroscopy so that analysts and researchers may determine whether the method is indeed appropriate for their needs.

PA infrared spectroscopy can be viewed from several perspectives. For example, it can be thought of as one of a large number of photoacoustic and photothermal methods currently used by physicists, materials scientists, and chemists to characterize both condensed phases and gases. It should be noted that both optical and thermal properties of matter can be investigated by these methods. From this viewpoint, PA infrared spectroscopy is a specialization within a much broader field, made possible by the development and application of a variety of detectors and optical instrumentation at progressively longer wavelengths. Some might suggest that this interpretation, while obviously valid, tends to ascribe a rather secondary status to PA infrared spectroscopy.

The vibrational (infrared and Raman) spectroscopist will almost certainly approach PA infrared spectroscopy in an entirely different way. For the spectroscopist, PA detection can be described as an enabling technology, significantly increasing the number and type of samples for which viable data can be obtained. The reader will soon recognize that the viewpoint of the infrared spectroscopist is adopted in this book. In fact, PA detection of infrared absorption spectra, using modern accessories and spectrometers, offers several well-known advantages. The most important of these are the following:

- Minimal sample preparation is required.
- The technique is suitable for opaque materials.
- Depth profiling can be effected for inhomogeneous or layered solids.
- PA spectroscopy is nondestructive (the sample is not consumed).

It is not an exaggeration to assert that these characteristics are critical in many circumstances: For example, samples of a difficult nature can be encountered in industrial laboratories on a daily basis. These include (but are not limited to) viscous liquids, semisolids, and dispersions; metal powders, carbonaceous solids, and granular materials; and polymers and layered samples, whose structures or chemical compositions may be altered by grinding. Traditional infrared sample preparation methods are often inappropriate or have deleterious effects on these substances. Hence the minimization of sample preparation in PA infrared spectroscopy can be considered as its most important attribute. A thorough reading of the PA literature affirms that the majority of spectroscopists using this technique implicitly agree with this statement. The use of PA infrared spectroscopy for the chemical analysis of problematic, even "intractable," samples is discussed throughout this book.

Notwithstanding these statements, the capacity of PA infrared spectroscopy for depth profiling is an almost equally important feature: This experimental technique has been utilized extensively to study layered polymer structures, adsorption on various substrates, and surface oxidation of hydrocarbon fuels or other species. Thus the PA spectroscopist also possesses the capability for the analysis of surface and subsurface layers (in this context, "surface" implies depths on the order of a few micrometres, and frequently extends to tens of micrometres), a goal that can surely be said to be the dream of many chemists and physicists.

PA spectroscopy—frequently referred to by the acronym PAS—is sometimes described as an "unconventional" infrared technique. The very significant number of publications in the primary scientific literature that report research-quality PA spectra belie this rather pejorative description. It is hoped that the present account adequately demonstrates the wide-ranging applicability of the technique, thereby making a convincing argument for its increased future use.

## 1.1. SINGLE- AND MULTIPLE-WAVELENGTH PA SPECTROSCOPIES

PA spectroscopy can be divided into two broad categories. The first can be described as single-wavelength spectroscopy since only one wavelength of light impinges on the sample of interest. Signal generation in a gas-microphone cell can be used to illustrate this technique. Three steps can be identified. First, modulated monochromatic light from a laser or other suitable source impinges on the sample; second, the absorbed radiation is converted to heat by radiationless processes; and finally, the heat generated within the sample is transferred to its immediate environment, where the resulting pressure wave is detected by a transducer (microphone). A variation of this method is scanning (sequential) spectroscopy in which a spectrum is built up by systematically changing the wavelength of the incident radiation. This can be accomplished by selecting different lines from a laser that is capable of emitting at several wavelengths, or by rotating the grating in a monochromator that is used to filter broadband radiation. These techniques were used to obtain PA infrared spectra by several research groups, particularly in the 1970s and early 1980s when modern PA spectroscopy was evolving. Currently, multiline gas ($CO_2$ and CO) and solid-state mid-infrared lasers are used for specific PA applications, the most important being trace gas detection. These implementations of single-wavelength PA infrared spectroscopy are discussed in later chapters.

The second category is multiple-wavelength (multiplex) PA spectroscopy, as practiced with modern Fourier transform infrared (FTIR) spectrometers. Most readers are, of course, aware of the fact that these spectrometers have achieved very wide acceptance in analytical, research, and teaching laboratories during the last three decades. In the present context, the most important feature of an FTIR spectrometer is its capability for simultaneous measurements at an entire range of wavelengths; spectral coverage is determined mainly by the optical characteristics of the beamsplitter, the window material in the sample accessory, and the detector. In conventional PA FTIR spectroscopy, no optical detector is required, and the accessible wavelength range depends only on the beamsplitter and the window of the gas-microphone cell. This technique has been used extensively for more than 20 years and is the source of the majority of the literature discussed in this book. Generation of the PA FTIR signal in the gas-microphone cell can be described in terms similar to those outlined in the previous paragraph, with the modulation being provided by the FTIR spectrometer itself. This is discussed in more detail in Chapter 3.

## 1.2. SCOPE

As noted in the previous section, PA infrared spectroscopy has long been practiced with lasers, scanning monochromators, and FTIR spectrometers. Indeed, the continuing use of numerous types of instrumentation in PA spectroscopy demonstrates the breadth of the subject field. While many workers today naturally tend to associate infrared spectroscopy with FTIR spectrometers, it should be emphasized at the outset that PA FTIR spectroscopy is, in fact, a subset of the broader discipline of PA spectroscopy at infrared wavelengths. This book adopts the wider definition of the field and examines a number of relevant PA infrared techniques from both historical and modern perspectives.

Spectroscopists are well aware that definitions of wavelength regions tend to differ for reasons that may be either historical, technological, or a combination of the two. Near-, mid-, and far-infrared PA spectroscopies are discussed in this book. Unless otherwise stated, these regions are herein defined as follows: near-infrared, 12,500–4000 $cm^{-1}$ (wavelengths 0.8–2.5 μm); mid-infrared, 4000–400 $cm^{-1}$ (2.5–25 μm); and far-infrared, 400–200 $cm^{-1}$ (25–50 μm). It should be noted that this division of the electromagnetic spectrum, although not uncommon, is somewhat arbitrary; for example, the low-frequency limit assumed for the far-infrared region is based on the few published PA far-infrared spectra, rather than the more conventional far-infrared limits that encompass much lower frequencies and longer wavelengths.

## 1.3. OTHER SOURCES OF INFORMATION

A significant number of review articles, conference proceedings, and texts on PA spectroscopy have been published during the last two decades. Many of the literature reviews provide an overview of the PA technique and emphasize work in shorter wavelength regions (ultraviolet and visible) that are not discussed in this book. Mid-infrared PA spectroscopy receives very limited attention in most of these reviews; some discuss near-infrared spectroscopy in more detail and are therefore relevant to specific sections of this text. The early work of Adams (1982) is a typical example. Several years later, Vargas and Miranda (1988) published a detailed summary of PA and photothermal techniques that contains a short section on PA spectroscopy in the near- and mid-infrared regions. Other reviews that discuss PA spectroscopy and its relationship to a series of PA and photothermal methods are listed among the references at the end of this chapter.

Early progress in mid-infrared PA spectroscopy was summarized by Vidrine (1982) and by Graham et al. (1985). McClelland (1983) discussed aspects of signal generation and instrumentation in an important survey of PA spectroscopy that emphasized infrared applications. The latter article is generally considered to be authoritative and continues to be cited by a number of workers who utilize PA infrared spectroscopy. Numerical methods, specifically phase correction and signal averaging, were discussed a few years later by the present author (Michaelian, 1990).

Two research groups that have made major contributions to PA infrared spectroscopy should be specifically mentioned with regard to review publications. R. A. Palmer of Duke University, together with a number of students and other collaborators, has long been active in research areas that include the PA infrared spectroscopy of polymers, phase analysis of PA spectra, and the development of step-scan instrumentation; a review on PA spectroscopy of polymers by Dittmar et al. (1994) contains a valuable summary of the history and principles of PA infrared spectroscopy. Numerous other publications by this research group are referred to in later chapters.

Similarly, J. F. McClelland and his co-workers at Iowa State University have made a very considerable contribution to the field of PA spectroscopy during the last three decades. These investigators have a substantive history in instrumentation that has culminated in the successful manufacture of commercial PA sample accessories for FTIR spectrometers. McClelland and his colleagues have published several review articles, including a summary of the PA FTIR technique that discusses signal generation and demonstrates a series of qualitative and quantitative applications (McClelland et al., 1992). Other reviews described sample handling in PA FTIR spectroscopy (McClelland et al., 1993), and the implementation of PA spectroscopy with step-scan and rapid-scan spectrometers (McClelland et al., 1998). Recently, this group reviewed PA FTIR spectroscopy in a chapter in *Handbook of Vibrational Spectroscopy* (McClelland et al., 2002), again with particular reference to signal generation, instrumentation, and sampling. These publications will be of considerable use to investigators who require an introduction to PA infrared spectroscopy. Their existence also affects (and restricts) the selection of subject matter in the present work.

## REFERENCES

Adams, M. J. (1982). Photoacoustic spectroscopy. *Prog. Analyt. Atom. Spectrosc.* **5**: 153–204.

Dittmar, R. M., Palmer, R. A., and Carter, R. O. (1994). Fourier transform photoacoustic spectroscopy of polymers. *Appl. Spectrosc. Rev.* **29**: 171–231.

Graham, J. A., Grim, W. M., and Fateley, W. G. (1985). Fourier transform infrared photoacoustic spectroscopy of condensed-phase samples. In: *Fourier Transform Infrared Spectroscopy*. J. R. Ferraro and L. J. Basile (eds.). Academic, New York, Vol. 4, pp. 345–392.

McClelland, J. F. (1983). Photoacoustic spectroscopy. *Anal. Chem.* **55**: 89A–105A.

McClelland, J. F., Luo, S., Jones, R. W., and Seaverson, L. M. (1992). A tutorial on the state-of-the-art of FTIR photoacoustic spectroscopy. In: *Photoacoustic and Photothermal Phenomena III.* D. Bićanić (ed.). Springer, Berlin, pp. 113–124.

McClelland, J. F., Jones, R. W., Luo, S., and Seaverson, L. M. (1993). A practical guide to FTIR photoacoustic spectroscopy. In: *Practical Sampling Techniques for Infrared Analysis.* P. B. Coleman (ed.). CRC Press, Boca Raton, FL, pp. 107–144.

McClelland, J. F., Bajic, S. J., Jones, R. W., and Seaverson, L. M. (1998). Photoacoustic spectroscopy. In: *Modern Techniques in Applied Molecular Spectroscopy.* F. M. Mirabella (ed.). Wiley, New York, pp. 221–265.

McClelland, J. F., Jones, R. W., and Bajic, S. J. (2002). Photoacoustic spectroscopy. In: *Handbook of Vibrational Spectroscopy.* J. M. Chalmers and P. R. Griffiths (eds.). Wiley, Chichester, Vol. 2, pp. 1231–1251.

Michaelian, K. H. (1990). Data treatment in photoacoustic FT-IR spectroscopy. In: *Vibrational Spectra and Structure.* J. R. Durig (ed.). Elsevier Science, Amsterdam, Vol. 18, pp. 81–126.

Vargas, H., and Miranda, L. C. M. (1988). Photoacoustic and related photothermal techniques. *Phys. Rep.* **161**: 43–101.

Vidrine, D. W. (1982). Photoacoustic Fourier transform infrared spectroscopy of solids and liquids. In: *Fourier Transform Infrared Spectroscopy.* J. R. Ferraro and L. J. Basile (eds.). Academic, New York, Vol. 3, pp. 125–148.

## RECOMMENDED READING

Almond, D., and Patel, P. (1996). *Photothermal Science and Techniques.* Chapman & Hall, London.

Griffiths, P. R., and de Haseth, J. A. (1986). *Fourier Transform Infrared Spectrometry*, Chem. Anal., Vol. 83, Wiley (Interscience), New York.

Pao, Y.-H. (Ed.) (1977). *Optoacoustic Spectroscopy and Detection.* Academic, New York.

Rosencwaig, A. (1978). Photoacoustic spectroscopy. *Adv. Electron. Electron Phys.* **46**: 207–311.

Rosencwaig, A. (1980). *Photoacoustics and Photoacoustic Spectroscopy*, Chem. Anal. Vol. 57, Wiley (Interscience), New York.

CHAPTER

2

# EVOLUTION OF PHOTOACOUSTIC INFRARED SPECTROSCOPY

This chapter is an account of the historical development of PA infrared spectroscopy. Relevant literature published in the period beginning in 1968 and ending in 1981 is reviewed in the following pages. The reader should appreciate that these dates are not entirely arbitrary: The former year marked the appearance of a prototypical article on PA detection of the absorption of infrared radiation in gases, while the latter appears to be the approximate point at which PA infrared spectroscopy first achieved a measure of maturity and acceptance by the vibrational spectroscopy community (see Fig. 2.1). This acceptance is, of course, signified by the publication of an appreciable number of studies on the subject; because of its size, this body of literature is not particularly amenable to the chronological perspective adopted here to place early work in context. Hence this second date is where a historical review ought to end, and discussions based on applications of PA infrared spectroscopy should logically begin.

Researchers who utilized PA infrared spectroscopy in these formative years may well question the criteria for inclusion of studies in this discussion. Inevitably, some articles have been excluded. There are at least three reasons for this. First, a few publications may not be widely available or have not come to the attention of the present author; and second, the current selection of studies was deliberately chosen so as to illustrate the evolution of the field. Finally, it should be noted that some works were intentionally omitted from this chapter and are discussed elsewhere in this book.

This chapter is divided into three principal sections. The first deals with work published between 1968 and 1978; the latter date can be considered a sort of milestone since it corresponds to publication of the first study on PA FTIR spectroscopy. The second section describes early multiple-wavelength PA infrared spectroscopy, which can be thought of as a precursor to multiplex (FTIR) PA infrared spectroscopy. Finally, the third section discusses the emergence of PA FTIR spectroscopy, which took place between 1978 and 1981. Additional work published in 1981 and beyond is reviewed in subsequent chapters, which are organized primarily according to subject matter, rather than year of publication.

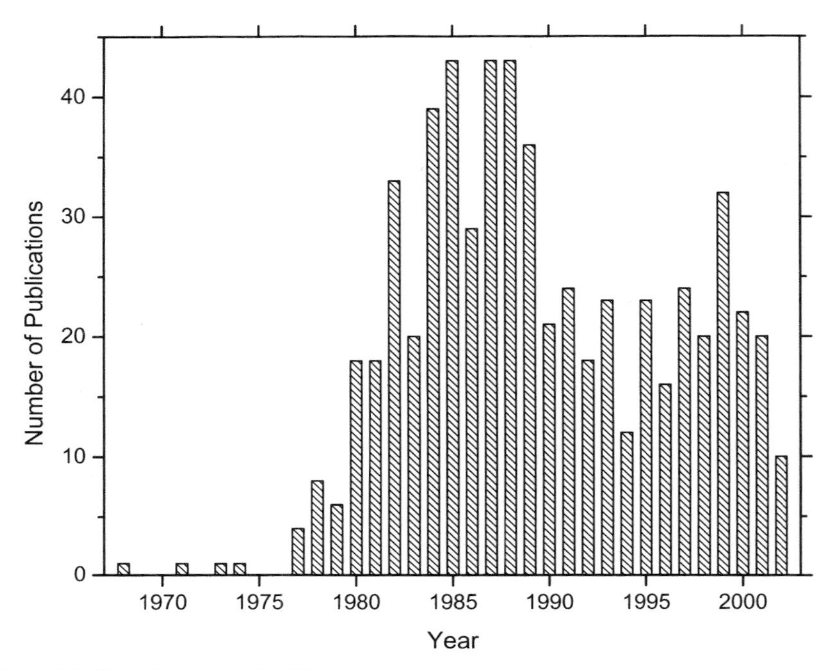


**Figure 2.1.** Plot of number of publications on PA infrared spectroscopy as function of year of publication.

## 2.1. EARLY HISTORY

Photoacoustic infrared spectroscopy is sometimes said to originate with the work of Busse and Bullemer (1978), who obtained a PA spectrum of methanol vapor using a commercial FTIR spectrometer and an absorption cell fitted with a microphone. This apparatus allowed comparison of a conventional mid-infrared absorption spectrum with a PA spectrum obtained under the same conditions. To put this work in proper context and describe the evolution of the field more accurately, it should be emphasized that this was, in fact, the first successful PA FTIR experiment described in the primary scientific literature; nevertheless, a number of articles on PA infrared spectroscopy using other instrumental methods were actually published prior to this work. These works are briefly summarized in this section.

Table 2.1 lists some of these early publications on PA infrared spectroscopy. All of these studies appeared before that of Busse and Bullemer (1978), that is, prior to 1978; it can be noted that the majority describe PA

**Table 2.1. Photoacoustic Infrared Spectroscopy: Early Publications**

| Author | Subject |
|---|---|
| Kerr and Atwood (1968) | $CO_2$ laser absorptivity spectrophone |
| Kreuzer (1971) | PA spectroscopy of $CH_4$, $CH_3OH$ |
| Rosengren (1973) | Theory of optoacoustic gas detector |
| Max and Rosengren (1974) | Optoacoustic detection of $NH_3$ in air |
| Kanstad and Nordal (1977) | Grazing incidence PA spectroscopy of metal surfaces |
| Nordal and Kanstad (1977a) | PA spectroscopy of $(NH_4)_2SO_4$, glucose |
| Nordal and Kanstad (1977b) | PA spectroscopy of $(NH_4)_2SO_4$, hexachlorobenzene |
| Patel et al. (1977) | Optoacoustic spectroscopy of excited-state NO |

spectra of gases. Some of these references are mentioned again in Chapter 6, which deals with specific applications of PA infrared spectroscopy.

In 1968, Kerr and Atwood described an "absorptivity spectrophone" designed to measure weak absorptivities at visible and infrared wavelengths. In the latter experiment a $CO_2$ laser, operating at 9.6 μm (1042 $cm^{-1}$) instead of the more familiar wavelength of 10.6 μm, was used to analyze $CO_2$ in $CO_2/N_2$ mixtures. The resulting signal from a differential pressure transducer was observed to vary linearly with $CO_2$ concentration. An absorptivity of $5.8 \times 10^{-7}$ was measured at 300 ppm, the standard concentration of $CO_2$ in air. Moreover, the authors predicted that an absorptivity as low as $1.2 \times 10^{-7}$ could be measured with the use of a 10-W $CO_2$ laser. Thus the high sensitivity achievable in PA infrared spectroscopy of gases was demonstrated at a very early date.

Three years later, Kreuzer (1971) discussed the theory of the absorption of infrared radiation by a gas, with particular reference to coherent sources of radiation. PA spectra of $CH_4$ and $CH_3OH$ vapor in a narrow section of the C—H stretching region were obtained by tuning a HeNe laser through a series of emission lines between 2947.87 and 2947.93 $cm^{-1}$. Despite the extremely limited frequency range that was covered, this experiment allowed the observation of a few absorption bands. The calculated sensitivity was about $10^{-8}$, although Kreuzer estimated that this might eventually be improved by a factor as great as $10^5$. A subsequent theoretical model for this experiment (Rosengren, 1973) rationalized and confirmed the minimum detectable molecule density for $CH_4$ reported by Kreuzer. This theoretical analysis concluded that the highest sensitivity predicted in the experimental study, which is equivalent to $10^{-4}$ ppm, was indeed achievable.

Trace gas detection was studied further by Max and Rosengren (1974), who constructed a resonant PA gas cell. These authors showed that the minimum concentration of gas that can be detected is determined by microphone and preamplifier noise; in particular, acoustic noise detected by the microphone was found to be the most important factor. In this investigation, a $CO_2$ laser was used to study trace amounts of $NH_3$ in air. For a laser power of 280 mW, the lower concentration limit giving rise to a measurable signal was about 0.003 ppm. It was again suggested that a detection limit of about $10^{-4}$ ppm should be achievable with this PA technique.

Early PA infrared spectra were also measured for condensed phases. Nordal and Kanstad (1977a,b) noted recent successes in PA spectroscopy at ultraviolet, visible, and near-infrared wavelengths and successfully extended the technique into the mid-infrared. A tunable $CO_2$ laser with lines in three wavelength intervals (approximately 9.2–9.7 μm, 10.1–10.3 μm, and 10.5–10.7 μm) was used for excitation; the PA cell was constructed in the laboratory of the authors. Ammonium sulfate (both solid and in aqueous solution), glucose, and hexachlorobenzene were successfully studied in these investigations. Insensitivity to morphology of solid samples was noted as a particular advantage; indeed, the appearance of the totally symmetric $\nu_1$ band of the sulfate ion, normally observed only in Raman spectra, was thought to arise from reduced symmetry in crystallite surface layers that would have been destroyed by grinding. Although this reasoning was subsequently modified by the authors, it nicely demonstrates that a very real advantage can be created by the minimal sample preparation required in PA spectroscopy. The suitability of PA infrared spectroscopy for quantitative analysis was also shown in this work: The intensity of the PA signal of $(NH_4)_2SO_4$ in solution was observed to vary linearly with concentration over a wide range. However, this signal tended to saturate at very high concentrations, a phenomenon that had already been observed in PA spectra at shorter wavelengths.

Another study published at about the same time by these authors (Kanstad and Nordal, 1977) described grazing incidence PA spectroscopy on metal surfaces, an application of a technique termed PA reflection-absorption spectroscopy (PARAS) by these researchers. This experiment also utilized a tunable $CO_2$ laser and was designed to study optical absorption in thin surface layers. The layout of the PA apparatus is shown schematically in Figure 2.2. The principle of the measurement is that light polarized in the plane of incidence may be absorbed by the sample, whereas light polarized perpendicular to this plane may not; the geometry of the experiment was carefully controlled to ensure that only the former polarization was available. Measurements were performed on aluminum oxide layers and on iron surfaces exposed to $SO_2$. For the former group of samples, a strong absorption near 10.55 μm (950 $cm^{-1}$) was observed and attributed to a longitudinal Al–O vibration. The results for the iron surface were less satisfactory but tentatively ascribed to the presence of $FeSO_4 \cdot 7H_2O$. Impor-

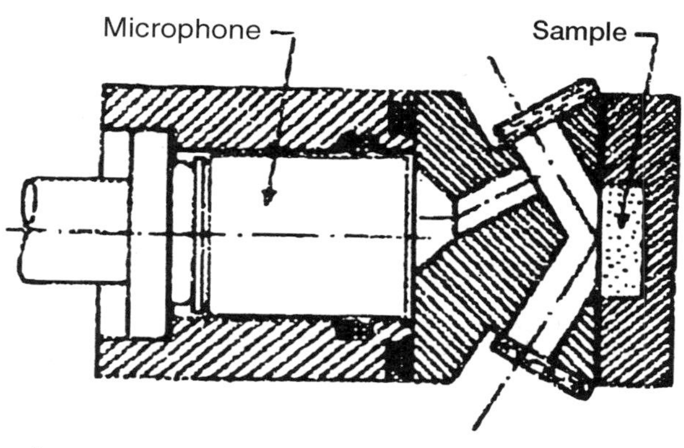


**Figure 2.2.** Experimental layout for PA reflection-absorption spectroscopy (PARAS). (Reproduced from Kanstad, S. O. and Nordal, P.-E., *Int. J. Quantum Chem.* **12** (Suppl. 2): 123–130, by permission of John Wiley & Sons, Inc.; copyright © 1977.)

tantly, Kanstad and Nordal concluded that aluminum oxide could be detected at surface coverages of only a fraction of a monolayer; the putative sulfate species was thought to have a detection limit of several layers. Thus very high sensitivity was also demonstrated in PA infrared spectroscopy of solids, in the region made accessible by the $CO_2$ laser.

The final study to be mentioned in this short review also involves PA spectroscopy of gases, but its perspective is significantly different from those in the works mentioned above. Patel et al. (1977) obtained high-resolution excited-state PA spectra of NO gas, specifically choosing PA detection because of its high sensitivity. In this experiment, NO was first excited with a CO laser line at 1917.8611 $cm^{-1}$ and then probed with tunable radiation from a spin-flip Raman laser. The PA signal was measured as a function of the magnetic field of this laser or, equivalently, of its infrared frequency. Zeeman spectra of excited-state NO were also obtained using PA detection. While this investigation provides a different type of information than that in the previously mentioned studies, it should still be considered relevant because it shows the versatility of PA spectroscopy in an interesting application at infrared wavelengths.

## 2.2. MULTIPLE-WAVELENGTH PA INFRARED SPECTROSCOPY

The second phase in the evolution of PA infrared spectroscopy can be thought of as the period in which the pioneering single-wavelength measurements described above gave way to multiple-wavelength (sequential) experiments, performed with scanning spectrometers or through the use of a series of infrared laser lines. The broader spectral coverage in these later PA experiments greatly increased the quantity and diversity of information that was obtainable, making it possible to use PA spectroscopy as a tool for the analysis of samples of greater interest to chemists; thus it was no longer strictly necessary to restrict PA studies to systems that absorb radiation at particular wavelengths dictated by the available monochromatic sources. Both mid- and near-infrared dispersive PA spectra are included in these more recent experiments. Results obtained using $CO_2$ laser radiation by Nordal and Kanstad, as well as Perlmutter and co-workers, are also discussed in this section. All of this multiple-wavelength work was published between 1978 and 1980, an interval during which several significant developments in PA FTIR spectroscopy concomitantly occurred. The latter topic is discussed separately in the next section of this chapter.

Dispersive PA spectra are discussed first in this section. In the mid-infrared region, Low and Parodi (1978) successfully used PA spectroscopy as a means to characterize surfaces and adsorbates. The goal of their work was to observe species adsorbed on catalysts or silica powder. The spectrometer available to these researchers utilized a LiF prism, making it possible to record infrared spectra from 2000 to 4000 $cm^{-1}$, but precluding observation of lower frequency bands due to heavier atoms such as silicon and oxygen. In their first experiment, PA spectra of a chromia-alumina catalyst exhibited bands due to $OCH_3$ and OH groups, the former appearing after exposure of the catalyst to $(CH_3O)_3CH$ vapor and the latter arising from surface hydroxyls and adsorbed water. Similarly, the PA infrared spectrum of treated silica exhibited bands due to $CH_3$ vibrations in Si–$OCH_3$ groups; after exposure to $HSiCl_3$ vapor, a new band characteristic of the chemisorbed silane was observed. Low and Parodi noted that similar results could, in principle, have been observed with thin pellets using transmission spectroscopy. However, PA spectroscopy offers a decisive advantage in the study of whole catalyst pellets, for which transmittance is obviously prohibitively low.

This initial demonstration of dispersive mid-infrared PA spectroscopy was soon followed by an entire series of publications by these authors (Low and Parodi, 1980a–e). This work is described in more detail in the first section of Chapter 3. In these investigations, the authors pursued their interest in the characterization of surfaces and opaque solids (Low and Parodi, 1980c,d) and of various carbonaceous samples (Low and Parodi, 1980a,b). With regard to the latter subject, it is relevant to note that PA spectra of carbons are frequently utilized to normalize spectra of less strongly absorbing samples; this topic is taken up in Chapter 5.

Several works on dispersive near-infrared PA spectroscopy are also relevant to this discussion. Three of these publications describe the work of

Kirkbright and his colleagues. In the first, Adams et al. (1978) described instrumentation and applications of PA spectroscopy at wavelengths between approximately 0.8 and 2.7 μm (12,500 and 3700 $cm^{-1}$). Experimental apparatus included a quartz–halogen lamp, chopper, monochromator, home-built PA cell, and a lock-in amplifier. Spectra were reported for two rare-earth oxides and for aliphatic and aromatic hydrocarbons; the hydrocarbon spectra are discussed in another section of this book. The bands observed for the rare-earth oxides arise from low-energy electronic transitions and were sufficiently well defined for the authors to suggest their use as wavelength standards.

In a related investigation, Kirkbright (1978) presented a brief overview of PA spectroscopy, mentioning an entire series of advantages and applications of the technique. These include the following: freedom from light-scattering interference; the capability for depth profiling; complementary measurements of thermal conductivity and sample thickness; the wide spectral range accessible by this technique; determination of luminescence quantum efficiencies; the capability for study of very weakly absorbing materials; and, finally, the use of an inert atmosphere, which facilitates the study of air-sensitive materials. This discussion was followed by near-infrared PA results for a wide variety of samples, including blood, proteins, clays and minerals, and the aforementioned hydrocarbons. Hence this study serves to illustrate the considerable diversity of PA near-infrared spectroscopy and the detailed information obtainable by this technique.

In subsequent research carried out by this group (Castleden et al., 1980), near-infrared photoacoustic spectroscopy was used to determine the moisture content in single-cell protein. The spectra reported in this work covered the wavelength range from about 1.3 to 2.3 μm (roughly 7700 to 4350 $cm^{-1}$). A band at 1.9 μm (5265 $cm^{-1}$) was found to vary with the amount of moisture present and was therefore employed for quantitation. The intensity of this band was first corrected by ratioing it against the intensity of an —NH band at 1.55 μm (6450 $cm^{-1}$); the latter feature was due to protein, which was known not to vary significantly from one sample to the next. The authors found that quantitative analysis was not feasible until samples were separated into series of like size fractions, since PA intensities were greater for smaller particle sizes. When spectra for specific size ranges were compared, it was observed that corrected PA intensities at 1.9 μm increased linearly with moisture content. These results were superior to those obtained by low-resolution nuclear magnetic resonance (NMR) spectroscopy: NMR was less selective than PA spectroscopy because it tended to detect all protons, rather than those in particular functional groups.

Other research groups were also active in dispersive near-infrared PA spectroscopy during this period. For example, Blank and Wakefield (1979), working at Gilford Instrument Laboratories, described a double-beam PA spectrometer that covered the ultraviolet, visible, and near-infrared spectral regions. This work contains near-infrared PA spectra of the rare-earth oxides $Ho_2O_3$, $Nd_2O_3$, and $Pr_6O_{11}$, which show a number of well-defined bands. A wide variety of other samples was also mentioned by these authors, although corresponding spectra were not shown. Thus, as was the case in the work of Kirkbright's group, the results of Blank and Wakefield demonstrated that PA near-infrared spectroscopy is a particularly versatile technique for chemical analysis. It should also be noted that the divided PA cell described in this work was subsequently adapted for mid-infrared PA spectroscopy.

The final example of dispersive near-infrared PA spectroscopy to be mentioned in this section is the work of Lochmüller and Wilder (1980a,b). These researchers used a Princeton Applied Research spectrometer equipped with a Xe lamp to obtain PA spectra at ultraviolet, visible, and near-infrared wavelengths. The first study by these authors (1980a) was concerned with qualitative characterization of chemically modified silica surfaces, while the second (1980b) emphasized quantitation of the degree of surface coverage for these surfaces. In the former study samples were prepared by reacting microparticulate silica gel with various organomethyldichlorosilanes so as to introduce hydrocarbonaceous cyclic systems onto the silica surface. Spectra of a series of silicas in which cyclohexyl, cyclohexenyl, or phenyl groups are bonded to the surface contained features due to the second overtone of the strongest aliphatic C—H stretching band at 1.186 μm (8430 $cm^{-1}$) and a shoulder at 1.173 μm (8525 $cm^{-1}$). These were attributed to cyclohexyl methylene groups and the adjacent methyl group on the organomethyldichlorosilane, respectively. Another band at 1.14 μm (8770 $cm^{-1}$) arises from cis hydrogens on $sp^2$ carbons and consequently was most intense for the phenyl system. These results demonstrate the capability of the technique for qualitative analysis. In the second investigation, it was observed that coverage with various alkyl, phenyl, and aminoalkyl compounds could be monitored using either the first overtone of the strongest aliphatic C—H stretching band, which occurs at 1.71 μm (5850 $cm^{-1}$), or the second overtone that was mentioned above. Because the first overtone—which, of course, is the stronger of the two—is superimposed on a sloping baseline, the authors chose to use the other band for quantitation. It was found that a linear relationship exists between PA signal amplitude and carbon content.

Three works by Nordal and Kanstad were described in the first section of this chapter. These authors used several $CO_2$ laser lines in their PA studies and were able to observe a number of important infrared bands. Five additional studies by this group, published between 1978 and 1980, are pertinent to the current discussion of multiple-wavelength PA spectroscopy. This later research was based on the use of emission lines from both $^{12}CO_2$ and $^{13}CO_2$ lasers, which effectively increased the number of wavelengths at which PA observations could be made.

In the first of these studies, Kanstad and Nordal (1978) described an "open-membrane" PA cell in which the sample was outside the acoustic chamber, suspended on a membrane that was used in place of the more familiar infrared-transparent window. In this experiment, heat generated as a consequence of light absorption in the sample was transported through the membrane into the chamber, where the resulting acoustic wave was detected by a microphone. This approach, a variation of the so-called open PA cell, was eventually adapted by other workers, and is mentioned in Chapter 3. Kanstad and Nordal used this technique to obtain PA spectra of an $Al_2O_3$ film, a polychlorotrifluoroethylene oil film, and hexachlorobenzene crystallites. Measurements were made at more than 20 wavelengths in four regions between 9.2 and 10.7 μm. Although the $CO_2$ laser radiation utilized in this work was not continuously tunable, the number of lines available was sufficient to characterize several absorption bands of interest to the authors.

PA infrared spectra of ammonium sulfate were obtained in the initial investigations by these authors described in the previous section. This work was extended to include the study of single crystals of $(NH_4)_2SO_4$ in two more recent publications (Kanstad and Nordal, 1979, 1980a). These later results are noteworthy because they include one of the first demonstrations of polarized PA infrared spectra in the literature. Indeed, the single-crystal study revealed that the $\nu_1$ sulfate band at 975 $cm^{-1}$ is relatively strong when the electric vector of the infrared radiation is parallel to the $\mathbf{a}_0$ crystal axis, but practically absent when the vector is polarized along $\mathbf{b}_0$. This result led to the conclusion that the appearance of the $\nu_1$ peak in the PA spectra of $(NH_4)_2SO_4$ powder obtained by these authors (Nordal and Kanstad, 1977a,b) was not due to surface properties, since the band was also observed for single crystals. The PA infrared spectrum of ammonium sulfate is discussed again in the following section of this chapter.

PA reflection-absorption spectroscopy (PARAS), which was briefly mentioned in the previous section, was considered in greater detail by these authors (Nordal and Kanstad, 1978; Kanstad and Nordal, 1979, 1980b). This technique was developed for studying thin films, particularly oxides, on metal surfaces. In addition to the open-membrane spectrophone discussed above, the authors described two other PARAS techniques, using either a modified closed-chamber spectrophone or piezoelectric detection. In the former experiment, the $CO_2$ laser radiation impinges on the sample at an angle of 60°, with the reflected beam being allowed to exit the spectrophone chamber (see Fig. 2.2). A microphone detects the acoustic wave that arises from absorption of infrared radiation by the sample. Spectra are plotted as the quantity $\Delta = \delta R_{\parallel} - \delta R_{\perp}$, where $\delta R_{\parallel}$ and $\delta R_{\perp}$ are the fractional changes in reflectivity for light polarized parallel and perpendicular, respectively, to the plane of incidence. For an oxide layer on aluminum, the 10.55-μm (948-$cm^{-1}$) band appeared only in the parallel-polarized spectrum. Moreover, the authors noted that this band shifts by a few reciprocal centimeters ($cm^{-1}$) when the oxide layer is thicker, and that the band frequencies observed by PARAS differ slightly from published values obtained by other techniques.

In the latter experiment, the piezoelectric transducer was affixed directly to the back surface of the sample, with the front surface being illuminated by the laser beam. This removed any requirement for sample modification or the use of a gas-microphone cell. This technique was utilized by Kanstad and Nordal (1979, 1980b) to obtain spectra of oxide layers on aluminum. Piezoelectric detection is discussed in more detail in Chapter 3.

The next publication to be discussed with regard to multiple-wavelength PA infrared spectroscopy also describes the use of $CO_2$ lasers. Perlmutter et al. (1979) described the PA detection of trace amounts of ethylene in the presence of interfering gases. The authors pointed out three limiting factors in these measurements: (1) additive noise, that is, microphone electrical noise; (2) modulation noise, which primarily arises from variations in laser power; and (3) the limited accuracies of the absorption coefficients of the interfering gases. The third factor is by far the most important, typically amounting to levels of a few percent. Perlmutter et al. (1979) used both $^{12}CO_2$ and $^{13}CO_2$ lasers in their work and detected $C_2H_4$ levels of 130 ppb in an urban environment. They estimated a detection limit of approximately 5 ppb but noted that the presence of 1% $CO_2$ would increase this limit by a factor of about 10. PA infrared detection of ethylene is discussed further in Chapter 6.

The final example in this section utilized a different laser as the source of radiation. Monchalin et al. (1979) obtained PA infrared spectra of chrysotile asbestos in the hydroxyl stretching region between about 3400 and 3800 $cm^{-1}$, using a series of emission lines near 2.7 μm from an HF laser. This form of asbestos, a hydrated magnesium silicate, was of interest to these authors as an environmental pollutant. About a dozen HF laser lines were available in this experiment; these were sufficient for observation of a doublet at 3645 and 3688 $cm^{-1}$, the second peak being the stronger of the two. These bands, which arise from hydroxyl groups in two different environments, displayed an intensity ratio consistent with their expected relative abundances. In the second part of this investigation, Monchalin et al. (1979) measured PA spectra for two orientations of asbestos fibers relative to the polarization of the laser; as expected, the PA signal was much smaller when the polarization was parallel to the fibers than when it was perpendicular to them.

## 2.3. ARRIVAL OF PA FTIR SPECTROSCOPY

The third stage in the development of PA infrared spectroscopy lasted between approximately 1978 and 1981. During this period, the enhanced capabilities and increasingly widespread availability of FTIR spectrometers led more and more vibrational spectroscopists to utilize these instruments, rather than traditional dispersive infrared spectrometers, on a routine basis. At the same time, recent advances in both dispersive near-infrared and multiple-wavelength mid-infrared PA spectroscopies helped to create and define a need for a multiplex technique for mid-infrared PA spectroscopy. These parallel developments made the emergence of PA FTIR spectroscopy in the late 1970s and early 1980s logical and, perhaps, inevitable.

As mentioned above, Busse and Bullemer (1978) are generally recognized as the first workers to describe a viable PA infrared spectrum obtained with an FTIR spectrometer. Several other groups developed productive research programs in PA FTIR spectroscopy during the next 2–3 years. Interestingly, virtually all of the latter efforts pertained to condensed matter (primarily solids) rather than gases; hence the investigators in these projects had to deal with PA signals that were, in principle, weaker than those in the first successful PA FTIR experiment. The most relevant publications of several of these groups in the years up to and including 1981 are reviewed in this section. Work published after this date is discussed in later chapters, which are organized according to subject matter rather than the date of publication.

Table 2.2 lists some of the publications on PA infrared spectroscopy that appeared between 1979 and 1981. These studies present much of the historically important early work in PA FTIR spectroscopy. The research groups mentioned are those of Rockley and his collaborators at Oklahoma State University; Mead, Vidrine, and their colleagues at Nicolet Instrument Corporation; and finally, Royce, Teng, and co-workers at Princeton University. The publication references from each of these three sources are grouped together in Table 2.2 and discussed in the same sequence in the following paragraphs.

This history of PA FTIR spectroscopy begins with the work of Rockley and collaborators, who had previously utilized PA spectroscopy at much shorter wavelengths. In what was purported to be the first publication illustrating a PA infrared spectrum of a solid obtained with an FTIR spectrometer, Rockley (1979) presented a single result for a polystyrene film in the

**Table 2.2. Photoacoustic FTIR Spectroscopy: Early Publications**

| Author | Subject |
|---|---|
| Rockley (1979) | Polystyrene |
| Rockley (1980a) | $(NH_4)_2SO_4$ |
| Rockley (1980b) | Charcoal, polystyrene, aspirin |
| Rockley and Devlin (1980) | Coal |
| Rockley et al. (1980) | Naphthalene, benzophenone, $KNO_3$, silica |
| Mead et al. (1979) | Coal |
| Mead et al. (1981) | LiF, Ge, Si, GaAs, near-, mid-, far-infrared |
| Vidrine (1980) | Polyvinyl chloride, polyurethane, lecithin, plastic, coal |
| Vidrine (1981) | Catalysts, conducting polymers, depth profiling |
| Laufer et al. (1980) | AgCN |
| Royce et al. (1980) | Polystyrene, acrylates, coal, zeolites |

spectral region extending from 4000 to 800 $cm^{-1}$. The PA cell used in this work, originally developed for use at ultraviolet and visible wavelengths, had been fitted with an NaCl window for this initial infrared study. This approach was not uncommon in the early days of PA infrared spectroscopy and illustrates how PA spectroscopy was gradually adapted to longer wavelengths. Several C—H stretching bands are visible in the uncorrected polystyrene spectrum reported by Rockley, with absorption by atmospheric $CO_2$ and $H_2O$ also being quite significant. Although this PA spectrum is noisy and also contains spikes due to ground loops, it can be described as noteworthy simply because it is the first of its kind. The author correctly predicted that the technique would eventually be used to study samples with rough surfaces, catalysts, and gases. Examples of these applications are discussed in later chapters of this book.

The use of PA FTIR spectroscopy for characterization of ammonium sulfate, which is of interest because it is an atmospheric aerosol, was described one year after the initial work on polystyrene (Rockley, 1980a). As discussed above, Nordal and Kanstad (1977a,b) had already identified the $\nu_1$ band of the sulfate ion in PA infrared spectra of solid $(NH_4)_2SO_4$ obtained with a $CO_2$ laser; Rockley (1980a) confirmed the existence of this band in the analogous PA FTIR spectrum. However, the more recent work showed that this band occurred as a shoulder rather than a well-defined peak in the FTIR spectrum of this compound. Rockley attributed this observation to local field-induced distortions that broaden the stronger $\nu_3$ band near 1100 $cm^{-1}$, causing the $\nu_1$ peak to appear as a shoulder. Figure 2.3 shows that the $\nu_1$ sulfate band is, in fact, barely visible in the PA infrared spectrum of polycrystalline ammonium sulfate obtained with modern instrumentation.

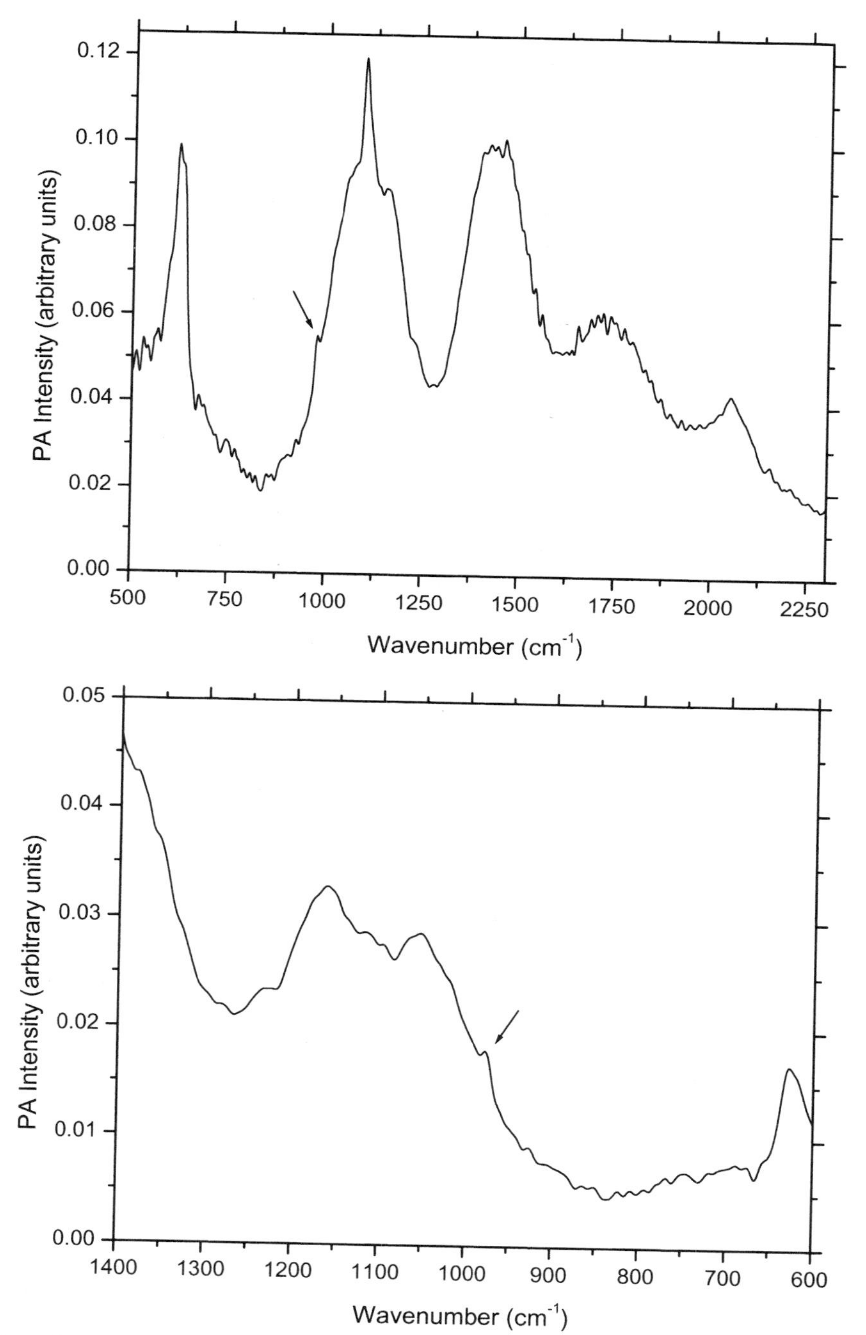


**Figure 2.3.** (*a*) PA FTIR spectrum of $(NH_4)_2SO_4$, measured at a resolution of 6 cm$^{-1}$ using a Bruker IFS 88 FTIR spectrometer and an MTEC 200 PA cell. Arrow indicates a frequency of 975 cm$^{-1}$. (*b*) Expanded plot, with a frequency scale similar to that used by Rockley (1980a).

Rockley (1980b) next described PA FTIR spectra of several opaque solids. In this work, charcoal, polystyrene beads used as catalyst supports, and an aspirin tablet were studied. Carbon black and finely divided lead were considered as reference materials. The author observed that absorption by atmospheric $H_2O$ and $CO_2$ tends to complicate PA infrared spectra: When these gases are present in the optical path between the interferometer and the PA cell, they absorb part of the incident radiation and give rise to transmissionlike features in the spectra. Further, Rockley found that these bands are only approximately canceled by ratioing the PA spectrum of a sample to that of a reference material, since the amounts of $CO_2$ and water vapor in the instrument are frequently different from one experiment to the next. Despite this complication, useful PA spectra of polystyrene and aspirin were obtained in this investigation.

PA infrared spectroscopy has been used extensively to characterize coals during the last 20 years. This subject is discussed in detail in the second section of Chapter 6. Rockley and Devlin (1980) were among the first to report PA FTIR spectra of coals, describing spectra of aged and freshly exposed coal surfaces. These authors obtained research-quality PA spectra of subbituminous, bituminous, and anthracitic coals at a time when the use of FTIR spectrometers in PA spectroscopy was barely underway—a rather impressive accomplishment. As shown in Figure 2.4, their investigation revealed noteworthy differences among the spectra of the three coals, the most striking being the relative weakness of the $CH_2$ and $CH_3$ bands in the anthracite spectrum. Significant differences were also observed among the PA spectra of the aged coal surfaces. These results are discussed in more detail in the second section of Chapter 6.

The final publication by Rockley's group to be discussed in this historical review reports early work on PA FTIR spectroscopy of solids, with particular reference to the capabilities of the technique for quantitative analysis (Rockley et al., 1980). The first attempt at quantitation by these authors utilized previously melted binary mixtures of naphthalene and benzophenone; this system did not display band intensities that were proportional to the concentrations of the components, possibly because of the formation of eutectic mixtures. However, the second example considered in this work (mixtures of $K^{14}NO_3$ and $K^{15}NO_3$) did exhibit a linear relationship between intensity and concentration. This led Rockley et al. to conclude that PA spectroscopy is, in fact, well suited for quantitative analysis. This topic is considered in more detail in Chapter 7.

In the same publication, Rockley et al. (1980) carried out one of the first investigations of the effect of particle size on PA infrared spectra. Silica powder with particle sizes ranging up to 60 μm was separated into five different size fractions. For silica particles up to about 25 μm, the ratio of

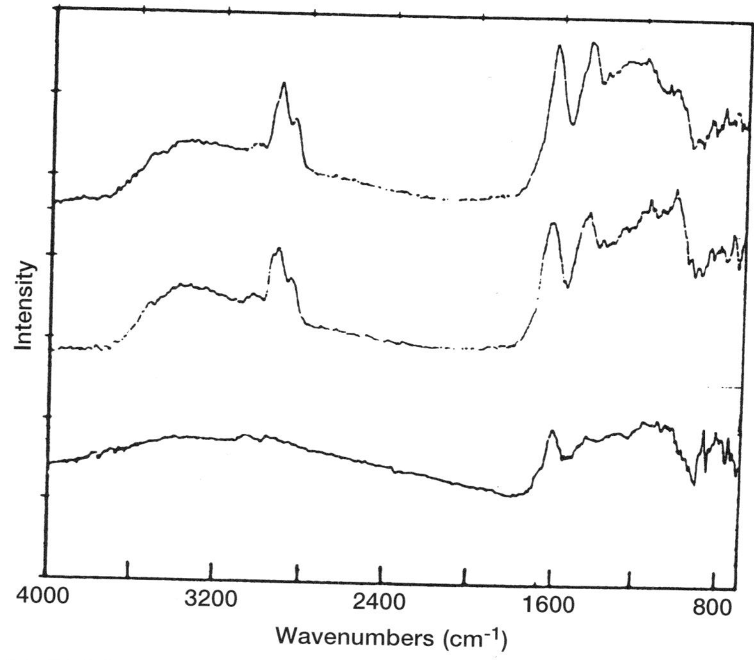


**Figure 2.4.** PA infrared spectra of three coals. Upper curve, Illinois No. 6; middle curve, Pittsburgh bituminous; bottom curve, Reading anthracite. (Reproduced from Rockley, M. G. and Devlin, J. P., *Appl. Spectrosc.* **34**: 407–408, by permission of the Society for Applied Spectroscopy; copyright © 1980.)

PA intensity at 1100 $cm^{-1}$ with respect to that at 2000 $cm^{-1}$ decreased as particle size increased; on the other hand, between 25 and 60 μm, PA intensity was approximately constant. The results were interpreted as an indication that analyte bands become visible in PA spectra only when particle sizes are similar to or less than the wavelength of the incident light—a model that does not seem to have been invoked in subsequent research.

Important exploratory work in PA infrared spectroscopy was carried out at Nicolet Instrument Corporation in the period between 1979 and 1981 by Mead, Vidrine, Lowry, and other researchers. Some of the earliest studies written by this group are listed in Table 2.2. A review by Vidrine that also dates from this period has already been mentioned in Chapter 1.

In a brief—but important—conference abstract Mead et al. (1979) compared the PA and transmission (KBr pellet) FTIR spectra of a Japanese coal. There are significant differences between the two spectra. For example, the authors noted that the PA spectrum clearly displayed bands from aliphatic and aromatic C—H groups, whereas the transmission spectrum was especially sensitive to hydroxyl oxygen content. In the latter case, the spectrum is affected by the interaction between the KBr diluent and the coal; this well-known phenomenon, which arises from the tendency for small quantities of water retained by KBr to form hydrogen bonds with oxygen-containing functional groups in coals, is completely obviated in PA infrared spectra of coal where neat (undiluted) samples are utilized. Hence this preliminary result serves to illustrate an important advantage of PA spectroscopy with regard to the characterization of coals. Other aspects of this analytical problem are discussed elsewhere in this book.

Mead et al. (1981) next described much more extensive PA FTIR studies of a variety of samples, including LiF, Ge, Si, and GaAs. Significantly, near-, mid-, and far-infrared data were presented in this work; both powders and crystals were studied. For the latter three (semiconducting) samples, the effects of dopants on the near-infrared band edges were considered. On the other hand, mid- and far-infrared PA spectra were recorded for LiF. A broad peak near 600 $cm^{-1}$ in the spectrum of LiF powder exhibited possible fine structure, which the authors attributed to surface vibrational modes and a change in the usual selection rules that govern the infrared spectrum of LiF. This, in turn, led to the suggestion that PA spectroscopy might be used to study surface vibrational modes, a point discussed above with regard to the work of Nordal and Kanstad (1977a,b).

Two additional articles written by Vidrine during this period are relevant to this discussion. In the first, Vidrine (1980) discussed PA FTIR spectroscopy of solids, with particular emphasis on the fact that traditional sample preparation is not required in this technique. To demonstrate the advantages that arise from the use of PA spectroscopy for the analysis of difficult samples, the author acquired PA FTIR spectra for a series of commercial materials, including plastic, polyurethane chips, phenoxy pellets, liquid lecithin, and a pharmaceutical tablet. The relative insensitivity of PA spectroscopy to sample morphology is demonstrated by the high quality of these spectra; a comparison of PA spectra of a nitrile plastic—examined as a powder, with sawn or smooth surfaces, and as pellets—affirmed this conclusion. PA spectra of coals were again investigated in this work. In fact, PA spectra of both low- and high-aromatic coals were shown to display better detail and signal-to-noise ratios than those observed in the diffuse reflectance spectra of the same coals.

This study (Vidrine, 1980) also discussed the use of a PA spectrum of carbon black as a reference for normalizing the PA FTIR spectrum of a sample measured under similar conditions, a procedure that continues to be widely used in PA infrared spectroscopy. Noise levels in PA spectra were characterized; it was observed that the root-mean-square (rms) noise in a PA spectrum is independent of acoustic frequency, and too small to be significant in most work. Coherent noise was also found to be relatively weak, except at very low wavenumbers where throughput is limited. The PA cell used in this work exhibited typical frequency response, with signal intensity diminishing as modulation frequency increased. Although cell resonance is not discussed in detail, the results in this work show that the PA cell had a resonant frequency of about 1.1 kHz.

In a study based on a lecture given at the 1981 International Conference on Fourier Transform Infrared Spectroscopy, Vidrine (1981) presented a general description of PA FTIR spectroscopy of solids, stressing the surface sensitivity of the technique. The phenomenology of the PA effect is explained in this work, and the relative weakness of the PA signal for solids—typically at least an order of magnitude smaller than that observed with an ordinary infrared detector—is stressed. The concept of depth of penetration (sampling depth) and its dependence on the thermal properties of the sample is discussed, although no values for this distance are given. The author stated that the PA and attenuated total reflectance (ATR) sampling methods are the only two surface-sensitive techniques for infrared spectroscopy, a generalization that would probably be disputed by contemporary workers who utilize other reflectance techniques.

The PA spectra illustrated by Vidrine (1981) in this article are noteworthy. For example, spectra of a catalyst were obtained at a series of different mirror velocities and provided one of the earliest examples of depth profiling in rapid-scan PA FTIR spectroscopy. This experiment is discussed again in Chapter 4. PA infrared spectra were also shown for conducting polymers, which can be difficult to analyze because of their high absorptivity in the infrared region. The PA spectroscopy of gases is also mentioned in this work, as is the use of He as a carrier gas in the gas-microphone PA cell. Thus Vidrine raised a number of significant issues in PA FTIR spectroscopy in this early work, many of which retain their importance in current work.

The Princeton University group of Royce, Teng, and colleagues was also among the first to utilize PA FTIR spectroscopy to study solids. In their first work, Laufer et al. (1980) described PA spectra of powdered AgCN. This material did not yield satisfactory transmission spectra when either pellets or undiluted powder were examined: The Christiansen effect caused the spectra to have sloping baselines, which depend on the refractive index of the matrix material. By contrast, PA spectra were obtained for a neat sample and were therefore free from this undesirable effect. The authors observed that the C≡N stretching band in the PA spectrum of AgCN is much wider than the corresponding Raman band; an increase in modulation frequency did not cause the width or shape of the PA band to change, leading to the conclusion that saturation was not the source of band broadening.

This work on PA infrared spectroscopy of solids was expanded in another investigation by these researchers (Royce et al., 1980). In addition to AgCN, polystyrene, acrylates, coal, and zeolites were included in this second work. Several features of FTIR spectrometers particularly relevant to PA infrared spectroscopy were noted; most important, of course, are the well-known multiplex (Fellgett) and throughput advantages. It was also pointed out that a PA interferogram qualitatively resembles an emission interferogram, rather than the more familiar transmission interferogram in which most of the information is confined to a narrow spatial region because of the large width of the blackbody spectrum. In other words, a PA interferogram contains more information in the "wings" farther from the center burst; the dynamic range of a typical PA interferogram is much smaller than that of an interferogram recorded in a transmission experiment. This implies that a single gain setting can be used in a PA FTIR measurement, in contrast with the situation in many other applications of FTIR spectroscopy.

## REFERENCES

Adams, M. J., Beadle, B. C., and Kirkbright, G. F. (1978). Optoacoustic spectrometry in the near-infrared region. *Anal. Chem.* **50**: 1371–1374.

Blank, R. E., and Wakefield, T. (1979). Double-beam photoacoustic spectrometer for use in the ultraviolet, visible, and near-infrared spectral regions. *Anal. Chem.* **51**: 50–54.

Busse, G., and Bullemer, B. (1978). Use of the opto-acoustic effect for rapid scan Fourier spectroscopy. *Infrared Phys.* **18**: 631–634.

Castleden, S. L., Kirkbright, G. F., and Menon, K. R. (1980). Determination of moisture in single-cell protein utilising photoacoustic spectroscopy in the near-infrared region. *Analyst* **105**: 1076–1081.

Kanstad, S. O., and Nordal, P.-E. (1977). Grazing incidence photoacoustic spectroscopy of species and complexes on metal surfaces. *Int. J. Quantum Chem.* **12**(Suppl. 2): 123–130.

Kanstad, S. O., and Nordal, P.-E. (1978). Open membrane spectrophone for photoacoustic spectroscopy. *Opt. Comm.* **26**: 367–371.

Kanstad, S. O., and Nordal, P.-E. (1979). Infrared photoacoustic spectroscopy of solids and liquids. *Infrared Phys.* **19**: 413–422.

Kanstad, S. O., and Nordal, P.-E. (1980a). Photoacoustic and photothermal spectroscopy. *Phys. Technol.* **11**: 142–147.

Kanstad, S. O., and Nordal, P.-E. (1980b). Photoacoustic reflection-absorption spectroscopy (PARAS) for infrared analysis of surface species. *Appl. Surf. Sci.* **5**: 286–295.

Kerr, E. L., and Atwood, J. G. (1968). The laser illuminated absorptivity spectrophone: A method for measurement of weak absorptivity in gases at laser wavelengths. *Appl. Opt.* **7**: 915–921.

Kirkbright, G. F. (1978). Analytical optoacoustic spectrometry. *Optica Pura y Aplicada* **11**: 125–136.

Kreuzer, L. B. (1971). Ultralow gas concentration infrared absorption spectroscopy. *J. Appl. Phys.* **42**: 2934–2943.

Laufer, G., Huneke, J. T., Royce, B. S. H., and Teng, Y. C. (1980). Elimination of dispersion-induced distortion in infrared absorption spectra by use of photoacoustic spectroscopy. *Appl. Phys. Lett.* **37**: 517–519.

Lochmüller, C. H., and Wilder, D. R. (1980a). Qualitative examination of chemically-modified silica by near-infrared photoacoustic spectroscopy. *Anal. Chim. Acta* **116**: 19–24.

Lochmüller, C. H., and Wilder, D. R. (1980b). Quantitative photoacoustic spectroscopy of chemically-modified silica surfaces. *Anal. Chim. Acta* **118**: 101–108.

Low, M. J. D., and Parodi, G. A. (1978). Infrared photoacoustic spectra of surface species in the 4000–2000 $cm^{-1}$ region using a broad band source. *Spectrosc. Lett.* **11**: 581–588.

Low, M. J. D., and Parodi, G. A. (1980a). Infrared photoacoustic spectra of solids. *Spectrosc. Lett.* **13**: 151–158.

Low, M. J. D., and Parodi, G. A. (1980b). Carbon as reference for normalizing infrared photoacoustic spectra. *Spectrosc. Lett.* **13**: 663–669.

Low, M. J. D., and Parodi, G. A. (1980c). Infrared photoacoustic spectroscopy of surfaces. *J. Mol. Struct.* **61**: 119–124.

Low, M. J. D., and Parodi, G. A. (1980d). Infrared photoacoustic spectroscopy of solids and surface species. *Appl. Spectrosc.* **34**: 76–80.

Low, M. J. D., and Parodi, G. A. (1980e). An infrared photoacoustic spectrometer. *Infrared Phys.* **20**: 333–340.

Max, E., and Rosengren, L.-G. (1974). Characteristics of a resonant opto-acoustic gas concentration detector. *Opt. Comm.* **11**: 422–426.

Mead, D. G., Lowry, S. R., Vidrine, D. W., and Mattson, D. R. (1979). Infrared spectroscopy using a photoacoustic cell. Fourth International Conference on Infrared and Millimeter Waves and Their Applications, p. 231. Miami Beach, FL, Dec. 10–15, 1979.

Mead, D. G., Lowry, S. R., and Anderson, C. R. (1981). Photoacoustic infrared spectroscopy of some solids. *Int. J. Infrared Millimeter Waves* **2**: 23–34.

Monchalin, J.-P., Gagné, J.-M., Parpal, J.-L., and Bertrand, L. (1979). Photoacoustic spectroscopy of chrysotile asbestos using a cw HF laser. *Appl. Phys. Lett.* **35**: 360–363.

Nordal, P.-E., and Kanstad, S. O. (1977a). Photoacoustic spectroscopy on ammonium sulphate and glucose powders and their aqueous solutions using a $CO_2$ laser. *Opt. Comm.* **22**: 185–189.

Nordal, P.-E., and Kanstad, S. O. (1977b). Infrared photoacoustic spectroscopy for studying surfaces and surface-related effects. *Int. J. Quantum Chem.* **12**(Suppl. 2): 115–121.

Nordal, P.-E., and Kanstad, S. O. (1978). Photoacoustic reflection-absorption spectroscopy (PARAS) of thin oxide films on aluminium. *Opt. Comm.* **24**: 95–99.

Patel, C. K. N., Kerl, R. J., and Burkhardt, E. G. (1977). Excited-state spectroscopy of molecules using opto-acoustic detection. *Phys. Rev. Lett.* **38**: 1204–1207.

Perlmutter, P., Shtrikman, S., and Slatkine, M. (1979). Optoacoustic detection of ethylene in the presence of interfering gases. *Appl. Opt.* **18**: 2267–2274.

Rockley, M. G. (1979). Fourier-transformed infrared photoacoustic spectroscopy of polystyrene film. *Chem. Phys. Lett.* **68**: 455–456.

Rockley, M. G. (1980a). Reasons for the distortion of the Fourier-transformed infrared photoacoustic spectroscopy of ammonium sulfate powder. *Chem. Phys. Lett.* **75**: 370–372.

Rockley, M. G. (1980b). Fourier-transformed infrared photoacoustic spectroscopy of solids. *Appl. Spectrosc.* **34**: 405–406.

Rockley, M. G., and Devlin, J. P. (1980). Photoacoustic infrared spectra (IR-PAS) of aged and fresh-cleaved coal surfaces. *Appl. Spectrosc.* **34**: 407–408.

Rockley, M. G., Richardson, H. H., and Davis, D. M. (1980). Fourier-transformed infrared photoacoustic spectroscopy, the technique and its applications. *Ultrasonics Symp. Proceed.* **2**: 649–651.

Rosengren, L.-G. (1973). A new theoretical model of the opto-acoustic gas concentration detector. *Infrared Phys.* **13**: 109–121.

Royce, B. S. H., Teng, Y. C., and Enns, J. (1980). Fourier transform infrared photoacoustic spectroscopy of solids. *Ultrasonics Symp. Proceed.* **2**: 652–657.

Vidrine, D. W. (1980). Photoacoustic Fourier transform infrared spectroscopy of solid samples. *Appl. Spectrosc.* **34**: 314–319.

Vidrine, D. W. (1981). Photoacoustic Fourier transform infrared (FTIR) spectroscopy of solids. *SPIE* **289**: 355–360.

## RECOMMENDED READING

Avramides, E., and Hunter, T. F. (1979). Analysis of energy transfer for thermally available vibrational energy levels as measured by photoacoustic and infrared emission techniques. *J. Chem. Soc., Faraday Trans. 2* **75**: 515–527.

Patel, C. K. N. (1978). Use of vibrational energy transfer for excited-state opto-acoustic spectroscopy of molecules. *Phys. Rev. Lett.* **40**: 535–538.

Roark, J. C., Palmer, R. A., and Hutchison, J. S. (1978). Quantitative absorption spectra via photoacoustic phase angle spectroscopy (ΦAS). *Chem. Phys. Lett.* **60**: 112–116.

CHAPTER

3

# EXPERIMENTAL METHODS

## 3.1. PA INFRARED SPECTROSCOPY WITH DISPERSIVE SPECTROMETERS

The discussion of multiple-wavelength PA infrared spectroscopy in Chapter 2 mentioned a number of early investigations that were carried out with dispersive spectrometers. This work, which can be regarded as the logical progression of previous research in ultraviolet and visible PA spectroscopy, began in the near-infrared region and eventually expanded to include mid-infrared wavelengths as well. Dispersive PA mid-infrared spectroscopy, in turn, was eventually displaced by multiplex (FTIR) PA spectroscopy. The use of dispersive spectrometers for the acquisition of PA infrared spectra is now therefore primarily of historical interest. This topic is briefly reviewed in this section, both for completeness and to partially establish the context for the discussion of other experimental PA methods in later sections of this chapter.

### 3.1.1. Near-Infrared PA Spectroscopy

The rapid growth in interest in PA and photothermal techniques in the 1970s prompted a number of workers to initiate research programs that involved dispersive near-infrared PA spectroscopy. Representative publications from four different groups are listed among the references at the end of this discussion. The reader should appreciate that this list is not meant to be comprehensive: The articles were chosen to briefly illustrate a variety of different applications of the technique and to indicate some of the most well-known research groups that were active in this area.

G. F. Kirkbright and his colleagues published several articles describing their work in dispersive PA near-infrared spectroscopy before eventually turning their efforts to multiplex spectroscopy at longer wavelengths. For example, Kirkbright (1978) presented a very useful general description of dispersive PA spectroscopy, with examples that were drawn from a variety of different fields. This work is mentioned in several places in this book. Some of the results that it describes were taken from relevant articles by this group, which are discussed in detail in Chapter 6. Related subsequent publications on dispersive PA near-infrared spectroscopy by these researchers discussed aromatic hydrocarbons (Adams et al., 1978), single-cell protein (Castleden et al., 1980), and pharmaceuticals (Castleden et al., 1982). These investigators made a number of important early contributions to PA infrared spectroscopy, although they generally did not specialize in a particular analytical application.

Several PA near-infrared spectrometers were commercially available during this period. Blank and Wakefield (1979) described the design and construction of a dual-beam spectrometer at Gilford Instrument Laboratories. Both solid and liquid samples were successfully analyzed with this apparatus. Similarly, Lochmüller and Wilder (1980a,b) utilized a Princeton Applied Research spectrometer to study the bonding of hydrocarbons to silica gel. The results of their work are described more fully in Chapters 2 and 6. In addition, a number of other groups successfully utilized a dispersive PA spectrometer that was manufactured by EDT Research.

Finally, the research of P. S. Belton and his collaborators on the PA near- and mid-infrared spectra of foods should be specifically mentioned. This group incorporated both dispersive and FT spectrometers in its research. Examples of the group's findings in dispersive near-infrared PA spectroscopy are given in works by Belton and Tanner (1983), as well as Belton et al. (1987). Belton's group made a number of important contributions with regard to the PA infrared spectroscopy of foods and to quantitative analysis of solid mixtures; their work in these areas is summarized in Chapters 6 and 7, respectively.

### 3.1.2. Mid-Infrared PA Spectroscopy

As mentioned above, dispersive mid-infrared PA spectroscopy was utilized for a limited period before the evolution and general acceptance of PA FTIR spectroscopy. The many well-known advantages of FTIR spectrometers made this progression inevitable. Nevertheless, viable research was conducted in dispersive mid-infrared PA spectroscopy for a number of years. Some relevant aspects of this work are mentioned in the following paragraphs.

The work of M. J. D. Low and his group is discussed in several places in this book. These researchers published an extensive literature on the PA infrared spectra of solids and surface species, one of their major contributions being a long series of studies of carbons and chars. In fact, many of the early studies of this group were performed using dispersive spectrometers. Some representative publications are listed among the references at the end of this chapter. It should be noted that studies carried out by this group

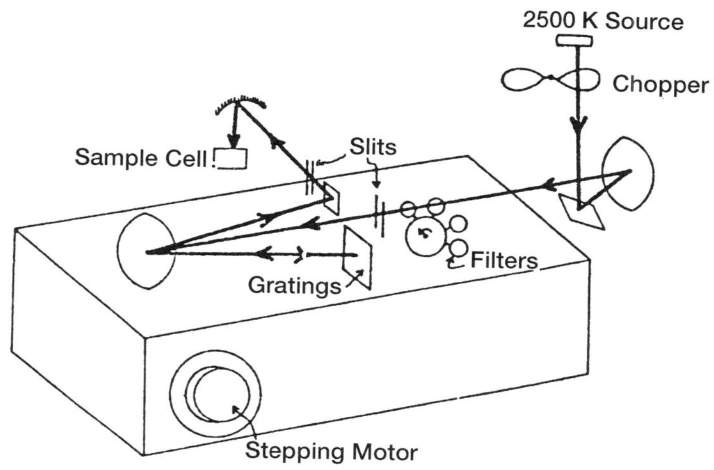


**Figure 3.1.** Optical system for dispersive mid-infrared PA spectroscopy. (Reprinted from *Infrared Phys.* **20**, Low, M. J. D. and Parodi, G. A., an infrared photoacoustic spectrometer, 333–340, copyright © 1980, with permission from Elsevier Science.)

using photothermal beam deflection spectroscopy are discussed separately later in this chapter.

Several relevant publications by this group were already mentioned in the historical review in Chapter 2. Among these, an important series of studies by Low and Parodi (1978, 1980a–e) describe the dispersive PA mid-infrared spectra of various solids. A drawing of the apparatus used in some of these experiments is shown in Figure 3.1 (Low and Parodi, 1980c). In another configuration of this equipment, LiF and $CaF_2$ prisms were used instead of a grating to separate the various wavelengths of infrared radiation (Low and Parodi, 1978, 1980a,d). A total of three dispersive infrared spectrometers were successfully modified for PA spectroscopy in this project (Low and Parodi, 1982). A wide variety of surface species, including adsorbents, catalysts, coals, carbons, and corrosion products, were studied in this way. However, despite the success of these pioneering studies, this group increasingly relied on FTIR spectrometers in its later PA work.

It appears that very little, if any, research in PA mid-infrared spectroscopy has been performed with dispersive spectrometers in recent years. Although dispersive mid-infrared spectrometers are still manufactured today, there are many features of both rapid- and step-scan FTIR spectrometers that make these instruments the logical choice for the present-day PA spectroscopist. Hence the vast majority of contemporary PA infrared spectra of solids and liquids are obtained using FTIR spectrometers. This situation is reflected in the published scientific literature that is discussed throughout the rest of this book.

### 3.2. RAPID-SCAN PA FTIR SPECTROSCOPY

The initial development of rapid-scan PA FTIR spectroscopy was outlined in Chapter 2. At the time of this writing, this technique is about 25 years old and has been the subject of more than 500 publications (see Appendix 2). Much of this literature is discussed in the rest of this book, particularly in Chapter 6. Accordingly, the present section is restricted to the mention of a few general points with regard to rapid-scan PA spectroscopy.

The major difference between rapid-scan PA infrared spectroscopy and the dispersive method described in the preceding section exists in the fact that no external modulation is required in the rapid-scan experiment. In simple terms, the moving mirror in the interferometer provides the audio-frequency modulation necessary for the generation of a PA signal. Gas-microphone PA cells that are compatible with rapid-scan FTIR spectrometers have been readily available for more than two decades; the acquisition of a PA spectrum with one of these instruments is a straightforward task in many circumstances. Therefore the nonspecialist can approach rapid-scan PA FTIR spectroscopy in a manner not unlike that which is appropriate for a variety of other infrared measurement techniques (e.g., diffuse reflectance, attenuated total reflectance (ATR), transmission). Qualitative—and sometimes quantitative—PA infrared spectra of a wide variety of solids and liquids are readily obtainable in this way.

The analyst will eventually be faced with another set of less obvious issues. Some are specific to the PA gas-microphone cell: These include effects due to limited sample quantities, the sizes of the sample cups, and even their method of filling. A related question, which has been discussed by Carter and Wright (1991), pertains to the position of the sample within the PA cell. As shown in Figure 3.2, these authors found that the PA signal was maximal when the distance between the sample surface and the cell window was smaller than the boundary layer (the thermal diffusion length of the carrier gas). They also concluded that powdered carbon black was the most suitable reference material for normalizing PA infrared spectra of finely divided solids, whereas carbon-filled rubber was more appropriate for nonporous materials (Carter and Paputa Peck, 1989). The normalization of PA infrared spectra is discussed in more detail in Chapter 5.

The dependence of the PA signal on sample position has recently been studied much more thoroughly (Jones and McClelland, 2001). The magnitude of the signal from the reference (glassy carbon), its dependence on modulation frequency, and its phase all vary with the separation between the

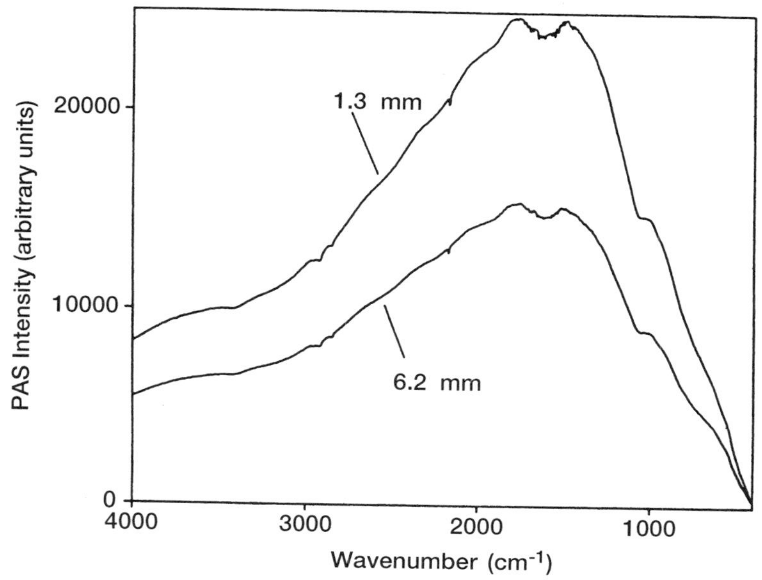


**Figure 3.2.** PA infrared spectra of carbon-black-filled rubber at two distances between the sample and the window of PA cell. Mirror velocity was 0.08 cm/s, corresponding to a modulation frequency range of 64–640 Hz in the mid-infrared. (Reproduced from Carter, R. O. and Wright, S. L., *Appl. Spectrosc.* **45**: 1101–1103, by permission of the Society for Applied Spectroscopy; copyright © 1991.)

reference material and the cell window. The values of the magnitude and phase at lower modulation frequencies (lower wavenumbers in rapid-scan experiments) differ significantly from those observed at higher frequencies. These observations were put down to the fact that the walls of the cell act as a heat sink in these experiments. The cell window and the microphone also contribute to the observed nonideal behavior. For accurate work, the sample position must be kept constant so as to reduce the number of variables that affect PA intensities.

When modern FTIR spectrometers became widely available in the 1970s, many infrared spectroscopists were pleasantly surprised at the very high signal-to-noise ratios that these instruments afforded under many circumstances. The excellent performance of these spectrometers led to greatly reduced acquisition times and also facilitated the implementation of numerical techniques such as deconvolution, curve-fitting, and chemometric methods, all of which are adversely affected by noise in the spectra of interest.

Despite the high performance of these FTIR instruments, many early rapid-scan PA FTIR spectra were corrupted by noise. In fact, typical signals from early PA cells were weaker than those obtained using optical detectors by about one to two orders of magnitude; for some samples, PA signals were even weaker. Random noise in these spectra could be eliminated by signal averaging, although at the expense of greatly increased measurement times. The reduction of environmental noise, for example, by interrupting the flow of the purge gas during data acquisition, was found to reduce the noise in rapid-scan PA spectra (Donini and Michaelian, 1988). Moreover, considerable success was achieved by isolating the PA cell in an insulated chamber (Duerst and Mahmoodi, 1984; Mahmoodi et al., 1984; Duerst et al., 1987). It should be noted that the environmental isolation required for older PA cells is usually not necessary with modern accessories.

PA signal generation in a rapid-scan FTIR experiment has been discussed in detail by McClelland, Jones, and their collaborators in several of the reviews cited in Chapter 1. The interested reader is encouraged to study these publications in order to gain greater familiarity with this subject. The phenomenology of the experiment has been outlined by McClelland et al. (1992) and can be briefly summarized as follows. Modulated infrared radiation emerges from the interferometer and impinges on the PA gas-microphone cell containing the sample of interest. Part of this radiation is absorbed at wavelengths that are dictated by the optical properties of the sample. Absorption induces temperature oscillations in a layer at a depth $x$ and a thickness $dx$ that are proportional to $P_o(1 - R)\beta e^{-\beta x}$ where $P_o$ is the incident power, $R$ is the fraction of incident light reflected by the surface of the sample, and $\beta$ is the absorption coefficient. The temperature oscillations generate thermal waves that decay according to $\mu_s = (\alpha/\pi f)^{1/2}$, where $\mu_s$ is the thermal diffusion length and $\alpha$ is the thermal diffusivity of the sample (see Appendix 1). A small fraction of each thermal wave is transferred to the layer of carrier gas immediately adjacent to the sample. The resulting pressure wave is detected by the microphone (transducer) in the PA cell. The microphone produces a voltage that is amplified and then input to the FTIR electronics for further processing.

## 3.3. STEP-SCAN PA FTIR SPECTROSCOPY

The many applications of step-scan PA infrared spectroscopy described in Chapters 4, 6, and elsewhere in this book attest to the considerable importance of this method. The reader will soon recognize that the predominant use of step-scan spectroscopy documented in these pages has been with regard to depth profiling experiments. Moreover, the step-scan technique also allows the PA spectroscopist to establish a constant modulation frequency throughout the entire infrared region within a particular spectrum.

As is discussed in Chapter 4, initial step-scan PA depth profiling results were obtained using amplitude modulation in the late 1980s. Phase modulation was successfully implemented in PA step-scan measurements at about the same time and has retained its popularity with PA spectroscopists during the intervening years. Consequently, the published literature on step-scan PA FTIR spectroscopy is too extensive to permit its discussion as a separate topic; instead, several references of a more general nature are mentioned below.

It is relevant to mention that the capability for rapid-scan movement of the mobile interferometer mirror was a key feature of the new generation of FTIR spectrometers that became readily available in the 1970s. In fact, many older interferometers utilized step-scan operation: The detector signal was digitized and recorded for later processing at each resting position of the movable mirror. This method of data acquisition was reliable but relatively slow; hence, most infrared spectroscopists welcomed the availability of rapid, continuous scanning. However, researchers interested in PA detection soon recognized that this new approach presented its own set of problems because each spectral ordinate is associated with a different modulation frequency. It is well known that the characterization of saturation effects and depth profiling experiments are both complicated by the existence of this range of frequencies. The subsequent reintroduction of the step-scan spectrometer obviated these problems and was therefore eagerly accepted by many PA infrared spectroscopists.

The Bruker IFS 88 FTIR spectrometer was introduced as the first modern commercially available spectrometer with both rapid- and step-scan capabilities in 1987. During the next few years, other instrument manufacturers developed similar instrumentation, and today's PA spectroscopist can choose from several options. Signal demodulation has also undergone major advances; whereas a lock-in amplifier was originally required for signal recovery in a step-scan PA experiment, modern electronics have made it possible to demodulate signals entirely in software. The simultaneous detection of PA signals at different frequencies or phase angles further facilitates depth profiling measurements.

The research group led by R. A. Palmer of Duke University modified several conventional rapid-scan FTIR spectrometers to enable their alternative use in step-scan operation. Smith et al. (1988) described the modification of an IBM Instruments IR/44 to permit step-scan operation. The position of the movable mirror was controlled by a feedback loop that monitored the interference pattern of the HeNe laser. Each step size was one-quarter of the HeNe wavelength, that is, about 0.158 μm; this increased the free spectral range to include the near-infrared and visible regions in addition to the far- and mid-infrared. Both amplitude modulation and phase modulation PA experiments were carried out with this instrument. The authors noted that phase modulation measures only the alternating current (ac) component of the interferogram, whereas amplitude modulation also detects the large direct current (dc) background. The latter point is discussed further in Chapter 4.

This instrument was described in greater detail in a subsequent publication (Manning et al., 1991). Particular emphasis was placed on the instrumentation and the mirror position control circuit in this second article. Step-scan PA spectra were obtained for carbon-black-filled elastomers and adhesive tape using phase modulation. The potential application of the step-scan technique to the study of various time-dependent phenomena was also mentioned in this work.

This group next modified a Nicolet 800 FTIR spectrometer to accommodate step-scan operation (Gregoriou et al., 1993). The performance of this second instrument was found to be generally superior to that of the original spectrometer described above; importantly, the amplitude of the phase modulation was potentially much greater in the second case. The authors went on to utilize this instrument for the measurement of PA spectra of polymer films, demonstrating the phase angle rotation method discussed in the next chapter.

Two relevant review articles were written by this group at about the same time that this second instrumental modification was completed. In the first, Palmer et al. (1993) summarized dynamic vibrational spectroscopy and several applications of step-scan FTIR spectrometers. The following advantages were attributed to step-scan PA spectroscopy: (a) near-infrared and visible wavelength results are improved because the high Fourier frequencies associated with these regions are replaced by much lower modulation frequencies; (b) the use of the phase spectrum extends the dynamic range, making it possible to detect weakly absorbing components within a strongly absorbing matrix; and (c) the depth profiling capability of step-scan PA spectroscopy is superior to that in rapid-scan PA spectroscopy. In fact, a depth resolution on the order of 1 μm is attainable in step-scan operation under favorable circumstances. The second review (Palmer, 1994) surveyed time- and phase-resolved techniques in vibrational spectroscopy. PA spectroscopy is mentioned briefly, together with a number of other infrared experiments. No new PA data are presented in this work.

## 3.4. PHOTOTHERMAL BEAM DEFLECTION SPECTROSCOPY

An alternative to the popular gas-microphone PA technique is discussed in this section. Photothermal beam deflection spectroscopy (PBDS)—which is based on the so-called mirage effect—has been used by a number of research groups to acquire near- or mid-infrared spectra of solids, liquids, and gases during the last two decades. In fact this detection scheme actually predates the development of modern PA FTIR spectroscopy, having originally been developed for studies in the visible and near-infrared wavelength regions. The adaptation of the method to the mid-infrared occurred in the early 1980s, at about the same time as various other experimental advances described throughout this book. Unlike the gas-microphone cell, which has been commercially available for many years, PBDS has generally been implemented with purpose-built instrumentation. This factor has undoubtedly tended to limit the number of infrared spectroscopists who utilize the technique.

The basis of PBD is the thermally induced refractive index gradient that exists near a surface as a consequence of the absorption of modulated light. In contrast with the gas-microphone method, however, the sample need not be confined to an enclosed cell in PBDS; when the sample is exposed to the environment, the heating of the gas does not produce a detectable acoustic wave. Instead, a beam of light from a probe laser is arranged to pass immediately above the sample, through the heated gas. Because the refractive index of this gas is modulated, the beam is periodically deflected from its original path. This phenomenon is reminiscent of the familiar mirage effect, after which it has been named by some authors.

The deflection of the beam can be monitored in various ways, a simple example being the use of a knife edge and a position detector. Many workers in this field have utilized bicells or various types of linear position detectors. The resulting signal can be demodulated with a lock-in amplifier and then analyzed in a manner similar to that for other PA spectra. The major advantage of this technique is that the sample does not have to be enclosed: This makes it possible to obtain spectra of relatively large or irregularly shaped objects.

The research of M. J. D. Low and his colleagues on the PA infrared spectra of carbons, catalysts, and other solids is discussed in several places in this book. This group used the PBD technique in many of its studies. A majority of the group's publications tended to emphasize the interpretation of the infrared spectra rather than the technique by which they were acquired; these articles are not discussed here. On the other hand, several studies that describe the PBD method in some detail were also published by these researchers (Low and Lacroix, 1982; Low et al., 1982a; DeBellis and Low, 1987, 1988). Some additional pertinent references are included in the list at the end of this chapter and mentioned in the next paragraph.

A simple drawing of an experimental arrangement for PBDS is reproduced in Figure 3.3 (Low and Tascon, 1985). As mentioned above, the

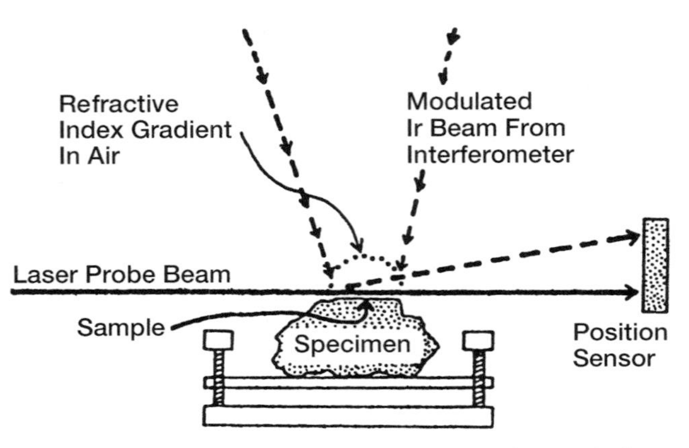


**Figure 3.3.** Schematic drawing of simple experimental layout for PBDS. (Reproduced from Low, M. J. D. and Tascon, J. M. D., *Phys. Chem. Minerals* **12**: 19–22, Fig. 1, by permission of Springer-Verlag; copyright © 1985.)

position sensor in this experiment can be used in conjunction with a knife edge, arranged to partially block the probe laser beam; a dual detector [measuring $(A+B)$ or $(A-B)/(A+B)$, where $A$ and $B$ are the intensities detected in the two channels] or a double-beam dual detector arrangement can also be employed (Low et al., 1982b). The effects of vibrational noise on this experiment can, of course, be quite deleterious. Low (1986) demonstrated that not all sources of noise in the spectra were random; extensive averaging therefore reduced, but did not eliminate, the noise in the spectra. Although irregularly shaped samples can be analyzed by this technique, powders were found to yield spectra superior to those for coarse solids (Low and Morterra, 1987). This result was put down to the increased surface area and scattering caused by the smaller particles. The geometry of the experiment is also better defined for a well-packed finely divided solid, or a film, than for a coarser sample.

Some of the most important early PBD experiments were carried out by Fournier, Boccara, and their collaborators. Two publications from this group will be mentioned here. In the first, Fournier et al. (1980) described the measurement of gas-phase spectra by means of the photothermal deflection technique. A $CO_2$ laser, with emission lines in the 9.4-μm (~1080-cm$^{-1}$) and 10.4-μm (~970-cm$^{-1}$) regions served as the infrared source in this experiment. The optical layout was such that the probe beam from a HeNe laser was nearly collinear with the $CO_2$ laser beam. The photothermal signal detected for ethylene varied linearly with concentration over four orders of magnitude. An impressive detection limit of 5 ppb was estimated in this work.

Another elegant experiment by this group demonstrated the high sensitivity of PBDS in two additional applications (Fournier et al., 1982). This technique was shown to be more sensitive than conventional PA spectroscopy by three orders of magnitude at near-infrared and visible wavelengths. Hence the authors were able to obtain absorption and circular dichroism data that were completely inaccessible by previously established techniques.

Several authors have also utilized the PBD technique in recent years. Bain et al. (1992) studied the C—H stretching region in PBD spectra of organic monolayers on both smooth and rough silicon surfaces. The authors described an enclosed cell that could be filled with perfluorodecalin or $CCl_4$ to increase the magnitude of the beam deflection. While this technique can increase the sensitivity of the method, it is suitable only for infrared-transparent liquids; this explains the authors' choices for the deflecting medium.

Near-infrared PBDS was used to characterize nonoxidized polythiophenes with different alkyl side chains by Einsiedel et al. (1998). These materials are of interest with regard to nonlinear optical waveguide applications. PBDS was used to detect several weak near-infrared absorption bands, even though the optical density varied by as much as five units. Finally, Bouzerar et al. (2001) obtained visible and near-infrared PBD spectra of amorphous hydrogenated carbon thin films (a-C:H). No well-defined absorption bands are visible in the spectra of these samples. Experimental details were not given.

Deng et al. (1996) designed an electrochemical cell that permitted the measurement of PBD spectra. The principle of the experiment is illustrated in Figure 3.4, which depicts a test cell used for setup and optimization. Infrared illumination was provided by tunable diode lasers that emit in the 8–16 μm (1250–660 $cm^{-1}$) region. The probe beam was arranged to pass through the acrylic block that supported the sample. Infrared absorption caused the deflection of the probe beam, which was monitored using a position detector. This system was used to obtain infrared spectra between 600 and 1000 $cm^{-1}$ for lithium surface films after exposure to air, and an electrolyte after both discharging and recharging. Although the spectral coverage afforded by the lasers was relatively narrow, the results demonstrated both the sensitivity and the versatility of this novel technique.

## 3.5. REVERSE MIRAGE PA SPECTROSCOPY

An arrangement of the PBDS experiment that incorporates a liquid deflection medium was described above. This concept is also utilized in a related technique referred to as reverse mirage PA spectroscopy. The motivation for

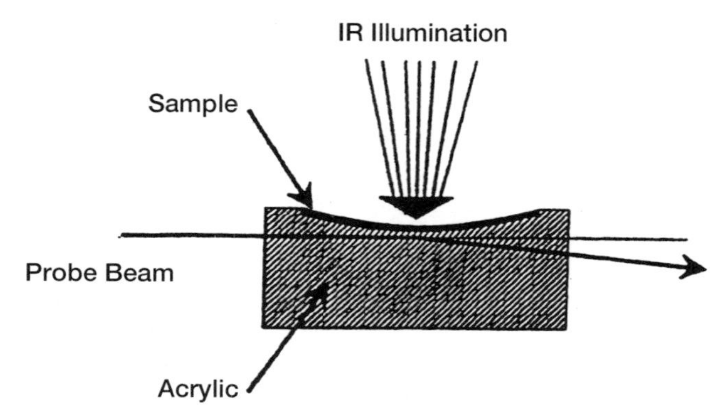


**Figure 3.4.** PBD test cell. Electrochemical cell was also constructed using this design. (Reproduced from Deng, Z. et al., *J. Electrochem. Soc.* **143**: 1514–1521, by permission of the Electrochemical Society, Inc.; copyright © 1996.)

this change is rather obvious. As mentioned with regard to PBDS, submersion of the sample in a liquid increases the magnitude of beam deflection relative to that obtained in air, thereby increasing the sensitivity of the method; however, the accessible spectral region is generally limited due to absorption of the infrared beam by the liquid. Because this beam does not traverse the liquid in the reverse mirage arrangement, this problem is obviated.

Figure 3.5 depicts the experimental setup. An infrared-transparent window, such as Si, Ge, or GaAs, is partially submerged in the liquid, which

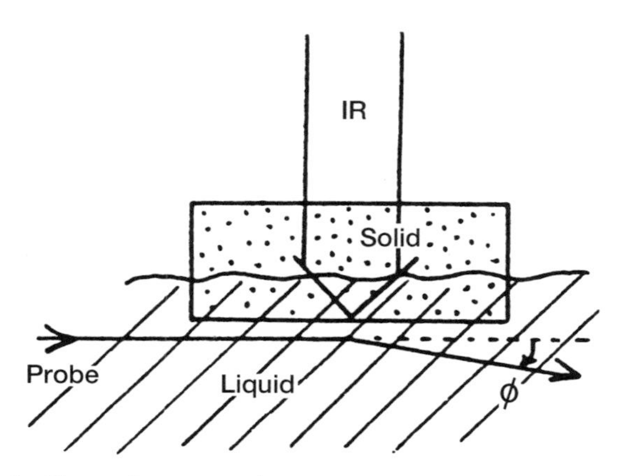


**Figure 3.5.** Sketch of layout for reverse mirage spectroscopy. Probe beam is deflected by angle $\phi$ due to absorption of infrared radiation by liquid. (Reproduced from Palmer, R. A. and Smith, M. J., *Can. J. Phys.* **64**: 1086–1092, by permission of NRC Research Press; copyright © 1986.)

acts as both sample and deflection medium. Modulated infrared radiation impinges on the window and is transmitted into the liquid. The transverse probe beam passes through the liquid immediately adjacent to the window. When the liquid absorbs the incident radiation, part of the energy is converted to heat by radiationless transitions. The resulting change in refractive index causes a deflection of the probe beam that can be monitored with a position-sensitive photodiode. The primary advantage of this reverse mirage arrangement is that the liquid need transmit only the probe beam (usually from a HeNe laser), not the infrared radiation. The method is suitable for studying absorption spectra of liquids and species at the solid–liquid interface.

The reverse mirage technique can be implemented in an FTIR spectrometer, or alternatively it can be employed to detect absorption of monochromatic infrared (laser) radiation. Pioneering FTIR experiments were described by Palmer and Smith (1986), who obtained a research-quality spectrum of acetonitrile using this method. These authors noted similarities between reverse mirage spectroscopy and the phenomenon of thermal lensing, the latter being the basis of the conventional mirage effect in air. Moreover, they also pointed out that reverse mirage spectroscopy is somewhat analogous to the well-known technique of ATR infrared spectroscopy, insofar as it yields a spectrum of a liquid or a solution at the solid–liquid interface. In fact, optical materials used as ATR crystals are also suitable as windows for reverse mirage spectroscopy.

Smith and Palmer (1987) continued their reverse mirage investigations of acetonitrile and observed a somewhat disconcerting result: The distance between the probe beam and the surface of the Si window had a dramatic effect on the spectra. In particular, they found that strong bands tended to be distorted in the reverse mirage spectra, whereas weaker features behaved in a more conventional manner. This distortion was greater when the probe beam was farther from the window. These results were rationalized as follows. Unlike the situation in the ordinary mirage effect, the thermal gradient in the reverse mirage effect does not arise from a localized source; instead, it is distributed within the absorbing medium. When absorption is strong, the optical penetration depth is shallow and the heat source is closer to the solid–liquid interface. On the other hand, the heat source is deeper within the sample for weaker bands. Thus the optimum location of the probe laser beam is not the same in both cases. Smith and Palmer (1987) showed that the thermal gradient is proportional to absorbance only within the first 15 μm of the liquid; at greater distances from the window, the thermal gradient passes through a maximum and tends toward zero with increasing absorptivity. Hence the probe beam must be arranged so that it literally grazes the window if strong bands are to be observed reliably by reverse mirage spectroscopy. Even when this is done, the proportionality between thermal gradient and absorbance extends only to absorption coefficients of about 1000 $cm^{-1}$.

A few years after these initial investigations, Palmer's research group published several additional studies dealing with reverse mirage spectroscopy. In a study of propylene carbonate, Manning et al. (1992a) again noted distortions of stronger absorption bands, even though the magnitude of the deflection signal varied monotonically at lower absorptivities. The distortion of the stronger bands, which is due to attenuation of the thermal gradient in the liquid before it reaches the probe beam, leads to the appearance of notched bands.

Because this investigation was carried out with a step-scan instrument, it was possible to calculate both phase and magnitude spectra. Manning et al. (1992a) observed that the phase spectrum was actually a close approximation to the "true" spectrum (obtained with a cylindrical ATR cell). In fact the reverse mirage phase signal saturates at higher absorptivity than the reverse mirage magnitude spectrum: This effectively extends the dynamic range of reverse mirage PA spectroscopy. The authors noted that it is the phase, rather than the in-phase and quadrature spectra used to calculate this quantity, that varies monotonically with absorptivity. However, caution must still be used when interpreting the phase data because calculation of the phase spectrum can be adversely affected when the phase angle changes from one quadrant to the next in a region of strong absorption.

Another, more theoretical, work by Manning et al. (1992b) presented a detailed mathematical model of the reverse mirage effect. This treatment explicitly included the thermal influence of the window and derived expressions for the thermal gradient at various distances within this influence (about three times its thermal diffusion length) and at greater distances where the effect of the window can be ignored. The Gaussian profile of the probe beam was incorporated through the use of numerical integration. One of the important results of this treatment is the prediction that the natural logarithm of the magnitude of the reverse mirage signal varies linearly with probe distance: The proportionality constant that relates these quantities is equal to the absorptivity. Thus it is possible to determine absolute absorptivity at infrared wavelengths using reverse mirage spectroscopy. It was estimated that absorptivity values as high as 3000 $cm^{-1}$ could be measured using the phase response obtained from step-scan reverse mirage data. Absorptivities were calculated for 11 different bands of acetonitrile and found to agree fairly well with those obtained by ATR spectroscopy (Manning et al., 1992c).

As mentioned above, reverse mirage spectroscopy can also be used to study absorption at discrete wavelengths. An example of this implementation of the technique occurs in the work of Bićanić et al. (1989), who studied reverse mirage spectra of liquid methanol at $CO_2$ laser wavelengths. According to these authors, the amplitude of the deflection signal is given by

$$S = \eta(L/n)(dn/d\theta)(d\theta/dx) \tag{3.1}$$

where $\eta$ is the transducer conversion factor (voltage per unit angle), $L$ is the interaction length, $n$ is the refractive index of the liquid, $\theta$ is the amplitude of the ac temperature rise, and $x$ is the distance between the probe beam and the window. The absorption coefficient is obtained from

$$\beta = (\Delta)^{-1} \ln[S_n(x)/S_n(x + \Delta)] \tag{3.2}$$

where $\Delta$ is the difference between two $x$ values and the subscript $n$ indicates that the signals are normalized. This relationship is expected to hold at probing distances up to approximately 20 μm for absorption coefficients as large as $10^5$ $m^{-1}$. These limiting values are similar to those reported by Smith and Palmer (1987) and discussed earlier in this section.

To demonstrate this method, Bićanić et al. (1989) carried out reverse mirage measurements on methanol using two strong $CO_2$ laser lines. For the 10P(20) and 10R(20) lines at 10.59 μm (944 $cm^{-1}$) and 10.26 μm (975 $cm^{-1}$), they obtained absorption coefficients of $0.122 \times 10^5$ $m^{-1}$ and $0.162 \times 10^5$ $m^{-1}$, respectively. These results are in good agreement with the expected values of $0.12 \times 10^5$ and $0.175 \times 10^5$ $m^{-1}$.

## 3.6. PIEZOELECTRIC DETECTION

The absorption of modulated infrared radiation can be detected photoacoustically in several ways. The most common, of course, is by means of the sound waves in the carrier gas above the sample, usually in a gas-microphone cell; as an alternative, it is possible to detect the elastic wave induced in a piezoelectric substrate in intimate contact with the sample. The reader is well aware of the fact that the majority of the applications discussed in this book are based on the former technique. On the other hand, it is important to recognize that piezoelectric detection of PA infrared spectra is also viable—indeed, piezoelectric PA infrared spectroscopy has been known for approximately the same length of time as its gas-microphone counterpart. One could speculate that the current rather limited use of piezoelectric detection is at least partially due to the fact that it was never developed and commercialized in the same manner as the gas-microphone cell, various models of which have been widely available for more than two decades. Piezoelectric detection in PA infrared spectroscopy is briefly summarized in this section.

An important series of papers by Kanstad and Nordal was discussed in Chapter 2. These authors used $CO_2$ laser radiation as an infrared source in their PA experiments, together with several means of detection. Some of their publications deal with piezoelectric PA infrared spectroscopy and are briefly summarized in the following paragraphs.

A scheme for the measurement of both piezoelectric and gas-microphone PA spectra that was utilized by Kanstad and Nordal (1980a,b) is depicted in Figure 3.6. It can be seen that the piezoelectric measurement is particularly simple: The piezoelectric transducer (PZT, also referred to as a piezotransducer) is affixed to the rear surface of the sample, with no need for a sample chamber or other similar enclosure. Absorption of the $CO_2$ infrared radiation by the sample leads to the production of a small voltage by the

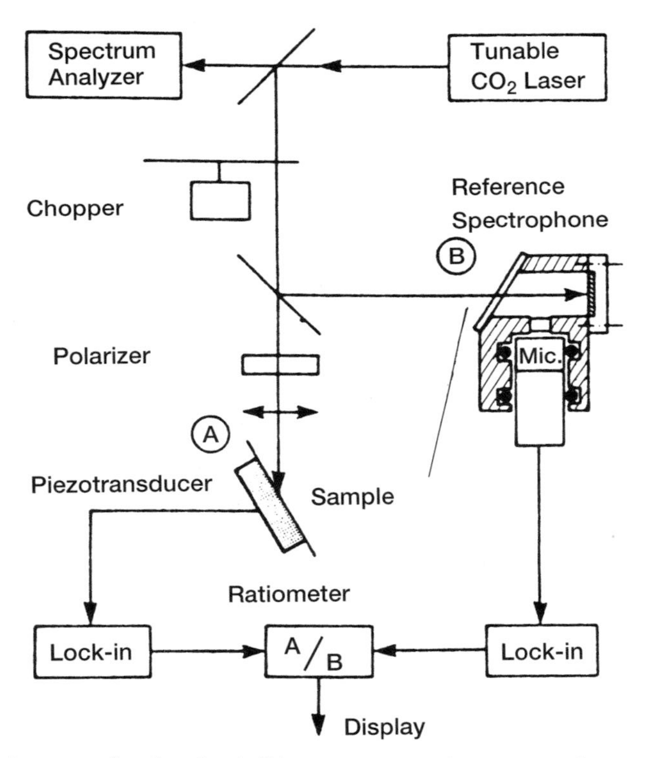


**Figure 3.6.** Apparatus for piezoelectric PA spectroscopy, using a spectrophone as a reference for source compensation. (Reprinted from *Appl. Surf. Sci.* **5**, Kanstad, S. O. and Nordal, P.-E., Photoacoustic reflection-absorption spectroscopy (PARAS) for infrared analysis of surface species, 286–295, copyright © 1980, with permission from Elsevier Science.)

PZT, which can be detected using a lock-in amplifier; in the configuration shown in Figure 3.6, the PZT signal is ratioed against that from a spectrophone or a gas-microphone cell. Kanstad and Nordal (1980b) obtained a PA spectrum of an aluminum oxide layer on a 0.04-mm-thick Al foil in this way and observed the prominent longitudinal optic mode of the Al—O stretching vibration at 10.49 μm (953 $cm^{-1}$). An uncompensated single-beam experiment would have involved the measurement of the PZT signal alone.

The piezoelectric PA technique described in the previous paragraph was subsequently used to obtain spectra of skin lipids by Kanstad et al. (1981). Source compensation was achieved with the use of a blackened reference transducer in this investigation. This work is discussed in more detail in the section on medical applications of PA infrared spectroscopy in Chapter 6.

One of the first demonstrations of piezoelectric PA infrared spectroscopy that utilized an FTIR spectrometer was carried out by Lloyd et al. (1982). These authors obtained both gas-microphone and piezoelectric PA spectra of samples such as tetraphenylcyclopentadienone on thin-layer chromatography (TLC) plates. The objective of this work was to develop methods for acquiring spectra of separated components without removing them from the plates. Neat pellets of the sample were also examined in this work. In the piezoelectric PA FTIR experiments, the PZT signals were preamplified and then detected with a lock-in amplifier, whose output was directed to the FTIR electronics.

Several noteworthy results were obtained in this investigation. For example, a pellet of the ketone that was attached to the PZT yielded a transmissionlike spectrum, rather than the expected (absorptive) spectrum. This result was put down to the fact that the PZT was not coupled rigidly to the sample, so that absorption of the incident light did not cause deformation of the PZT; instead, radiation transmitted by the sample could have reached the detector, essentially producing a piezoelectrically detected transmission spectrum. By contrast, coupling between an alumina TLC plate and the PZT was more efficient, and the expected absorption spectrum was obtained. Another, more general, observation from this study was that the elimination of a carrier gas in piezoelectric PA spectroscopy offered an important simplification with respect to the familiar gas-microphone technique.

Piezoelectric detection of PA spectra can, of course, be employed at wavelengths much shorter than the mid-infrared. Moreover, there may be no need to employ an FT spectrometer in these experiments. This was the case in the work of Manzanares et al. (1993), who studied both fundamental and overtone spectra with regard to the C—H stretching vibrations of *cis*- and *trans*-3-hexene. For the fourth ($\Delta v = 5$) and fifth ($\Delta v = 6$) overtones, a dye laser, acousto-optic (AO) modulator, and a piezoelectric detector were used. The AO modulator interrupted the $Ar^+$ laser used to pump the dye laser at a frequency of about 80 kHz, which is about two orders of magnitude higher than the typical modulation frequencies that are used in PA FTIR spectroscopy. The wavelength of the incident light was scanned by means of a birefringent filter in the dye laser in this experiment, while the signal from the PZT was detected with a lock-in amplifier. This study is primarily concerned with the curve fitting of the spectra and the assignment of the recovered bands. Local mode frequencies and anharmonicities were also calculated. In other words, the use of PA detection was somewhat incidental to this work.

All of the publications described so far in this section refer to experiments in which a lock-in amplifier was used to recover the signal from a piezoelectric transducer. In rapid-scan PA FTIR spectroscopy modulation is, of course, achieved by means of the moving mirror in the interferometer. The analyst may anticipate that this modulation—albeit at frequencies much lower than those normally employed with PZT detectors—could be utilized directly in a piezoelectric PA FTIR experiment. It turns out that rather weak PA infrared spectra are, indeed, obtainable in this way.

Zhang et al. (1997) recorded piezoelectric PA infrared spectra of mica, as well as stearic acid adsorbed on mica. The PZT detector was in direct contact with the rear surface of the mica sheets in these experiments, while the infrared radiation impinged on the front surface. The signal from the PZT was input directly to the electronics of the FTIR spectrometer. The relatively small voltage generated by the PZT yielded weak but stable interferograms. However, at the low mirror velocities normally used with gas-microphone cells, these interferograms lacked center bursts at or near the zero-path-difference point: This caused the standard FTIR software to misidentify this location and calculate unrealistic spectra. Correct spectra were obtained only when the interferogram phase was constrained to the first and fourth quadrants (Fig. 3.7). It is interesting to note that these piezoelectric PA spectra qualitatively resemble conventional transmission spectra, a result reminiscent of that observed by Lloyd et al. (1982) and discussed above. It should also be pointed out that the phase correction problem encountered with these PA data is somewhat similar to that which occurs in some types of differential spectroscopy, where both positive and negative bands can appear in the spectra. Improper phase correction wrongly forces the negative-going bands in these spectra to become positive features.

Thin films are particularly amenable to the measurement of piezoelectric PA spectra because it is possible to ensure good contact between the sample and the detector. Some applications of this technique may be of greater interest to materials scientists and physicists than to chemists: This is the case in the examples from the recent literature that are reviewed in the following paragraphs.

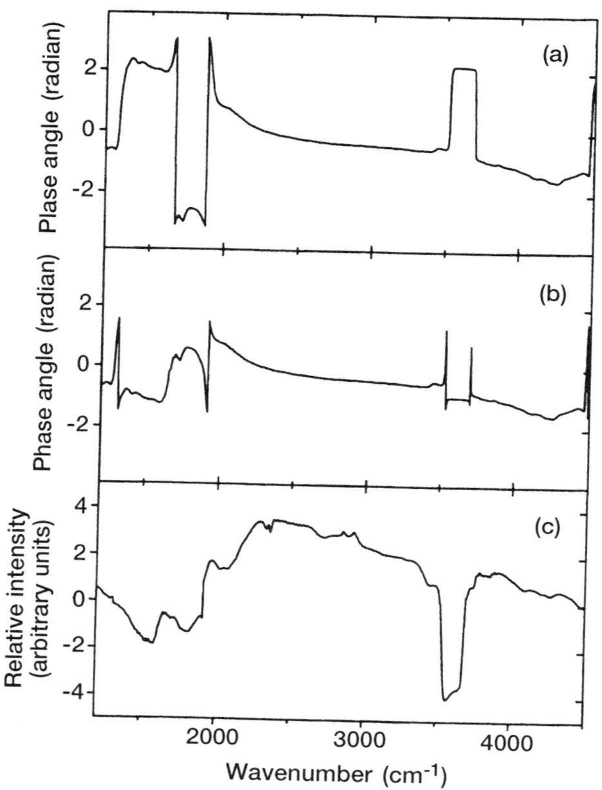


**Figure 3.7.** Effect of phase calculation on piezoelectric PA infrared spectrum of mica. (*a*) Phase calculated in all four quadrants; (*b*) phase calculated in first and fourth quadrants; and (*c*) PA spectrum corresponding to the phase in (*b*). (Reproduced from Zhang, S. L. et al., Phase correction in piezoelectric photoacoustic Fourier transform infrared spectroscopy of mica, *Opt. Eng.* **36**: 321–325, by permission of the International Society for Optical Engineering; copyright © 1997.)

Yoshino et al. (1999) reported piezoelectric PA spectra of the I–III–$VI_2$ chalcopyrite semiconductor $CuInSe_2$, grown on an oriented GaAs substrate by molecular beam epitaxy. PA spectra were obtained in the near-infrared region (0.6–1.8 eV, or 4840–14,520 $cm^{-1}$) at both liquid nitrogen and room temperatures using a disk-shaped PZT attached to the rear surface of the sample. The objective of this work was to investigate defect levels and the bandgap energy of $CuInSe_2$, which is of interest for possible solar cell applications. Spectra were obtained before and after quenching by illumination with a photon energy greater than 1.1 eV (8870 $cm^{-1}$). As shown in

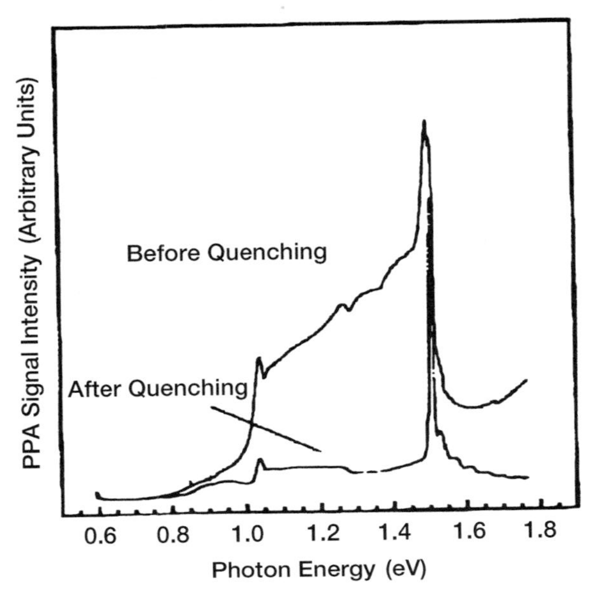


**Figure 3.8.** Piezoelectric PA spectra of $CuInSe_2$ film on GaAs substrate, before (upper curve) and after (lower curve) quenching at 77 K by irradiation at 1.1 eV. (Reprinted from *Thin Solid Films* **343–344**, Piezoelectric photoacoustic spectra of $CuInSe_2$ thin film grown by molecular beam epitaxy, 591–593, copyright © 1999 with permission from Elsevier Science.)

Figure 3.8, the spectra change dramatically as a consequence of quenching. The positions of the bands give the activation energies of the lattice defects.

In related work, Fukuyama et al. (2000) investigated near-infrared piezoelectric PA spectra of GaAs. This detection method was chosen because it directly monitors nonradiative relaxation and is sensitive to very weak absorption in highly transparent samples. The PA spectra, which are rather featureless except for two peaks superimposed on a continuum that extends between about 0.8 and 1.5 eV (6450 and 12,100 $cm^{-1}$), changed as the sample was quenched by 1.12-eV (9035-$cm^{-1}$) illumination at 80 K. This suggested the contribution of an acceptorlike deep level that changed the activation energy during the photoquenching process.

Piezoelectric PA spectra of *n*-type phosphorus-doped and *p*-type boron-doped silicon were measured near the energy gap by Kuwahata et al. (2000). The effect of carrier concentration was investigated for these spectra, which extended from visible to near-infrared wavelengths and exhibited two very broad bands in the 500–1200 nm region. PA intensities at energies higher than the energy gap decreased for the *n*-type samples and increased for the *p*-type samples, due to the increase in free electrons at the bottom of the conduction band and the increase in holes at the top of the valence band, respectively. Variation of the carrier concentration suggested that free electrons suppress thermoelastic transfer. Heat generated by light absorption is diffused by these electrons, causing suppression of the elastic wave and a decrease in PA signal intensity. At carrier concentrations above $10^{17}$ cm$^{-3}$, PA spectra were not observed.

## 3.7. OPTOTHERMAL WINDOW SPECTROSCOPY

The final experimental technique to be described in this chapter is referred to as optothermal window (OW) spectroscopy. In contrast with many of the methods discussed so far, the OW technique has received somewhat limited attention in the literature. Nevertheless, it is important to include it in this survey of experimental methods: It bears a clear relationship to the other PA techniques mentioned and, importantly, has been utilized in both medical applications and the study of food products.

The use of the term *optothermal window* is, in fact, quite instructive: This experiment is based on the detection of the temperature increase in an optical window that is in intimate contact with an absorbing sample. As is the case in other PA experiments, the absorption of light results in the production of a thermal wave; in OW spectroscopy the heat is transferred to the window, where it is detected piezoelectrically or by means of a thermistor. Hence the OW scheme is essentially a photothermal experiment, although an important aspect of the technique is also reminiscent of PA spectroscopy with PZT detection (see above).

The basis of OW spectroscopy has been given by Bićanić et al. (1995) and Helander (1993). For a thermally thick, optically opaque sample and a thermally thick window, the amplitude $\psi$ of the optothermal signal is given by

$$\psi = \beta_s\mu_s[(1+\beta_s\mu_s)^2+1]^{-1/2}(1+e_s/e_w)^{-1} \tag{3.3}$$

where the subscripts *s* and *w* denote sample and window, respectively, and *e* is the thermal effusivity (see Appendix 1). The phase is given by

$$\phi = -\tan^{-1}(1+\beta_s\mu_s)^{-1} \tag{3.4}$$

Equation (3.1) can be rearranged to yield an expression for the dimensionless optothermal signal amplitude *S*:

$$\begin{aligned} S &= \beta_s\mu_s[(1+\beta_s\mu_s)^2+1]^{-1/2} \\ S &= \psi(1+e_s/e_w) \end{aligned} \tag{3.5}$$

For the important case where $\beta_s\mu_s \ll 1$, the simple expressions $S = \beta_s\mu_s/2^{1/2}$

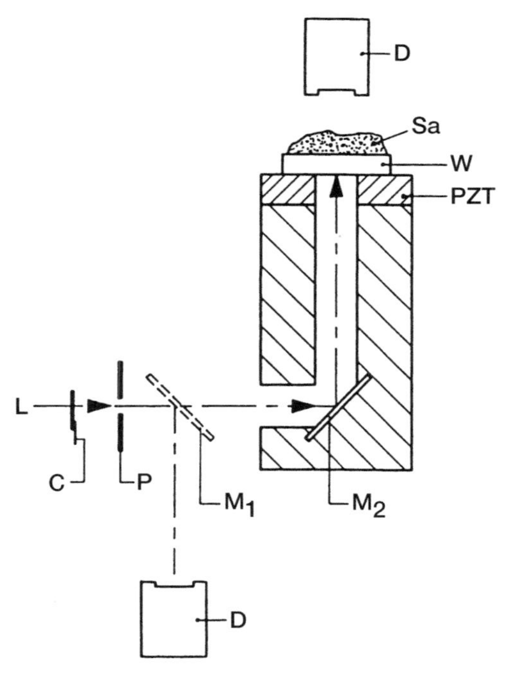


**Figure 3.9.** Experimental arrangement for OW spectroscopy. L, laser beam; C, chopper; P, diaphragm; $M_1$, $M_2$, mirrors; D, pyroelectric detector; PZT, annular piezoelectric detector; W, sapphire window; Sa, sample. A CO laser was used in this experiment. (Reproduced from Bićanić, D. et al., *Appl. Spectrosc.* **49**: 1485–1489, by permission of the Society for Applied Spectroscopy; copyright © 1995.)

and $\phi = -45°$ obtain. Thus the optothermal signal is proportional to the sample absorbance, i.e., an absorption spectrum can be obtained using this technique. On the other hand, the phase is saturated and does not provide any useful information under most conditions. For strongly absorbing samples, $\mu_s$ is constrained to low values, which in turn implies that a high modulation frequency must be used. This naturally implies reduced signal strengths, for reasons analogous to those that affect both rapid-scan and step-scan PA FTIR spectra.

A simplified view of the experimental apparatus used by Bićanić et al. (1995) for the observation of OW spectra of liquids is shown in Figure 3.9. In this arrangement, a CO laser was used as the source of infrared radiation; obviously, one could instead utilize a broadband near- and/or mid-infrared thermal source and appropriate optical filters to select light in the wavelength region of interest. The $CO_2$ laser is another suitable infrared source. The imaginative reader might even suggest that this experiment be performed using an FTIR spectrometer, so that an entire mid-infrared OW spectrum could be obtained. The output voltage from the PZT would be preamplified and directed to the electronics of the spectrometer, rather than a lock-in amplifier, in that case.

### 3.7.1. Research of Bićanić's Group

Bićanić et al. (1995) used the OW apparatus shown in Figure 3.9 to obtain multiple-wavelength (sequential) PA infrared spectra of oleic acid ($C_{18}H_{34}O_2$) and water. A CO laser, with about 30 emission lines between 1710 and 1890 $cm^{-1}$, was used as the infrared source in these experiments. This interval was sufficiently wide to allow the characterization of a band near 1720 $cm^{-1}$ in the spectrum of oleic acid; by contrast, only the high-frequency wing of the water $\nu_2$ band was detectable in this experiment. The data were calibrated using the previously known absorption coefficient of water at the 1781-$cm^{-1}$ CO laser frequency. This permitted the calculation of values for $\beta$ at each of the laser line frequencies and the construction of plots of $\beta$ vs. wavenumber for both liquids.

This initial investigation was followed by several studies on food products. These publications are discussed in greater detail in Chapter 6. It should be noted that a $CO_2$ laser, which emits near 1000 $cm^{-1}$, was used as the infrared source in these investigations.

The OW technique was compared with several other analytical methods with regard to the detection of trans fatty acids (TFAs) in margarines (Favier et al., 1996) and unsaturated vegetable oils (Bićanić et al., 1999). These acids exhibit a characteristic infrared absorption band at 966 $cm^{-1}$, a frequency that fortuitously coincides with a $CO_2$ laser line. TFA contents that ranged from about 4 to 60% were successfully measured in these studies. Similarly, Favier et al. (1998) used OW spectroscopy to detect the adulteration of extra-virgin olive oil by safflower and sunflower oils. These contaminants were analyzed at frequencies of 953 and 1041 $cm^{-1}$, respectively.

This group has also used OW spectroscopy recently in a medical application. Annyas et al. (1999) determined total body water in human blood serum samples by measuring absorption intensities at a wavelength of 4 μm. Heavy water was used as a tracer in this investigation. An impressive detection limit of 30 ppm was demonstrated; this could be further improved by replacing the thermal source (lamp) with a suitable laser.

### 3.7.2. Research of McQueen's Group

Optothermal spectroscopy has also been used to analyze foods by D. H. McQueen at the Chalmers University of Technology and his colleagues at several other institutions. The principal results of this work are summarized in Chapter 6. The present discussion is therefore restricted to a few salient points.

McQueen et al. (1995a) compared optothermal near-infrared spectroscopy and ATR mid-infrared spectroscopy with regard to their capabilities for the analysis of cheese samples. No spectra are illustrated in this study. In fact, measurements were carried out in just three relatively broad wavelength regions that were selected through the use of bandpass filters. These wavelengths were chosen so as to yield information on the contents of protein, fat, and moisture in the cheeses. The simplicity of this experiment and the quality of the results it provided both led the authors to conclude that the optothermal technique was superior to ATR spectroscopy for this particular application.

The authors also noted several limitations of the optothermal method as implemented in their laboratory. These include the following: (a) instrument reproducibility was about 1.5% of signal strength; (b) good thermal contact between the sample and the window was essential; and (c) temperature variations tended to affect spectral sensitivities and the transport of heat from the cheese samples to the window. These points were reiterated in a review article on the analysis of foods by near- and mid-infrared PA spectroscopy (McQueen et al., 1995b).

## REFERENCES

Adams, M. J., Beadle, B. C., and Kirkbright, G. F. (1978). Optoacoustic spectrometry in the near-infrared region. *Anal. Chem.* **50**: 1371–1374.

Annyas, J., Bićanić, D., and Schouten, F. (1999). Novel instrumental approach to the measurement of total body water: Optothermal detection of heavy water in the blood serum. *Appl. Spectrosc.* **53**: 339–343.

Bain, C. D., Davies, P. B., and Ong, T. H. (1992). Vibrational spectroscopy of monolayers by pulsed photothermal beam deflection (PBD). In: *Photoacoustic and Photothermal Phenomena III*. D. Bićanić (ed.). Springer, Berlin, pp. 158–160.

Belton, P. S., and Tanner, S. F. (1983). Determination of the moisture content of starch using near infrared photoacoustic spectroscopy. *Analyst* **108**: 591–596.

Belton, P. S., Wilson, R. H., and Saffa, A. M. (1987). Effects of particle size on quantitative photoacoustic spectroscopy using a gas-microphone cell. *Anal. Chem.* **59**: 2378–2382.

Bićanić, D., Krüger, S., Torfs, P., Bein, B., and Harren, F. (1989). The use of reverse mirage spectroscopy to determine the absorption coefficient of liquid methanol at $CO_2$ laser wavelengths. *Appl. Spectrosc.* **43**: 148–153.

Bićanić, D., Chirtoc, M., Chirtoc, I., Favier, J. P., and Helander, P. (1995). Photothermal determination of absorption coefficients in optically dense fluids: Application to oleic acid and water at CO laser wavelengths. *Appl. Spectrosc.* **49**: 1485–1489.

Bićanić, D., Fink, T., Franko, M., Močnik, G., van de Bovenkamp, P., van Veldhuizen, B., and Gerkema, E. (1999). Infrared photothermal spectroscopy in the science of human nutrition. *AIP Conf. Proc.* **463**: 637–639.

Blank, R. E., and Wakefield, T. (1979). Double-beam photoacoustic spectrometer for use in the ultraviolet, visible, and near-infrared spectral regions. *Anal. Chem.* **51**: 50–54.

Bouzerar, R., Amory, C., Zeinert, A., Benlahsen, M., Racine, B., Durand-Drouhin, O., and Clin, M. (2001). Optical properties of amorphous hydrogenated carbon thin films. *J. Non-Cryst. Sol.* **281**: 171–180.

Carter, R. O., and Paputa Peck, M. C. (1989). Photoacoustic detection of rapid-scan Fourier transform infrared spectra from low surface-area solid samples. *Appl. Spectrosc.* **43**: 468–473.

Carter, R. O., and Wright, S. L. (1991). Evaluation of the appropriate sample position in a PAS/FT-IR experiment. *Appl. Spectrosc.* **45**: 1101–1103.

Castleden, S. L., Kirkbright, G. F., and Menon, K. R. (1980). Determination of moisture in single-cell protein utilising photoacoustic spectroscopy in the near-infrared region. *Analyst* **105**: 1076–1081.

Castleden, S. L., Kirkbright, G. F., and Long, S. E. (1982). Quantitative assay of propranolol by photoacoustic spectroscopy. *Can. J. Spectrosc.* **27**: 245–248.

DeBellis, A. D., and Low, M. J. D. (1987). Dispersive infrared photothermal beam deflection spectroscopy. *Infrared Phys.* **27**: 181–191.

DeBellis, A. D., and Low, M. J. D. (1988). Dispersive infrared phase angle photothermal beam deflection spectroscopy. *Infrared Phys.* **28**: 225–237.

Deng, Z., Spear, J. D., Rudnicki, J. D., McLarnon, F. R., and Cairns, E. J. (1996). Infrared photothermal deflection spectroscopy: A new probe for the investigation of electrochemical interfaces. *J. Electrochem. Soc.* **143**: 1514–1521.

Donini, J. C., and Michaelian, K. H. (1988). Low-frequency photoacoustic spectroscopy of solids. *Appl. Spectrosc.* **42**: 289–292.

Duerst, R. W., and Mahmoodi, P. (1984). IR-PAS chamber for signal-to-noise enhancement. *Preprints, ACS Div. Polym. Chem.* **25**: 194–195.

Duerst, R. W., Mahmoodi, P., and Duerst, M. D. (1987). IR-PAS studies: Signal-to-noise enhancement and depth profile analysis. In: *Fourier Transform Infrared Characterization of Polymers*. H. Ishida (ed.). Plenum, New York, pp. 113–122.

Einsiedel, H., Kreiter, M., Leclerc, M., and Mittler-Neher, S. (1998). Photothermal beam deflection spectroscopy in the near IR on poly[3-alkylthiophene]s. *Opt. Mater.* **10**: 61–68.

Favier, J. P., Bićanić, D., van de Bovenkamp, P., Chirtoc, M., and Helander, P. (1996). Detection of total trans fatty acids content in margarine: An intercomparison study of GLC, GLC + TLC, FT-IR, and optothermal window (open photoacoustic cell). *Anal. Chem.* **68**: 729–733.

Favier, J. P., Bićanić, D., Cozijnsen, J., van Veldhuizen, B., and Helander, P. (1998). $CO_2$ laser infrared optothermal spectroscopy for quantitative adulteration studies in binary mixtures of extra-virgin olive oil. *J. Am. Oil Chem. Soc.* **75**: 359–362.

Fournier, D., Boccara, A. C., Amer, N. M., and Gerlach, R. (1980). Sensitive *in situ* trace-gas detection by photothermal deflection spectroscopy. *Appl. Phys. Lett.* **37**: 519–521.

Fournier, D., Boccara, A. C., and Badoz, J. (1982). Photothermal deflection Fourier transform spectroscopy: A tool for high-sensitivity absorption and dichroism measurements. *Appl. Opt.* **21**: 74–76.

Fukuyama, A., Akashi, Y., Suemitsu, M., and Ikari, T. (2000). Detailed observation of the photoquenching effect of EL2 in semi-insulating GaAs by the piezoelectric photoacoustic measurements. *J. Cryst. Growth* **210**: 255–259.

Gregoriou, V. G., Daun, M., Schauer, M. W., Chao, J. L., and Palmer, R. A. (1993). Modification of a research-grade FT-IR spectrometer for optional step-scan operation. *Appl. Spectrosc.* **47**: 1311–1316.

Helander, P. (1993). A method for the analysis of the optothermal and photoacoustic signals. *Meas. Sci. Technol.* **4**: 178–185.

Jones, R. W., and McClelland, J. F. (2001). Phase references and cell effects in photoacoustic spectroscopy. *Appl. Spectrosc.* **55**: 1360–1367.

Kanstad, S. O., and Nordal, P.-E. (1980a). Photoacoustic reflection-absorption spectroscopy (PARAS) for infrared analysis of surface species. *Appl. Surf. Sci.* **5**: 286–295.

Kanstad, S. O., and Nordal, P.-E. (1980b). Photoacoustic and photothermal spectroscopy. *Phys. Technol.* **11**: 142–147.

Kanstad, S. O., Nordal, P.-E., Hellgren, L., and Vincent, J. (1981). Infrared photoacoustic spectroscopy of skin lipids. *Naturwiss.* **68**: 47–48.

Kirkbright, G. F. (1978). Analytical optoacoustic spectrometry. *Optica Pura y Aplicada* **11**: 125–136.

Kuwahata, H., Muto, N., and Uehara, F. (2000). Carrier concentration dependence of photoacoustic spectra of silicon by a piezoelectric transducer method. *Jpn. J. Appl. Phys.* **39**: 3169–3171.

Lloyd, L. B., Yeates, R. C., and Eyring, E. M. (1982). Fourier transform infrared photoacoustic spectroscopy in thin-layer chromatography. *Anal. Chem.* **54**: 549–552.

Lochmüller, C. H., and Wilder, D. R. (1980a). Qualitative examination of chemically-modified silica surfaces by near-infrared photoacoustic spectroscopy. *Anal. Chim. Acta* **116**: 19–24.

Lochmüller, C. H., and Wilder, D. R. (1980b). Quantitative photoacoustic spectroscopy of chemically-modified silica surfaces. *Anal. Chim. Acta* **118**: 101–108.

Low, M. J. D. (1986). Some practical aspects of FT-IR/PBDS. Part I: vibrational noise. *Appl. Spectrosc.* **40**: 1011–1019.

Low, M. J. D., and Parodi, G. A. (1978). Infrared photoacoustic spectra of surface species in the 4000–2000 $cm^{-1}$ region using a broad band source. *Spectrosc. Lett.* **11**: 581–588.

Low, M. J. D., and Parodi, G. A. (1980a). Infrared photoacoustic spectroscopy of surfaces. *J. Mol. Struct.* **61**: 119–124.

Low, M. J. D., and Parodi, G. A. (1980b). Infrared photoacoustic spectra of solids. *Spectrosc. Lett.* **13**: 151–158.

Low, M. J. D., and Parodi, G. A. (1980c). An infrared photoacoustic spectrometer. *Infrared Phys.* **20**: 333–340.

Low, M. J. D., and Parodi, G. A. (1980d). Infrared photoacoustic spectroscopy of solids and surface species. *Appl. Spectrosc.* **34**: 76–80.

Low, M. J. D., and Parodi, G. A. (1980e). Carbon as reference for normalizing infrared photoacoustic spectra. *Spectrosc. Lett.* **13**: 663–669.

Low, M. J. D., and Lacroix, M. (1982). An infrared photothermal beam deflection Fourier transform spectrometer. *Infrared Phys.* **22**: 139–147.

Low, M. J. D., and Parodi, G. A. (1982). Dispersive photoacoustic spectroscopy of solids in the infrared range. *J. Photoacoustics* **1**: 131–144.

Low, M. J. D., and Tascon, J. M. D. (1985). An approach to the study of minerals using infrared photothermal beam deflection spectroscopy. *Phys. Chem. Minerals* **12**: 19–22.

Low, M. J. D., and Morterra, C. (1987). Some practical aspects of FT-IR/PBDS. Part II: sample handling procedures. *Appl. Spectrosc.* **41**: 280–287.

Low, M. J. D., Lacroix, M., and Morterra, C. (1982a). Infrared spectra of massive solids by photoacoustic beam deflection Fourier transform spectroscopy. *Spectrosc. Lett.* **15**: 57–64.

Low, M. J. D., Lacroix, M., and Morterra, C. (1982b). Infrared photothermal beam deflection Fourier transform spectroscopy of solids. *Appl. Spectrosc.* **36**: 582–584.

Mahmoodi, P., Duerst, R. W., and Meiklejohn, R. A. (1984). Effect of acoustic isolation chamber on the signal-to-noise ratio in infrared photoacoustic spectroscopy. *Appl. Spectrosc.* **38**: 437–438.

Manning, C. J., Palmer, R. A., and Chao, J. L. (1991). Step-scan Fourier-transform infrared spectrometer. *Rev. Sci. Instrum.* **62**: 1219–1229.

Manning, C. J., Dittmar, R. M., Palmer, R. A., and Chao, J. L. (1992a). Use of step-scan FT-IR to obtain the photoacoustic/photothermal response phase. *Infrared Phys.* **33**: 53–62.

Manning, C. J., Palmer, R. A., Chao, J. L., and Charbonnier, F. (1992b). Photothermal beam deflection using the reverse mirage geometry: Theory and experiment. *J. Appl. Phys.* **71**: 2433–2440.

Manning, C. J., Charbonnier, F., Chao, J. L., and Palmer, R. A. (1992c). Reverse mirage photothermal beam deflection: Theory and experiment. In: *Photoacoustic and Photothermal Phenomena III*. D. Bićanić (ed.). Springer, Berlin, pp. 161–164.

Manzanares, C., Blunt, V. M., and Peng, J. (1993). Vibrational spectroscopy of nonequivalent C—H bonds in liquid *cis*- and *trans*-3-hexene. *Spectrochim. Acta* **49A**: 1139–1152.

McClelland, J. F., Luo, S., Jones, R. W., and Seaverson, L. M. (1992). A tutorial on the state-of-the-art of FTIR photoacoustic spectroscopy. In: *Photoacoustic and Photothermal Phenomena III*. D. Bićanić (ed.). Springer, Berlin, pp. 113–124.

McQueen, D. H., Wilson, R., Kinnunen, A., and Jensen, E. P. (1995a). Comparison of two infrared spectroscopic methods for cheese analysis. *Talanta* **42**: 2007–2015.

McQueen, D. H., Wilson, R., and Kinnunen, A. (1995b). Near and mid-infrared photoacoustic analysis of principal components of foodstuffs. *Trends Anal. Chem.* **14**: 482–492.

Palmer, R. A. (1994). Time-resolved and phase-resolved vibrational spectroscopy by use of step-scan FT-IR. *SPIE* **2089**: 53–61.

Palmer, R. A., and Smith, M. J. (1986). Rapid-scanning Fourier-transform infrared spectroscopy with photothermal beam-deflection (mirage effect) detection at the solid-liquid interface. *Can. J. Phys.* **64**: 1086–1092.

Palmer, R. A., Chao, J. L., Dittmar, R. M., Gregoriou, V. G., and Plunkett, S. E. (1993). Investigation of time-dependent phenomena by use of step-scan FT-IR. *Appl. Spectrosc.* **47**: 1297–1310.

Smith, M. J., and Palmer, R. A. (1987). The reverse mirage effect: Catching the thermal wave at the solid/liquid interface. *Appl. Spectrosc.* **41**: 1106–1113.

Smith, M. J., Manning, C. J., Palmer, R. A., and Chao, J. L. (1988). Step scan interferometry in the mid-infrared with photothermal detection. *Appl. Spectrosc.* **42**: 546–555.

Yoshino, K., Fukuyama, A., Yokoyama, H., Meada, K., Fons, P. J., Yamada, A., Niki, S., and Ikari, T. (1999). Piezoelectric photoacoustic spectra of $CuInSe_2$ thin film grown by molecular beam epitaxy. *Thin Solid Films* **343–344**: 591–593.

Zhang, S. L., Michaelian, K. H., and Burt, J. A. (1997). Phase correction in piezoelectric photoacoustic Fourier transform infrared spectroscopy of mica. *Opt. Eng.* **36**: 321–325.

## RECOMMENDED READING

Bićanić, D., Jalink, H., Chirtoc, M., Sauren, H., Lubbers, M., Quist, J., Gerkema, E., van Asselt, K., Miklós, A., Sólyom, A., Angeli, Gy. Z., Helander, P., and Vargas, H. (1992). Interfacing photoacoustic and photothermal techniques for new hyphenated methodologies and instrumentation suitable for agricultural, environmental and medical applications. In: *Photoacoustic and Photothermal Phenomena III*. D. Bićanić (ed.). Springer, Berlin, pp. 20–27.

Débarre, D., Boccara, A. C., and Fournier, D. (1981). High-luminosity visible and near-IR Fourier-transform photoacoustic spectrometer. *Appl. Opt.* **20**: 4281–4286.

Lloyd, L. B., Riseman, S. M., Burnham, R. K., Eyring, E. M., and Farrow, M. M. (1980). Fourier transform photoacoustic spectrometer. *Rev Sci. Instrum.* **51**: 1488–1492.

CHAPTER

4

# DEPTH PROFILING

The ability to analyze solid samples at selected depths is generally recognized to be one of the most important attributes of PA spectroscopy. The present chapter is concerned with depth profiling in PA infrared spectroscopy, a topic that has now formed an active area of research for more than two decades. An examination of the relevant literature published in this period shows that some of the early attempts at characterization of layered or heterogeneous materials were adversely affected by low signal-to-noise ratios and various experimental limitations; however, the quality of the data acquired in these studies improved quickly, and excellent depth profiling results have been routinely obtained in many laboratories for a number of years.

As discussed elsewhere in this book, the PA signal normally arises from a surficial layer whose thickness is approximately equal to the thermal diffusion length $\mu_s$ of the sample. This dimension is given by the well-known expression $\mu_s = (\alpha/\pi f)^{1/2}$, where $\alpha$ denotes thermal diffusivity and $f$ is the modulation frequency. The mathematical definition of $\mu_s$ provides a basis for depth profiling: As $f$ is changed, $\mu_s$ varies accordingly and different layer thicknesses are analyzed. In general, this statement applies to both single- and multiple-wavelength PA experiments.

In a rapid-scan PA FTIR experiment with a Michelson interferometer, $f$ is determined by the mirror velocity $V$ according to the equation $f = 2V\nu$, where $\nu$ is the infrared frequency. (In a Genzel interferometer, the different optical layout implies that $f = 4V\nu$.) Consequently, the acquisition of PA spectra at different mirror velocities is equivalent to changing both $f$ and $\mu_s$. It should be pointed out that some FTIR spectrometers offer little or no operator control of mirror velocity; obviously, this depth profiling technique is not feasible with such instruments.

Vidrine (1981), as well as Vidrine and Lowry (1983), were among the first to demonstrate PA depth profiling by changing the mirror velocity. Numerous authors successfully applied this principle to a wide variety of samples in succeeding years (Donini and Michaelian, 1984; Muraishi, 1984; Ochiai, 1985; Yang et al., 1985; Urban and Koenig, 1986; Zerlia, 1986; Yang and Fateley, 1987; Yang et al., 1987; Urban, 1987). These and other representative publications that describe the results of PA infrared depth profiling experiments are listed in Table 4.1.

While the dependence of $\mu_s$ on mirror velocity allows the analyst to readily perform depth profiling studies in PA FTIR spectroscopy, the use of this technique must necessarily be accompanied by an important caveat: Since $\mu_s$ also varies inversely with the square root of the infrared frequency, the sampling depth varies considerably across the mid-infrared range in a rapid-scan PA experiment at a particular velocity. Specifically, the value of $\mu_s$ at the lower limit of this region (400 $cm^{-1}$) is greater than that at the upper boundary (4000 $cm^{-1}$) by a factor of $\sqrt{10} = 3.16$. Although this effect does not influence the interpretation of spectra of homogeneous samples, it is obviously important in other situations where composition is depth-dependent. On the other hand, if the comparison of a series of PA spectra obtained with different mirror velocities is confined to a relatively narrow spectral region, the slight wavenumber dependence of $\mu_s$ can be neglected.

The development of modern step-scan FTIR instruments in the late 1980s and early 1990s dramatically changed the situation with regard to depth profiling. In a step-scan spectrometer, the mirror movement in the interferometer is discrete rather than continuous; in effect, the scan velocity is reduced to zero, and some form of external modulation must be supplied to carry out the PA experiment. In the method referred to as amplitude modulation, the infrared beam is periodically interrupted by a chopper. Obviously, infrared radiation of all wavelengths is completely modulated (i.e., it is either "on" or "off") in this experiment. Alternatively, the moving mirror can be oscillated ("dithered") at each stopping position using a method known as phase modulation. Spectroscopists may recognize this as a derivative technique; in fact, the recorded phase modulation interferogram is equivalent to the first derivative of the amplitude modulation or rapid-scan interferograms. The depth of the modulation varies with wavelength in this approach. Phase modulation is favored by many PA researchers because all of the light that emerges from the interferometer is allowed to impinge on the PA cell, in contrast with the situation in an amplitude modulation experiment. Moreover, the amplitude of the phase modulation and its frequency are usually controlled by the FTIR software, reducing the complexity of these measurements.

The major depth profiling techniques used in PA infrared spectroscopy are summarized in Table 4.2. The main principles involved in each method, as well as several advantages and limitations, are also listed in this table. Depth profiling in rapid-scan PA FTIR spectroscopy constitutes the first method and was briefly described above. The other five techniques in Table 4.2 are normally implemented in step-scan experiments. These methods are discussed in the following sections of this chapter, with amplitude modula-

**Table 4.1. Examples of Depth Profiling in PA Infrared Spectroscopy**

| Reference | Application/Sample[a] |
|---|---|
| Dittmar et al. (1991) | EVAc/PP, polyimide |
| Donini and Michaelian (1984) | PE/ink |
| Drapcho et al. (1997) | Polymer films |
| Gonon et al. (1999) | Styrene/isoprene |
| Gonon et al. (2001) | Styrene/isoprene |
| Gregoriou and Hapanowicz (1996) | Polymer films |
| Gregoriou and Hapanowicz (1997) | Polymer laminate |
| Irudayaraj and Yang (2000) | Cheese wrapper |
| Irudayaraj and Yang (2002) | Starch/protein/polyethylene |
| Jiang et al. (1995) | Polymer films |
| Jiang and Palmer (1997) | EVAc/PP |
| Jiang et al. (1997a) | Polymer films |
| Jiang et al. (1997b) | Polymer films |
| Jiang et al. (1998) | Polymer films |
| Jiang (1999) | Coated bead, grease/fiber, hair gel |
| Jones and McClelland (1996) | PET/polycarbonate |
| Jones and McClelland (2002) | PET/polycarbonate |
| Lerner et al. (1989) | Polymer laminate |
| McClelland et al. (1992) | Polymer films |
| McClelland et al. (1994) | PE, PS |
| McClelland et al. (1997) | PET/polycarbonate |
| Michaelian (1989) | Coal |
| Michaelian (1991) | Coal |
| Muraishi (1984) | PVC/PVF |
| Noda et al. (1997) | Polymer films |
| Ochiai (1985) | PVC/PVF |
| Palmer and Dittmar (1993) | Polymer films |
| Palmer et al. (1993) | EVAc/PP, Teflon/polyimide |
| Palmer et al. (1994) | Polymer films |
| Palmer et al. (1997) | Polymer films |
| Sowa and Mantsch (1994a) | Human tooth |
| Sowa and Mantsch (1994b) | Human tooth |
| Sowa et al. (1995) | Finger nails |
| Sowa et al. (1996) | PE sulfonation |
| Story et al. (1994) | PDMS/PE/PS |
| Story and Marcott (1998) | Adhesive label |
| Stout and Crocombe (1994) | Polymer films |
| Szurkowski and Wartewig (1999a) | Olive oil/water |
| Szurkowski and Wartewig (1999b) | Olive oil/water |
| Szurkowski et al. (2000) | Olive oil/water |
| Teramae and Tanaka (1985) | Polymer films |
| Urban and Koenig (1986) | $PVF_2$/PET |

*Continued*

**Table 4.1.** *(Continued)*

| Reference | Application/Sample[a] |
|---|---|
| Urban (1987) | $PVF_2$/PET |
| Urban et al. (1999) | Butyl acrylate/polyurethane latexes |
| Vidrine (1981) | Catalyst |
| Vidrine and Lowry (1983) | Catalyst |
| Wahls and Leyte (1998a) | PP/PBA |
| Wahls and Leyte (1998b) | PP/PET |
| Yamada et al. (1996) | Coals |
| Yang et al. (1985) | PET |
| Yang and Fateley (1987) | PVC |
| Yang et al. (1987) | PET |
| Yang et al. (1990) | PET |
| Yang and Irudayaraj (2000) | Protein and fat in cheese packaging |
| Yang et al. (2001) | Coatings, microorganisms on foods |
| Yang and Irudayaraj (2001) | Beef, pork |
| Zerlia (1986) | Coals |

[a] EVAc, ethylene-vinylacetate; PDMS, polydimethylsiloxane; PE, polyethylene; PET, poly(ethylene terephthalate); PVC, polyvinyl chloride; PVF, polyvinyl fluoride; $PVF_2$, poly(vinylidene fluoride); PP, polypropylene; PS, polystyrene.

tion being mentioned first. After this, phase modulation techniques are considered, and the uses of phase spectra and phase rotation plots are reviewed. Generalized two-dimensional correlation (which also has many applications outside of PA spectroscopy) is described in the last section of this chapter.

## 4.1. AMPLITUDE MODULATION

The methodology utilized in depth profiling by amplitude modulation is relatively straightforward. As mentioned previously, the thermal diffusion length $\mu_s$ in a PA experiment is given by the expression

$$\mu_s = (\alpha/\pi f)^{1/2} \tag{4.1}$$

where $\alpha$ is the thermal diffusivity of the sample and $f$ denotes the modulation frequency; $\mu_s$ is the distance over which the thermal wave decays to a value equal to $1/e$ (where $e = 2.7183$) of its original amplitude; to a first approximation, $\mu_s$ is also equal to the sampling depth in PA spectroscopy. In an amplitude modulation depth profiling experiment, the infrared radiation falling on the sample is periodically interrupted using a mechanical device such as a chopper or shutter. Variation of the modulation frequency results

**Table 4.2. Principal Depth Profiling Techniques in PA Infrared Spectroscopy**

| Method | Salient Features | Comments |
|---|---|---|
| Rapid-scan | Sampling depth varies inversely with mirror velocity | Thermal diffusion length depends on wavenumber |
| Amplitude modulation | Plot of PA intensity vs. modulation frequency allows depth profiling | Incident intensity lost during modulation |
| Phase modulation | Sampling depth varies inversely with modulation frequency | Thermal diffusion length independent of wavenumber |
| Phase spectrum | Plot of relative PA phase vs. $\nu$ indicates spatial origin | High depth resolution<br>Data acquired in a single scan |
| Phase rotation | Plot of PA intensity vs. phase angle exhibits characteristic minimum | Complete extinction of bands possible<br>Can be used to determine layer thickness |
| G2D[a] correlation | Cross peaks in synchronous plot have similar spatial origins; peaks in asynchronous plot indicate phase differences, enabling depth profiling | Combines magnitude and phase in a single qualitative representation<br>Aids identification of overlapping bands |

[a] G2D, generalized two-dimensional.

in the detection of PA signals from surface layers having different thicknesses. Comparison of these data makes it possible to characterize layered or inhomogeneous samples.

The reader is obviously aware of the fact that the emphasis in this book is primarily on PA FTIR spectroscopy. On the other hand, it should be recognized that single-wavelength PA infrared depth profiling experiments may be adequate in some situations. For example, these measurements can be performed with pulsed or chopped laser radiation at a wavelength that corresponds to an infrared absorption band of specific interest. Alternatively, single or multiple-wavelength PA depth profiling experiments can be carried out using a broadband source such as a lamp, the wavelength(s) being selected by means of suitable optical filters.

For most infrared spectroscopists, the depth profiling experiments of greatest interest will undoubtedly be those carried out with modern FTIR spectrometers. In a typical amplitude modulation experiment of this sort, the beam that emerges from the interferometer is chopped before it impinges on the sample in the PA cell. The signal from the PA detector is demodulated, commonly through the use of a lock-in amplifier, and then processed by the FTIR electronics. Because the infrared radiation passes through the chopper, all wavelengths are modulated at the selected frequency. The spectroscopist controls $\mu_s$ by changing this chopping frequency. Depth profiling can thus be effected for a sample with a composition that varies with the distance from the surface.

Amplitude modulation depth profiling experiments were carried out on both fresh and oxidized coal samples shortly after the reintroduction of commercial step-scan FTIR spectrometers in the late 1980s (Michaelian, 1989a, 1991). Surface oxidation of coal, which has several negative technological consequences (see Chapter 6), creates relevant samples that are amenable to depth profiling by PA infrared spectroscopy. This oxidation produces several carbonylic functional groups that absorb in the spectral region near 1750 $cm^{-1}$. Choosing a simplistic model in which the infrared spectrum of oxidized coal is assumed to consist of a contribution from the corresponding fresh coal plus a generic carbonyl band, one can write

$$S_{ox}(\nu) = S_{fr}(\nu) + S_{C=O}(\nu) \tag{4.2}$$

where the subscripts ox, fr, and C=O denote oxidized, fresh, and carbonyl, respectively. The carbonyl band can be recovered through division of $S_{ox}(\nu)$ by $S_{fr}(\nu)$: This ratio is approximately equal to one, except in the region near 1750 $cm^{-1}$ where $S_{C=O}(\nu)$ also contributes to the observed intensity (Fig. 4.1).

In this experiment, the intensity of the retrieved carbonyl band was measured as a function of modulation frequency $f$. At higher frequencies the signal arises mainly from the surface layer of the oxidized coal and the band is more intense. As this frequency was reduced, $\mu_s$ increased and deeper layers of the coal particles contributed to the PA signal. The C=O band was weaker in these spectra. These results are consistent with a simple intuitive model in which the surface of the coal particles is more extensively oxidized than the interior and demonstrate successful depth profiling by amplitude modulation in PA infrared spectroscopy.

$S_{C=O}(\nu)$ can also be recovered by subtraction of $S_{fr}(\nu)$ from $S_{ox}(\nu)$. When this approach is taken, the observed intensities must be corrected for the usual $1/f$ dependence of the PA signal (Michaelian, 1991). Depth profiling is again accomplished by comparing carbonyl intensities at a series of modulation frequencies (Fig. 4.2). The results show that these intensities increase concomitantly with $f$ up to about 180 Hz and remain approximately constant thereafter. Using the previously known value of $0.1 \times 10^{-6}$ $m^2\ s^{-1}$ for the thermal diffusivity of coal, it can be concluded that the first 12 μm of the coal particles is oxidized to more or less the same extent. Depth

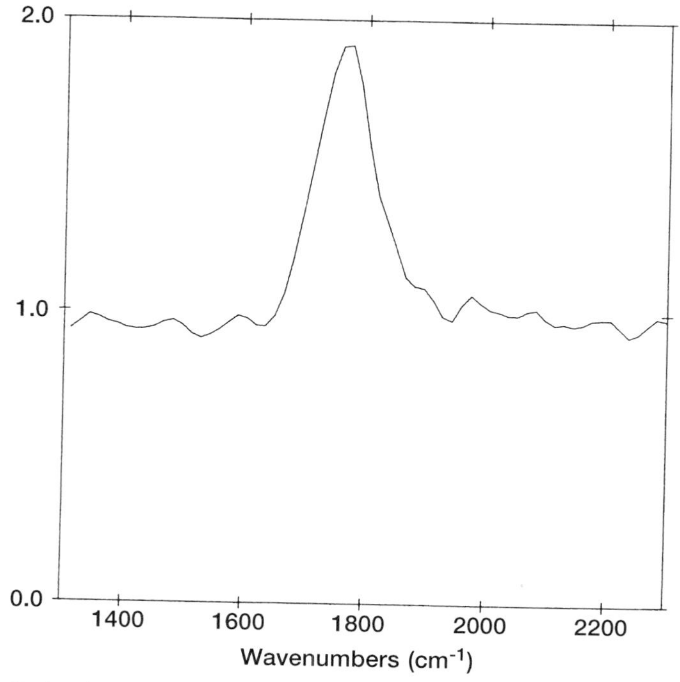


**Figure 4.1.** Ratio of step-scan PA spectra of oxidized and fresh coals in the carbonyl region. (Reproduced from Michaelian, K. H., *Appl. Spectrosc.* **43**: 185–190, by permission of the Society for Applied Spectroscopy; copyright © 1989.)

discrimination of this surface layer by PA infrared spectroscopy would be possible only for less extensively oxidized coals.

Because $\mu_s$ varies inversely with the square root of $f$, Figure 4.2 is not easily interpreted with regard to diffusion length. The data are presented in a different manner in Figure 4.3. The $\mu_s$ was calculated using the modulation frequencies and the value of $\alpha$ given in the previous paragraph. Figure 4.3 illustrates an approximately linear relationship between carbonyl intensity and sampling depth, that is, the extent of oxidation varies in direct proportion to the distance from the surface of the coal particles in the region from about 13 to 26 μm.

After these initial amplitude modulation depth profiling experiments, phase modulation techniques gained popularity with many investigators. Several aspects of amplitude modulation, which may be described as disadvantages, probably contributed to this trend. These include the following: (a) The method requires the use of an external chopper; (b) half of the available infrared radiation never reaches the sample, instead being sacri-

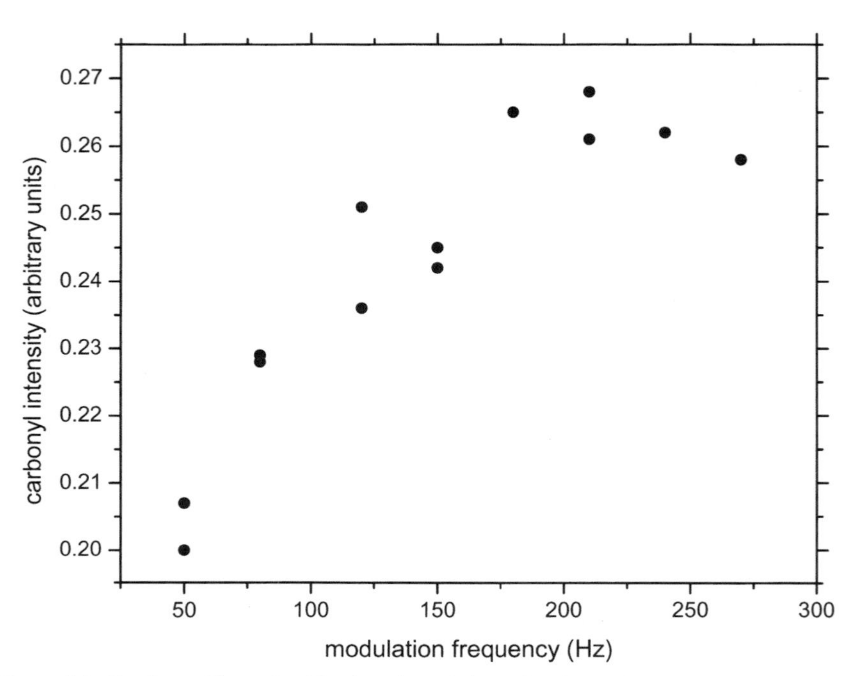


**Figure 4.2.** Depth profiling of oxidized coal: variation of carbonyl intensity with modulation frequency.

ficed on the chopper blades; and (c) the large dc background signal (which is independent of path difference, and can be observed in both rapid- and step-scan operation) reduces the dynamic range of the experiment. Moreover, it is essential that the output of the infrared source be very stable, insofar as a change in source intensity can produce a result similar to that caused by optical absorption. The last point obviously becomes more important in depth profiling studies that involve comparison of spectra obtained at significantly different times.

## 4.2. PHASE MODULATION

Depth profiling by phase modulation was demonstrated shortly after the initial amplitude modulation experiments described above. Phase modulation quickly became the technique of choice for most workers in this field for several substantive reasons: (a) No external modulator (chopper) is required; (b) the experiment is conveniently controlled by the FTIR software; and (c)

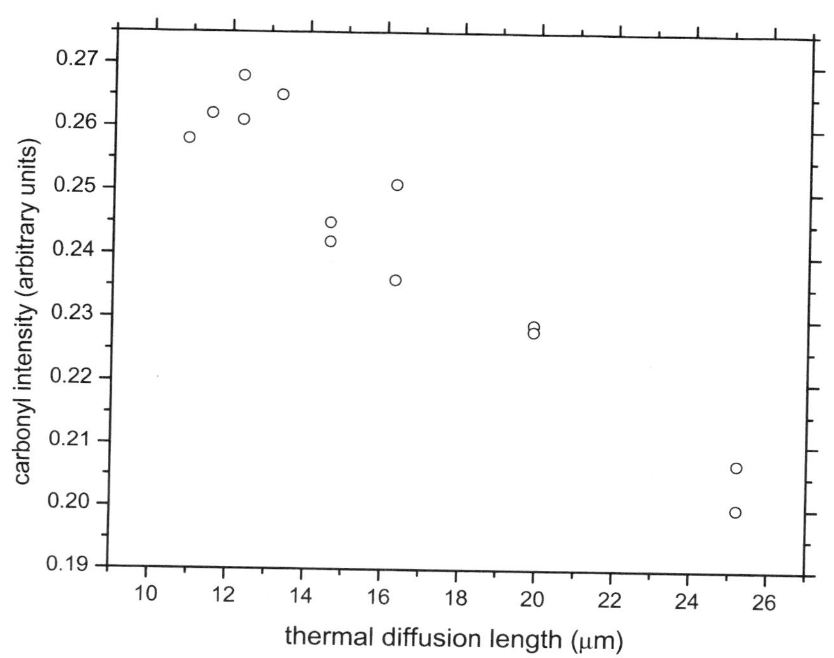


**Figure 4.3.** Depth profiling of oxidized coal: variation of carbonyl intensity with thermal diffusion length, $\mu_s$.

in-phase and in-quadrature spectra can be easily obtained in a single scan. (The last point also applies to amplitude modulation when a dual-phase lock-in amplifier is used.) Moreover, the specific limitations mentioned above with regard to the use of amplitude modulation do not affect phase modulation.

Both the magnitude and the phase of the PA signal in phase modulation experiments can be utilized in depth profiling studies. These quantities are related by

$$M = [Q^2 + I^2]^{1/2} \tag{4.3}$$

and

$$\theta = \tan^{-1}[Q/I] \tag{4.4}$$

where $M$ is the magnitude spectrum, $Q$ and $I$ are its quadrature and in-phase components, respectively, and $\theta$ is the phase angle. For simplicity, the wavenumber dependence of these quantities has not been shown explicitly in these equations.

The phase modulation magnitude spectrum (like its amplitude modulation counterpart) differs from rapid-scan magnitude spectra used for depth profiling in one important way. Because all infrared wavelengths are modulated at the same frequency in each experiment, the thermal diffusion length does not vary across the infrared region: Sampling depth is constant. Therefore, a change in modulation frequency $f$ has a simple and predictable effect on $\mu_s$, and depth profiling is comparatively straightforward. This has led to the use of phase modulation magnitude spectra in depth profiling studies by a significant number of research groups (Dittmar et al., 1991; Palmer and Dittmar, 1993; Szurkowski and Wartewig, 1999a, 1999b; Urban et al., 1999; Irudayaraj and Yang, 2000, 2002; Yang and Irudayaraj, 2000, 2001a; Szurkowski et al., 2000; Gonon et al., 2001; Yang et al., 2001; Jones and McClelland, 2002). The last reference in this list describes a numerical method for reducing the saturation in spectra recorded at low modulation frequency to the level that exists at higher frequency; this approach can also be applied in rapid-scan depth profiling experiments.

Another level of sophistication has been introduced into this experiment through the use of digital signal processing (DSP). This method allows simultaneous demodulation of the PA signal at several harmonics of the phase modulation frequency, thus providing all of the data necessary for depth profiling in one measurement. Obviously, a considerable amount of experimental time is saved in this way. Moreover, DSP detection has been found to yield signal-to-noise ratios that are superior to those obtained under similar conditions using a lock-in amplifier. While early experiments required a special DSP demodulator, the technique is now implemented entirely in software (Drapcho et al., 1997). DSP detection has been discussed further with regard to depth profiling in PA infrared spectroscopy by Story and Marcott (1998), and by Jiang (1999). As many as nine harmonic frequencies can be examined at the same time in this way.

### 4.2.1. Phase Spectrum

The depth profiling techniques discussed so far are based on the modulation frequency dependence of the PA signal. On the other hand, because of the significant time required for the thermal wave in each sample layer to propagate to the surface, the phase lag associated with the PA signal also conveys depth-related information. This is very conveniently shown by plotting the phase angle $\theta$ vs. wavenumber, a presentation that is referred to as a phase spectrum. It might be said that this spectrum is not very intuitive since it usually does not bear a close resemblance to absorbance or PA spectra;

however, the high phase resolution in such a plot is particularly useful in some specific applications, such as depth profiling studies that involve very thin films.

The use of phase spectra for depth profiling of layered polymer samples was pioneered in the early 1990s by R. A. Palmer and his collaborators. More recently, this method has been utilized by several research groups in a variety of applications, some of which are described below. It is also relevant to point out that the phase spectrum yields results that are mathematically equivalent to those from the phase rotation technique discussed in the following section. These two approaches have also been shown experimentally to yield equivalent results (Palmer et al., 1994; Jiang and Palmer, 1997).

The phase spectrum obtained for a two-layer polymer sample by Jiang and Palmer (1997) is shown in Figure 4.4. The letters in this figure denote the frequencies of magnitude-spectrum bands of particular interest. The larger phase angles of bands B, $C_2$, and $C_3$ imply deeper spatial origins, whereas the smaller angles for $A_1$, $A_2$, and $C_1$ indicate that the bands arise from a layer nearer to the surface. This demonstrates the use of phase angle spectra for depth profiling. In a series of related investigations, Palmer's group has described the application of this method to two-layer (Palmer et al., 1994, 1997), as well as three- and four-layer, polymer samples (Jiang et al., 1997a).

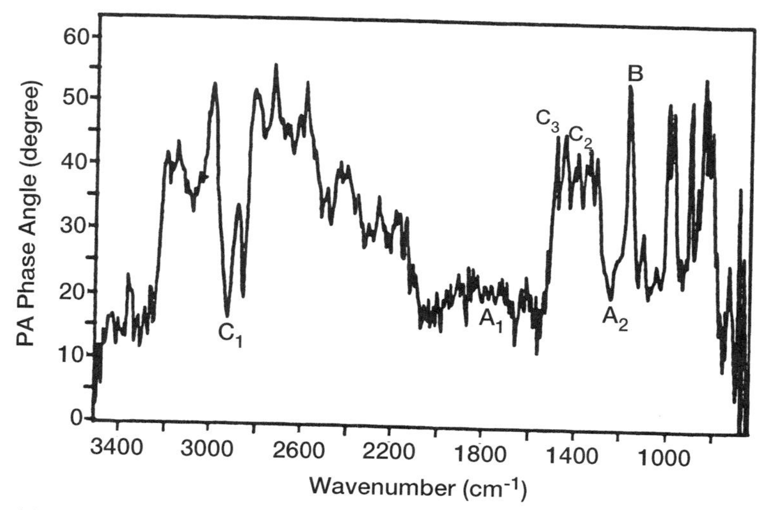


**Figure 4.4.** Phase angle spectrum of a two-layer polymer sample. The top layer was a 12-μm-thick ethylene-vinylacetate copolymer, while the substrate was a 60-μm thick layer of polypropylene. (Reprinted with permission from Jiang, E. Y. and Palmer, R. A., *Anal. Chem.* **69**: 1931–1935. Copyright © 1997 American Chemical Society.)

Phase angle spectra have also been used very recently in depth profiling studies on food products and food packaging materials. For example, the phase angles for a model three-layer (polyethylene/protein/starch) sample confirmed the ordering of the layers in this system: The angles were smallest for the polyethylene, intermediate for the protein, and greatest for the starch (Irudayaraj and Yang, 2002). Similar results were obtained for two-layer (protein/apple, starch/apple) and three-layer (protein/starch/apple) samples (Yang et al., 2001). These experiments are discussed in more detail in the section on generalized two-dimensional correlation at the end of this chapter. Phase angle spectra were also used to confirm the penetration of components from cheese into polymeric packaging material (Irudayaraj and Yang, 2000).

### 4.2.2. Phase Rotation

The phase rotation method was recently described in a series of studies by McClelland, Jones, and their colleagues in the PA research group at Iowa State University. This procedure has also been implemented in other laboratories. It is included in the present discussion because the corresponding experimental data are normally collected using phase modulation step-scan PA infrared spectroscopy. In fact, the information utilized in the phase rotation technique is essentially equivalent to that employed in the calculation of phase spectra, which were discussed in the preceding section.

As is demonstrated below, the phase rotation plot is a graphical representation of the variation of PA intensity with phase rotation angle. Data can be acquired by varying the detector phase angle (e.g., using a lock-in amplifier) so as to determine where the extinction of the bands of interest occurs; alternatively, band intensities can be monitored as this angle is systematically incremented in a series of related PA experiments. A third technique involves the measurement of PA spectra at two phase settings that are separated by 90°. The intensities at the other angles are then calculated by adding these two spectra vectorially.

A significant advantage of the method is that the phase angle plots are rather intuitive in appearance; this naturally tends to facilitate their interpretation. With regard to possible applications, phase rotation can be used in depth profiling studies that are undertaken to confirm the ordering of the layers in a sample. It also provides information regarding the sampling depth in a PA experiment. Moreover, this technique can be used to determine layer thickness in appropriate circumstances. All of these applications are demonstrated in the following paragraphs.

Some of the capabilities of phase rotation are shown in a brief report that describes phase modulation PA infrared spectra of chemically surface-treated polystyrene spheres (McClelland et al., 1994). The magnitude spec-

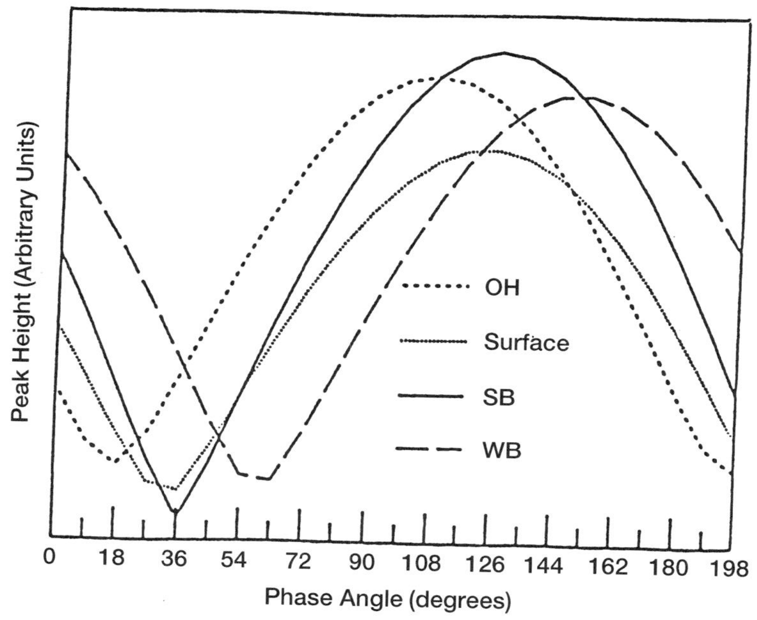


**Figure 4.5.** Phase rotation plot for chemically surface-treated polystyrene spheres. OH, hydroxyl groups; SB, strong band; WB, weak band (see text). (Reproduced from McClelland, J. F. et al., Depth profiling with step-scan FTIR photoacoustic spectroscopy, *SPIE* **2089**: 302–303, by permission of the International Society for Optical Engineering; copyright © 1994.)

trum for this sample contains a significant number of bands, among which are four of particular interest. These include a hydroxyl band, a feature due to the surface species, and another two bands that were simply described as weak and strong, respectively. Obviously, no information regarding the relative depths of the absorbing species can be derived from a single spectrum such as this. On the other hand, the variation of peak height with the phase angle (Fig. 4.5) is much more informative. In order of increasing phase angle minimum in the 18–63° region, the curves in this plot arise from (a) the hydroxyl band, (b) the peak from the surface species, (c) the strong band, and (d) the weak band. This sequence is consistent with the known ordering of the sample layers and confirms that depth profiling is indeed possible by the phase rotation approach. It should be noted that the weak band exhibits a larger phase lag than the strong feature because the signal arises from a layer that is nearer to the surface in the latter case (the infrared beam is attenuated more quickly for the stronger band).

A second application of the phase rotation technique was described by Bajic et al. (1995), who studied underground storage tank waste simulants.

Aqueous mixtures of sodium nitrate and disodium nickel ferrocyanide, $Na_2NiFe(CN)_6$, were used to model the stored waste from nuclear fuel processes in this work. Solid samples for PA infrared analysis were prepared in three different ways (freeze-drying, air-drying, and oven-drying), leading to the question as to whether the dried salts might form artificial layers with varying thermal properties. Indeed, the phase angle plots showed that the nitrate bands in the spectra of the air-dried and oven-dried samples exhibited earlier minima than those in the freeze-dried sample. This indicated that the first two drying techniques caused the migration of the soluble sample to the surface, whereas freeze-drying yielded a more homogenous sample. A similar phenomenon did not occur for $Na_2NiFe(CN)_6$, which is insoluble and therefore does not migrate during any of the drying processes. The phase rotation plots of the characteristic bands for the two salts also reflected the fact that the ferrocyanide absorption is the stronger of the two (its absorption coefficient maximizes at a smaller angle).

Jiang and Palmer (1997) used the phase rotation method to analyze PA data obtained for a two-layer sample that consisted of a 12-μm ethylene-vinyl acetate copolymer on a 60-μm polypropylene substrate. This experiment was also mentioned above in the discussion of the phase spectrum technique. Three bands due to the top layer exhibited a minimum in the phase rotation plot at an angle of about 80°; two peaks with contributions from both layers showed a distinctly different minimum near 100°; and finally, the characteristic angle for a weak substrate band was approximately 118°. These results nicely demonstrate that the phase rotation method can be used for depth profiling in two-layer polymer samples. Jiang and Palmer also reported that the three respective angles determined by phase rotation differ by practically the same amounts as the angles deduced from the corresponding phase spectrum.

As mentioned earlier, phase rotation can also be used to calculate layer thickness. For the particular case of a transparent layer and an absorbing substrate, Adams and Kirkbright (1977a,b) showed that the thermal wave at the surface lags behind that at the interface by

$$\Delta\theta = d/\mu_s \tag{4.5}$$

where $d$ is the thickness of the upper layer and $\mu_s$ is its thermal diffusion length. In simple terms, the phase lag $\Delta\theta$ exists because an interval of time is required for the thermal wave to propagate through the transparent layer and reach the surface of the two-layer sample.

Obviously, the upper layer is not transparent at all frequencies; in the example under discussion, it need only be transparent at the characteristic wavenumber where the substrate absorbs. In fact, the phase reference is established at a second frequency where the upper layer absorbs strongly. This is demonstrated below.

The phase lag $\Delta\theta$ is equal to the difference between the phase angle for the substrate band in the spectrum of a sample with a particular overlayer thickness and the reference angle, determined in a separate measurement for a strong band arising from the surface layer. It is important to note that the reference angle is expected to be essentially independent of film thickness.

This method has been demonstrated by Jones and McClelland (1996), McClelland et al. (1997), as well as Wahls and Leyte (1998a). In the first two studies, thin films of polethylene terephthalate (PET) on a much thicker polycarbonate substrate were examined. The reference phase angle was determined using the strong PET band at about 1725 $cm^{-1}$; as anticipated, this angle did not vary with film thickness. A polycarbonate band at 1771 $cm^{-1}$ was then studied in a series of spectra for samples having different PET film thicknesses. The differences between the phase angles at which the 1771-$cm^{-1}$ band exhibited maximal intensity and the reference angle (McClelland et al., 1997) are plotted as open circles in Figure 4.6. The filled

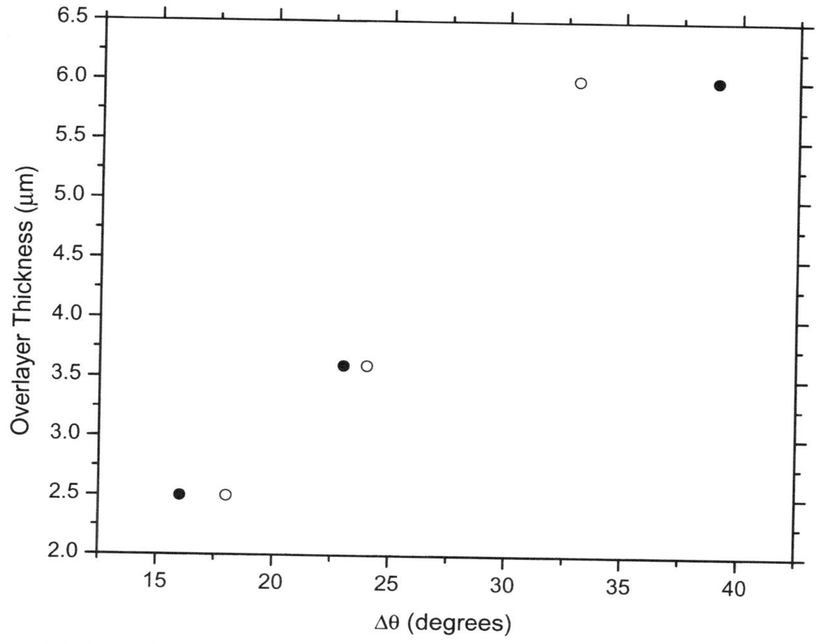


**Figure 4.6.** Determination of the thickness of PET films on a polycarbonate substrate by the phase rotation method (McClelland et al., 1997). Open circles, experimental results; filled circles, predicted values.

circles are the corresponding predicted values, calculated by assuming a thermal diffusivity of $1.0 \times 10^{-3}$ cm$^2$ s$^{-1}$ for PET. The good agreement in these results shows that this method can indeed be used to determine layer thickness. The data have been further analyzed using the numerical method known as expectation minimum analysis (Power and Prystay, 1995). Similar results were obtained in the other study of this two-layer system (Jones and McClelland, 1996).

It was mentioned above that PA spectra can be measured at a series of phase angles, with larger phase delays corresponding to greater sample depths. Alternatively, spectra can be calculated at various angles using the in-phase and in-quadrature components of the PA signal at a given modulation frequency. This approach was followed in several investigations by Sowa, Mantsch, and their colleagues at National Research Council Canada. Most of this work involved biomedical applications. Examples included an extracted but intact human tooth, where depth profiling revealed an increasing protein contribution relative to the mineral (hydroxyapatite and phosphate) component as depth increased. Despite the complexity of tooth enamel, the distribution of these two components resembles a simple two-layer system (Sowa and Mantsch, 1994a,b). Human finger nails were also analyzed successfully at selected depths using this principle (Sowa et al., 1995). This research is also discussed in the section on medical applications in Chapter 6.

### 4.3. GENERALIZED TWO-DIMENSIONAL (G2D) CORRELATION

The final depth profiling method to be discussed in this chapter is generalized two-dimensional (G2D) correlation. This procedure, originally developed by I. Noda and colleagues at the Procter and Gamble Company, has been used successfully by several groups in PA infrared experiments on layered samples—the most popular example being polymer laminates. Some relevant examples of the use of this technique are described below.

As explained by Noda et al. (1997), G2D is an extension of the original method of two-dimensional infrared spectroscopy, where fluctuations in infrared intensities induced by small-amplitude sample deformations are studied by cross-correlation analysis. Several key studies that describe the more general technique of 2D infrared spectroscopy are included as references in the recommended reading list at the end of this chapter. In G2D, the external perturbation applied to the sample does not have to be mechanical. Moreover, there is no requirement that the perturbation correspond to a sinusoidal variation; in fact, it need not be time dependent. The only requirement is that it must give rise to selective changes in the intensities of the bands in a spectrum.

The basis of G2D is that synchronous and asynchronous correlation intensities arise from similarities and dissimilarities, respectively, between the variations of the spectral intensities at wavenumbers $\nu_1$ and $\nu_2$. Peaks that occur along the diagonal of a synchronous correlation plot ("autopeaks") represent the magnitude of the changes in spectral intensities during the perturbation; those in off-diagonal positions ("cross peaks") indicate the simultaneous intensification of two bands (positive) or, alternatively, opposite variations in band intensities (negative). A synchronous correlation map is symmetric with respect to its diagonal.

In an asynchronous 2D correlation spectrum, cross peaks appear when the changes in band intensities differ: A positive feature occurs when $\nu_1$ is affected before $\nu_2$, and a negative feature occurs in the reverse case. An asynchronous correlation map is antisymmetric with regard to its diagonal.

Because all of the information contained in a conventional (one-dimensional) spectrum is reorganized in a two-dimensional display, G2D can also help in the identification of weak features that are obscured by stronger neighboring bands in ordinary circumstances. This is particularly true in asynchronous correlation. However, as pointed out by Jiang and Palmer (1997), artifactual peaks are occasionally created by this method. Hence the correlation results should be interpreted with care, as is the situation in any numerical analysis of spectra.

In the context of a PA step-scan depth profiling experiment, the above-mentioned perturbation corresponds to the external modulation. As discussed earlier, either amplitude or phase modulation can in principle be used. However, the latter technique is certainly the more popular of the two in practice. For both types of modulation, the bands arising from surface species exhibit greater relative intensities at higher modulation frequencies, while those due to the bulk sample are proportionally stronger at lower frequencies.

Synchronous cross peaks indicate the similarity of the time signatures of the PA signals at the two wavenumbers in question. Because a time lag is associated with the propagation of the thermal wave through the sample, synchronicity implies that the signals arise from the same depth. The contours are positive because all band intensities decrease as modulation frequency increases. On the other hand, in the asynchronous spectrum a positive peak at a given spectral coordinate implies that the first peak arises from a layer that is shallower than that for the second peak. The depth relationship is reversed for a negative asynchronous cross peak. This principle permits depth profiling of heterogeneous or layered samples in PA infrared spectroscopy by G2D correlation analysis.

### 4.3.1. G2D Correlation in the Study of Layered Polymer Systems

Several studies have presented straightforward illustrations of the G2D correlation technique for relatively simple layered polymer samples. For example, Gregoriou and Hapanowicz (1997) examined a two-layer system consisting of a micrometre-thick acrylic polymer structure coated on a PET substrate that was several tens of micrometres thick. The synchronous correlation map for this sample confirmed that the PET carbonyl overtone was correlated with the 3055-$cm^{-1}$ band, which was also known to originate from PET. Similarly, Jiang et al. (1998) used the G2D asynchronous correlation spectrum to analyze data for a sample consisting of a 1-μm polystyrene layer on a 100-μm mylar substrate. The signs of the contours in the correlation plot were consistent with the known ordering of the layers. Relatively high phase modulation frequencies were used in these experiments because the top layers were so thin.

Jiang and Palmer (1997) compared the phase rotation, phase spectrum, and G2D correlation methods with regard to the depth profiling of a two-layer polymer sample. In this study, the thicknesses of the layers were not as disparate as those mentioned above: The top (ethylene-vinylacetate copolymer) layer was 12 μm thick, whereas the substrate consisted of 60-μm polypropylene. The synchronous map (Fig. 4.7) shows that the bands labeled $A_1$, $A_2$, $C_2$, and $C_3$ have similar phases, that is, their spatial origins (depths) are about the same. Bands $A_1$ and $A_2$ are due to the ethylene-vinylacetate layer, while the other two (overlapping) bands arise mainly from the polypropylene. By contrast, the asynchronous map (Fig. 4.8) reveals that the cross peaks involving the much weaker band B (1169 $cm^{-1}$) and the other four bands are quite prominent: This confirms that band B arises from a greater sample depth (the polypropylene layer). In the interpretation of these data, it should be kept in mind that both PA intensities and phases affect the results.

The G2D technique has also been applied to more complex layered systems. Jiang et al. (1997b) reported results for a four-layer sample consisting of polyethylene (10 μm), polypropylene (10 μm), polyethylene terephthalate (6 μm), and polycarbonate (6 mm). The ordering of the layers was confirmed from the signs of a series of asynchronous correlation contours that involved the comparison of various band pairs. Some improvement in the definition of partly overlapping bands was also observed in the 2D plots.

The above results clearly demonstrate that G2D correlation can be successfully used to characterize polymer laminates. While the work carried out so far has involved samples of known compositions, the application of the technique to less well-characterized samples can also be confidently predicted to yield many useful results.

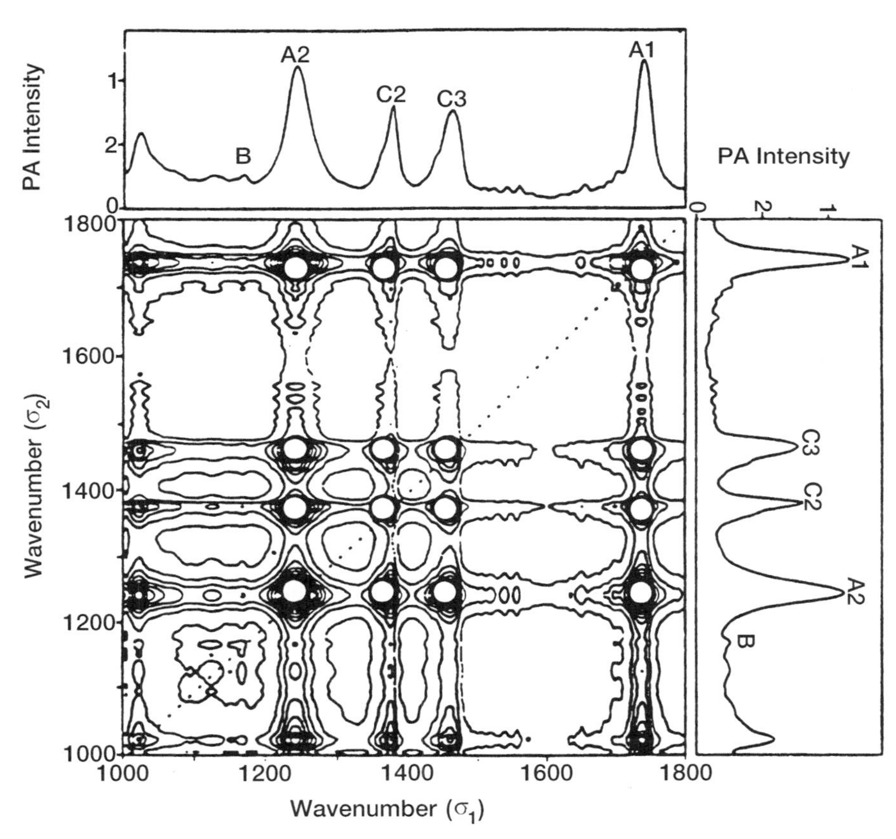


**Figure 4.7.** Synchronous 2D plot for two-layer polymer sample. The top layer was a 12-μm-thick ethylene-vinylacetate copolymer, and the substrate was 60-μm thick polypropylene. The modulation frequency was 200 Hz. (Reprinted with permission from Jiang, E. Y. and Palmer, R. A., *Anal. Chem.* **69**: 1931–1935. Copyright © 1997 American Chemical Society.)

### 4.3.2. G2D Correlation in the Study of Foods

PA infrared spectroscopy has been successfully used to study food products by a number of research groups. The literature on this subject is reviewed in detail in Chapter 6. In the context of the present discussion on depth profiling by G2D correlation analysis, it is important to mention several recent publications by Irudayaraj and his colleagues at Pennsylvania State University. This group has successfully utilized G2D correlation, along with other depth profiling methods, in the analysis of foods and food wrapping products. The principal results of this work are described in the next two paragraphs.

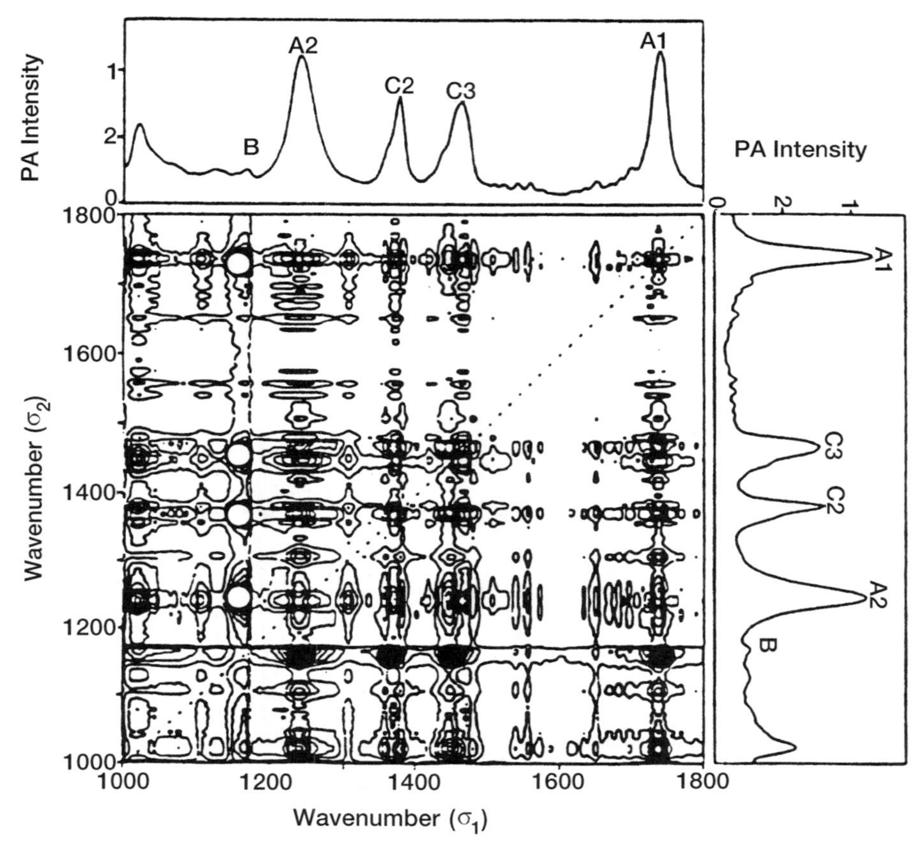


**Figure 4.8.** Asynchronous 2D plot for the two-layer polymer sample in Fig. 4.7. Open circles represent positive peaks, whereas filled circles denote negative peaks. (Reprinted with permission from Jiang, E. Y. and Palmer, R. A., *Anal. Chem.* **69**: 1931–1935. Copyright © 1997 American Chemical Society.)

Yang and Irudayaraj (2000) used G2D in a study of the migration of cheese components into polymeric packaging material. PA spectra of cheese package wrappers were obtained after various storage periods and then analyzed for evidence of penetration of the cheese into the films. The synchronous G2D spectrum did not yield any information on the adherence of the cheese components to the package film, although it did facilitate the identification of a C=C band at 1651 $cm^{-1}$. The asynchronous correlation spectrum was more informative, indicating that the amide I and II (protein) bands of the cheese originated from a shallower layer of the package than several bands arising from $CH_2$ and $CH_3$ groups in the polymeric packaging material. This result confirmed the presence of some components of the cheese near the surface of the package film. In addition, a 1461-cm$^{-1}$ C—H band that is not identifiable in the one-dimensional PA spectrum was clearly revealed in this G2D correlation spectrum.

This group also studied edible coatings on fruit surfaces with this technique (Yang et al., 2001). For a model two-layer sample consisting of a protein film on an apple skin, G2D was used to verify the known fact that the apple skin was below the protein film layer. Similarly, two different three-layer samples were examined: G2D confirmed the starch/protein/apple and protein/starch/apple ordering of these more complicated systems. Phase angle spectra yielded analogous results. A starch/protein/polyethylene three-layer system was also successfully analyzed by this technique (Irudayaraj and Yang, 2002). The ability of G2D to resolve overlapping peaks was noted as a particular advantage in this work. This research shows that G2D can be utilized to detect the penetration of foods into package materials and to study the interactions between foods and their coatings.

## REFERENCES

Adams, M. J., and Kirkbright, G. F. (1977a). Analytical optoacoustic spectrometry. Part III. The optoacoustic effect and thermal diffusivity. *Analyst* **102**: 281–292.

Adams, M. J., and Kirkbright, G. F. (1977b). Thermal diffusivity and thickness measurements for solid samples utilising the optoacoustic effect. *Analyst* **102**: 678–682.

Bajic, S. J., Luo, S., Jones, R. W., and McClelland, J. F. (1995). Analysis of underground storage tank waste by Fourier transform infrared photoacoustic spectroscopy. *Appl. Spectrosc.* **49**: 1000–1005.

Dittmar, R. M., Chao, J. L., and Palmer, R. A. (1991). Photoacoustic depth profiling of polymer laminates by step-scan Fourier transform infrared spectroscopy. *Appl. Spectrosc.* **45**: 1104–1110.

Donini, J. C., and Michaelian, K. H. (1984). Effect of cell resonance on depth profiling in photoacoustic FTIR spectra. *Infrared Phys.* **24**: 157–163.

Drapcho, D. L., Curbelo, R., Jiang, E. Y., Crocombe, R. A., and McCarthy, W. J. (1997). Digital signal processing for step-scan Fourier transform infrared photoacoustic spectroscopy. *Appl. Spectrosc.* **51**: 453–460.

Gonon, L., Vasseur, O. J., and Gardette, J.-L. (1999). Depth profiling of photooxidized styrene–isoprene copolymers by photoacoustic and micro-Fourier transform infrared spectroscopy. *Appl. Spectrosc.* **53**: 157–163.

Gonon, L., Mallegol, J., Commereuc, S., and Verney, V. (2001). Step-scan FTIR and photoacoustic detection to assess depth profile of photooxidized polymer. *Vibr. Spectrosc.* **26**: 43–49.

Gregoriou, V. G., and Hapanowicz, R. (1996). Sub-micron resolution depth profiling of thin coatings using step-scan photoacoustic FT-IR spectroscopy. *Prog. Nat. Sci.* **6**(Suppl.) S-10–S-13.

Gregoriou, V. G., and Hapanowicz, R. (1997). Applications of photoacoustic step-scan FT-IR spectroscopy to polymeric materials. *Macromol. Symp.* **119**: 101–111.

Irudayaraj, J., and Yang, H. (2000). Analysis of cheese using step-scan Fourier transform infrared photoacoustic spectroscopy. *Appl. Spectrosc.* **54**: 595–600.

Irudayaraj, J., and Yang, H. (2002). Depth profiling of a heterogeneous food-packaging model using step-scan Fourier transform infrared photoacoustic spectroscopy. *J. Food Eng.* **55**: 25–33.

Jiang, E. Y. (1999). Heterogeneity studies of a single particle/fiber by using Fourier transform infrared micro-sampling photoacoustic spectroscopy. *Appl. Spectrosc.* **53**: 583–587.

Jiang, E. Y., and Palmer, R. A. (1997). Comparison of phase rotation, phase spectrum, and two-dimensional correlation methods in step-scan Fourier transform infrared photoacoustic spectral depth profiling. *Anal. Chem.* **69**: 1931–1935.

Jiang, E. Y., Palmer, R. A., and Chao, J. L. (1995). Development and applications of a photoacoustic phase theory for multilayer materials: The phase difference approach. *J. Appl. Phys.* **78**: 460–469.

Jiang, E. Y., Palmer, R. A., Barr, N. E., and Morosoff, N. (1997a). Phase-resolved depth profiling of thin-layered plasma polymer films by step-scan Fourier transform infrared photoacoustic spectroscopy. *Appl. Spectrosc.* **51**: 1238–1244.

Jiang, E. Y., McCarthy, W. J., Drapcho, D. L., and Crocombe, R. A. (1997b). Generalized two-dimensional Fourier transform infrared photoacoustic spectral depth-profiling analysis. *Appl. Spectrosc.* **51**: 1736–1740.

Jiang, E. Y., Drapcho, D. L., McCarthy, W. J., and Crocombe, R. A. (1998). Frequency-resolved, phase-resolved and time-resolved step-scan Fourier transform infrared photoacoustic spectroscopy. *AIP Conf. Proc.* **430**: 381–384.

Jones, R. W., and McClelland, J. F. (1996). Quantitative depth profiling of layered samples using phase-modulation FT-IR photoacoustic spectroscopy. *Appl. Spectrosc.* **50**: 1258–1263.

Jones, R. W., and McClelland, J. F. (2002). Quantitative depth profiling using saturation-equalized photoacoustic spectra. *Appl. Spectrosc.* **56**: 409–418.

Lerner, B., Perkins, J. H., Pariente, G. L., and Griffiths, P. R. (1989). Sample depth profiling by PAS/step-scanning interferometry: Consideration of mirror positioning errors. *SPIE* **1145**: 476–477.

McClelland, J. F., Luo, S., Jones, R. W., and Seaverson, L. M. (1992). A tutorial on the state-of-the art of FTIR photoacoustic spectroscopy. In: *Photoacoustic and Photothermal Phenomena III.* D. Bićanić (ed.). Springer, Berlin, pp. 113–124.

McClelland, J. F., Jones, R. W., and Ochiai, S. (1994). Depth profiling with step-scan FT-IR photoacoustic spectroscopy. *SPIE* **2089**: 302–303.

McClelland, J. F., Jones, R. W., Bajic, S. J., and Power, J. F. (1997). Depth profiling by FT-IR photoacoustic spectroscopy. *Mikrochim. Acta [Suppl.]* **14**: 613–614.

Michaelian, K. H. (1989). Depth profiling and signal saturation in photoacoustic FT-IR spectra measured with a step-scan interferometer. *Appl. Spectrosc.* **43**: 185–190.

Michaelian, K. H. (1991). Depth profiling of oxidized coal by step-scan photoacoustic FT-IR spectroscopy. *Appl. Spectrosc.* **45**: 302–304.

Muraishi, S. (1984). Depth profile analysed by FT-IR/PAS spectrophotometry. *Bunko Kenkyu* **33**: 269–270.

Noda, I., Story, G. M., Dowrey, A. E., Reeder, R. C., and Marcott, C. (1997). Applications of two-dimensional correlation spectroscopy in depth-profiling photoacoustic spectroscopy, near-infrared dynamic rheo-optics, and spectroscopic imaging microscopy. *Macromol. Symp.* **119**: 1–13.

Ochiai, S. (1985). FT-IR spectroscopy analysis of depth profile of coating films. *Toso Kogaku* **20**: 192–195.

Palmer, R. A., and Dittmar, R. M. (1993). Step-scan FT-IR photothermal spectral depth profiling of polymer films. *Thin Solid Films* **223**: 31–38.

Palmer, R. A., Chao, J. L., Dittmar, R. M., Gregoriou, V. G., and Plunkett, S. E. (1993). Investigation of time-dependent phenomena by use of step-scan FT-IR. *Appl. Spectrosc.* **47**: 1297–1310.

Palmer, R. A., Jiang, E. Y., and Chao, J. L. (1994). Phase analysis and its application in step-scan FT-IR photoacoustic depth profiling. *SPIE* **2089**: 250–251.

Palmer, R. A., Jiang, E. Y., and Chao, J. L. (1997). Step-scan FT-IR photoacoustic ($S^2$ FT-IR PA) spectral depth profiling of layered materials. *Mikrochim. Acta [Suppl.]* **14**: 591–594.

Power, J. F., and Prystay, M. C. (1995). Expectation minimum (EM): A new principle for the solution of ill-posed problems in photothermal science. *Appl. Spectrosc.* **49**: 709–724.

Sowa, M. G., and Mantsch, H. H. (1994a). FT-IR step-scan photoacoustic phase analysis and depth profiling of calcified tissue. *Appl. Spectrosc.* **48**: 316–319.

Sowa, M. G., and Mantsch, H. H. (1994b). Phase modulated–phase resolved photoacoustic FT-IR study of calcified tissues. *SPIE* **2089**: 128–129.

Sowa, M. G., Wang, J., Schultz, C. P., Ahmed, M. K., and Mantsch, H. H. (1995). Infrared spectroscopic investigation of in vivo and ex vivo human nails. *Vibr. Spectrosc.* **10**: 49–56.

Sowa, M. G., Fischer, D., Eysel, H. H., and Mantsch, H. H. (1996). FT-IR PAS depth profiling investigation of polyethylene surface sulfonation. *J. Mol. Struct.* **379**: 77–85.

Story, G. M., and Marcott, C. (1998). Uniform depth profiling of multiple sample depth ranges in a single step-scanning FT-IR photoacoustic experiment. *AIP Conf. Proc.* **430**: 513–515.

Story, G. M., Marcott, C., and Noda, I. (1994). Phase correction and two-dimensional correlation analysis for depth-profiling photoacoustic step-scan FT-IR spectroscopy. *SPIE* **2089**: 242–243.

Stout, P. J., and Crocombe, R. A. (1994). PAS/FT-IR spectroscopy with a step-scan spectrometer. *SPIE* **2089**: 300–301.

Szurkowski, J., and Wartewig, S. (1999a). Application of photoacoustic spectroscopy in visible and infrared regions to studies of thin olive oil layers on water. *AIP Conf. Proc.* **463**: 618–620.

Szurkowski, J., and Wartewig, S. (1999b). Application of photoacoustic spectroscopy to studies of thin olive oil layers on water. *Instrument. Sci. Technol.* **27**: 311–317.

Szurkowski, J., Pawelska, I., Wartewig, S., and Pogorzelski, S. (2000). Photoacoustic study of the interaction between thin oil layers with water. *Acta Phys. Polon. A* **97**: 1073–1082.

Teramae, N., and Tanaka, S. (1985). Subsurface layer detection by Fourier transform infrared photoacoustic spectroscopy. *Appl. Spectrosc.* **39**: 797–799.

Urban, M. W. (1987). Photoacoustic Fourier transform infrared spectroscopy: A new method for characterization of coatings. *J. Coatings Technol.* **59**: 29–34.

Urban, M. W., and Koenig, J. L. (1986). Depth-profiling studies of double-layer $PVF_2$-on-PET films by Fourier transform infrared photoacoustic spectroscopy. *Appl. Spectrosc.* **40**: 994–998.

Urban, M. W., Allison, C. L., Johnson, G. L., and Di Stefano, F. (1999). Stratification of butyl acrylate/polyurethane (BA/PUR) latexes: ATR and step-scan photoacoustic studies. *Appl. Spectrosc.* **53**: 1520–1527.

Vidrine, D. W. (1981). Photoacoustic Fourier transform infrared (FTIR) spectroscopy of solids. *SPIE* **289**: 355–360.

Vidrine, D. W., and Lowry, S. R. (1983). Photoacoustic Fourier transform IR spectroscopy and its application to polymer analysis. *Adv. Chem. Ser. (Polym. Charact.)* **203**: 595–613.

Wahls, M. W. C., and Leyte, J. C. (1998a). Step-scan FT-IR photoacoustic studies of a double-layered polymer film on metal substrates. *Appl. Spectrosc.* **52**: 123–127.

Wahls, M. W. C., and Leyte, J. C. (1998b). Fourier transform infrared photoacoustic spectroscopy of polymeric laminates. *J. Appl. Phys.* **83**: 504–509.

Yamada, O., Yasuda, H., Soneda, Y., Kobayashi, M., Makino, M., and Kaiho, M. (1996). The use of step-scan FT-IR/PAS for the study of structural changes in coal and char particles during gasification. *Preprints, ACS Div. Fuel Chem.* **41**: 93–97.

Yang, C. Q., and Fateley, W. G. (1987). Fourier-transform infrared photoacoustic spectroscopy evaluated for near-surface characterization of polymeric materials. *Anal. Chim Acta* **194**: 303–309.

Yang, C. Q., Ellis, T. J., Bresee, R. R., and Fateley, W. G. (1985). Depth profiling of FT-IR photoacoustic spectroscopy and its applications for polymeric material studies. *Polym. Mater. Sci. Eng.* **53**: 169–175.

Yang, C. Q., Bresee, R. R., and Fateley, W. G. (1987). Near-surface analysis and depth profiling by FT-IR photoacoustic spectroscopy. *Appl. Spectrosc.* **41**: 889–896.

Yang, C. Q., Bresee, R. R., and Fateley, W. G. (1990). Studies of chemically modified poly(ethylene terephthalate) fibers by FT-IR photoacoustic spectroscopy and X-ray photoelectron spectroscopy. *Appl. Spectrosc.* **44**: 1035–1039.

Yang, H., and Irudayaraj, J. (2000). Depth profiling Fourier transform analysis of cheese package using generalized two-dimensional photoacoustic correlation spectroscopy. *Trans. Am. Soc. Agric. Eng.* **43**: 953–961.

Yang, H., and Irudayaraj, J. (2001). Characterization of beef and pork using Fourier transform infrared photoacoustic spectroscopy. *Lebensm.-Wiss. u.-Technol.* **34**: 402–409.

Yang, H., Irudayaraj, J., and Sakhamuri, S. (2001). Characterization of edible coatings and microorganisms on food surfaces using Fourier transform infrared photoacoustic spectroscopy. *Appl. Spectrosc.* **55**: 571–583.

Zerlia, T. (1986). Depth profile study of large-sized coal samples by Fourier transform infrared photoacoustic spectroscopy. *Appl. Spectrosc.* **40**: 214–217.

## RECOMMENDED READING

Curbelo, R. (1998). Digital signal processing (DSP) applications in FT-IR. Implementation examples for rapid and step scan systems. *AIP Conf. Proc.* **430**: 74–83.

Michaelian, K. H. (1989). Interferogram symmetrization and multiplicative phase correction of rapid-scan and step-scan photoacoustic FT-IR data. *Infrared Phys.* **29**: 87–100.

Michaelian, K. H. (1990). Step-scan photoacoustic infrared spectra of kaolinite. *Infrared Phys.* **30**: 181–186.

Noda, I. (1989). Two-dimensional infrared spectroscopy. *J. Am. Chem. Soc.* **111**: 8116–8118.

Noda, I. (1990). Two-dimensional infrared (2D IR) spectroscopy: theory and applications. *Appl. Spectrosc.* **44**: 550–561.

Noda, I. (1993). Generalized two-dimensional correlation method applicable to infrared, Raman, and other types of spectroscopy. *Appl. Spectrosc.* **47**: 1329–1336.

Palmer, R. A., Manning, C. J., Chao, J. L., Noda, I., Dowrey, A. E., and Marcott, C. (1991). Application of step-scan interferometry to two-dimensional Fourier transform infrared (2D FT-IR) correlation spectroscopy. *Appl. Spectrosc.* **45**: 12–17.

Wahls, M. W. C., Toutenhoofd, J. P., Leyte-Zuiderweg, L. H., de Bleijser, J., and Leyte, J. C. (1997). Step-scan FT-IR photoacoustic spectroscopy: the signal phase. *Appl. Spectrosc.* **51**: 552–557.

Wahls, M. W. C., Weisman, J. L., Jesse, W. J., and Leyte, J. C. (1998). Step-scan FTIR photoacoustic spectroscopy: The phase-response of the surface reference sample carbon black. *AIP Conf. Proc.* **430**: 392–394.

CHAPTER

5

# NUMERICAL METHODS

## 5.1. NORMALIZATION OF PA INFRARED SPECTRA

Most modern infrared spectra are measured in single-beam experiments and then corrected for wavenumber-dependent instrumental effects through division by a reference ("background") spectrum. This strategy implicitly assumes that the stability of the instrumentation used is adequate to ensure reliable results, even though the sample and reference spectra are acquired at different times. In the specific case where an FTIR spectrometer is used, the reference spectrum can be regarded as a representation of the optical throughput of the "empty instrument"; in other words, its profile is dictated by the optical and electronic characteristics of the spectrometer. This spectrum is measured in the absence of a chromophore. For example, a reference for transmission experiments on liquids is sometimes obtained using an empty sample cell. Similarly, a background that is suitable for correcting spectra of KBr pellets or diffuse reflectance data is often acquired by recording the corresponding spectrum of the neat diluent. Analogous strategies are used with a variety of other sample accessories.

The requirement for a reference spectrum naturally also exists in PA infrared spectroscopy. It is well known that the detection of a PA signal depends on the presence of an absorber; because no optical detector is employed in this experiment, an empty-cell spectrum similar to that described above does not exist. A different strategy is adopted instead: The reference PA spectrum is recorded using a substance with strong, ideally featureless, absorption. This spectrum is expected to be proportional to the instrumental curve mentioned above and, therefore, to be suitable for normalizing the PA spectra of a variety of samples. Typical PA reference spectra are measured using powdered carbon black, glassy carbon, carbon-filled polymers, or (occasionally) finely divided metals. The normalization of PA infrared spectra in this way is discussed and illustrated below.

### 5.1.1. Early Studies on Normalization

During the period when PA infrared spectroscopy was emerging as a modern analytical technique, several research groups examined the issue of normalization of the data. A number of studies published in the 1980s discussed this subject. Some important findings from this work are summarized in this section. For example, Chalmers et al. (1981) noted the similarity between PA FTIR spectra of powdered carbon black (particularly those recorded at lower mirror velocities) and the single-beam energy spectrum obtained with a mercury cadmium telluride (MCT) detector. PA spectra of carbon black were then used to correct the PA spectrum of polyvinyl chloride for instrumental effects. The result closely resembled the conventional absorption spectrum that was measured for this polymer, suggesting the validity of this approach.

Riseman and Eyring (1981) studied the PA spectra of a number of different carbons with regard to their suitability for normalizing FTIR spectra of solids. They observed that the spectra of carbons measured at low mirror velocities most closely resembled the empty-instrument spectrum obtained with a deuterated triglycine sulfate (DTGS) detector. However, the velocity dependence of the PA spectra of the carbons, and various discrepancies between the PA and DTGS data, led the authors to the somewhat pessimistic conclusion that no carbon was actually suitable for normalizing mid-infrared PA spectra. Riseman and Eyring instead proposed that double-beam experiments be utilized so as to record properly normalized PA spectra—a suggestion that was questioned by Low (1983) and, moreover, does not appear to have been received favorably by many spectroscopists.

As noted above, the material used to obtain a PA reference spectrum should not give rise to any well-defined absorption bands. This issue was taken up in several of these early studies, with various apparently anomalous results being put down to the existence of absorption bands in the PA spectra of carbons. Rockley et al. (1984) examined the spectra of hexane soot using attenuated total reflectance (ATR) diffuse reflectance, and PA infrared spectroscopies, observing several bands despite the fact that the carbon content of this sample was 95%. With regard to the PA data, Rockley et al. attributed several suspicious results to the response of the DTGS detector used to obtain the characteristic instrument spectrum in this investigation. This conclusion was subsequently disputed by Low (1985), who ascribed the results to structure in the PA spectrum of hexane soot. Similarly, Riseman and Eyring (1981) mentioned the occurrence of bands that might signify the presence of functional groups in some of their carbons. Low and Parodi (1980) likewise showed that sucrose chars yielded bands attributable to a variety of functional groups, making these carbons a rather poor choice when acquiring spectra to be used for normalization.

### 5.1.2. Sample Calculations

The normalization of a typical PA infrared spectrum of a clay mineral is illustrated in Figures 5.1–5.4. Figure 5.1 depicts the data used in the calcu-

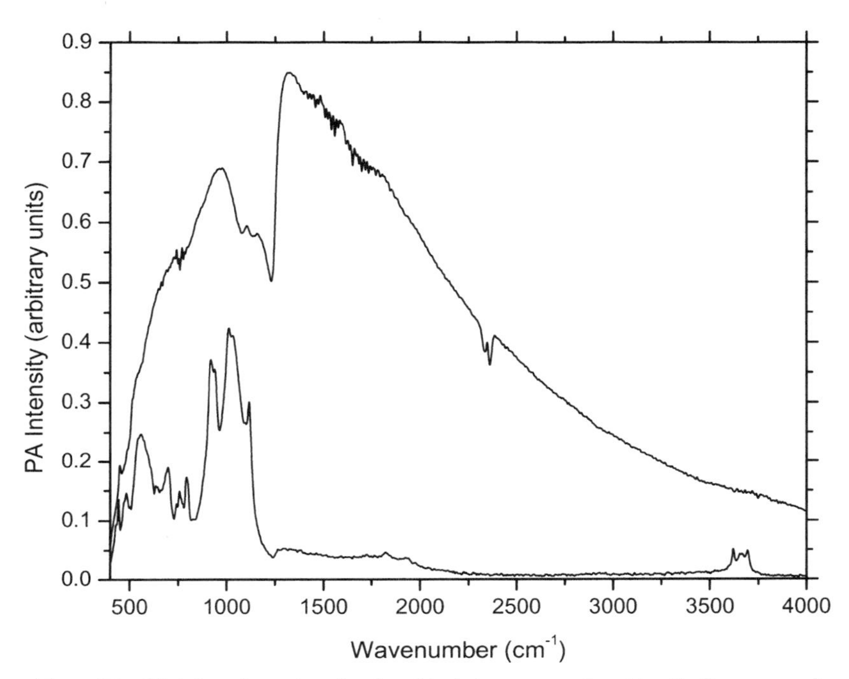


**Figure 5.1.** PA infrared spectra of carbon black (upper curve) and kaolin (lower curve).

lations, which were obtained with a Bruker IFS 113v FTIR and an MTEC 300 PA cell. The upper curve is the reference spectrum recorded for carbon black. It is featureless except for the broad beamsplitter absorption between about 1000 and 1300 $cm^{-1}$, as well as minor bands arising from the water vapor and $CO_2$ in the ambient atmosphere. The lower curve in Figure 5.1 is the PA spectrum measured under similar conditions for kaolin, a mixture of layer silicates with the empirical composition $Al_4Si_4O_{10}(OH)_8$. This is the spectrum to be corrected.

A simple division of the kaolin spectrum by the carbon black reference spectrum yields the result in Figure 5.2. It is evident that this correction has its most noticeable effect at both ends of the mid-infrared region, where the intensity of the radiation incident on the sample is much lower than in the central part (roughly 700–2500 $cm^{-1}$). A ratioed spectrum such as this is generally considered adequate for most qualitative and semiquantitative work.

The observant reader may have already noticed that the hydroxyl stretching bands between 3600 and 3700 $cm^{-1}$ are somewhat weaker relative to the low-frequency bands in Figure 5.2 than in published infrared spectra of this clay, most of which were obtained by transmission measurements on KBr

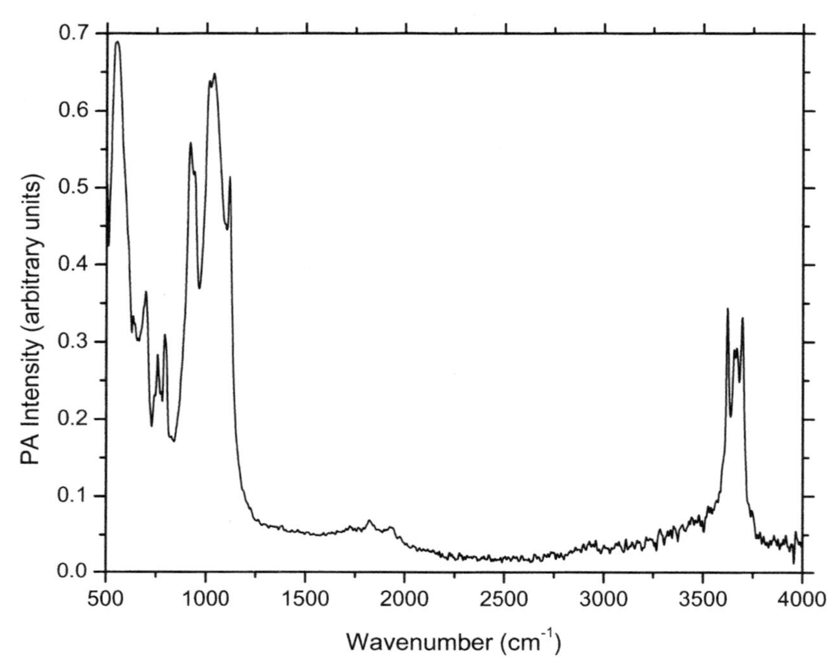


**Figure 5.2.** Normalized PA infrared spectrum of kaolin, obtained through division of lower spectrum in Fig. 5.1 by upper spectrum.

pellets. There is one major reason for this discrepancy. For an optically opaque, thermally thick solid PA intensity is proportional to $\omega^{-3/2}$, where $\omega$ denotes modulation frequency; this fact was mentioned in early investigations on the subject of normalization by Chalmers et al. (1981) and by Teramae et al. (1982). On the other hand, for an optically opaque, thermally thin solid such as carbon black, PA intensity is proportional to $\omega^{-1}$. The second relationship is demonstrated in Figure 5.3, which shows the frequency dependence of the PA signal intensity for carbon black in a simple experiment carried out with a HeNe laser, a mechanical chopper and a lock-in amplifier.

In rapid-scan PA FTIR spectroscopy, $\omega$ is proportional to the infrared wavenumber $\nu$. Therefore, to correct for this difference between the frequency dependences of the sample and reference intensities, the ratioed intensity must be multiplied by $\nu^{1/2}$ at each spectral ordinate. This procedure was performed for the current data and yielded the result plotted in Figure 5.4. Indeed, the redistribution of intensities effected by this calculation produces a modified PA spectrum that agrees much more closely with pre-

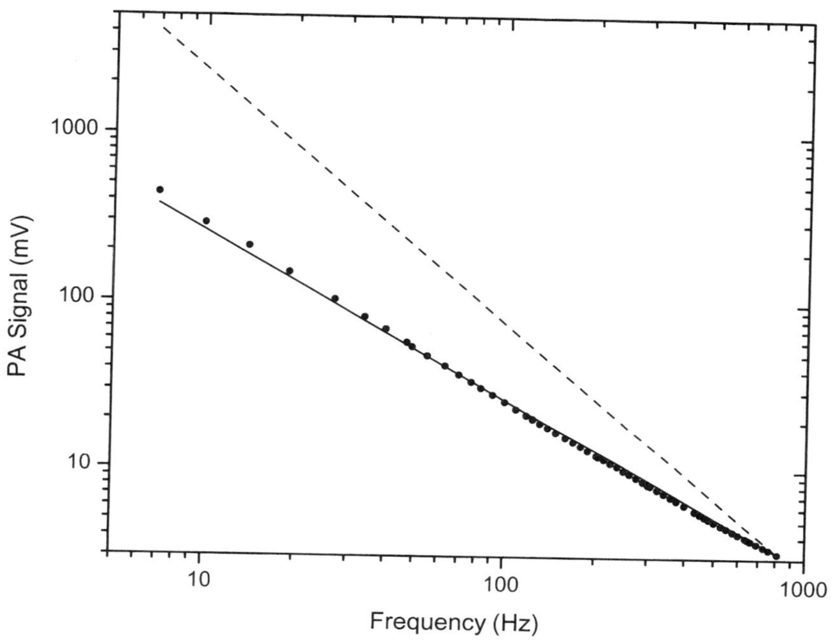


**Figure 5.3.** Frequency dependence of PA signal observed for carbon black using MTEC 300 cell, HeNe laser, mechanical chopper, and lock-in amplifier. Solid line has slope of −1, while dashed line has slope of −1.5.

viously published absorbance spectra of this clay. Although this approach to the correction of PA infrared spectra is not utilized very widely, its simplicity and effectiveness in some circumstances suggest that it should be kept in mind whenever a qualitative comparison between PA and other types of infrared spectra is required.

### 5.1.3. Other Reference Materials

Glassy carbon is now frequently used as an alternative to carbon black powder so as to obtain a PA reference spectrum. The single-wavelength (632.8-nm) experiment mentioned above was repeated for the glassy carbon reference provided with the MTEC 300 PA cell, yielding the results shown in Figure 5.5. It is evident that the frequency dependence of the PA intensity is quite close to $\omega^{-3/2}$ in this case. This implies that the correction of ratioed PA spectra through multiplication by $\nu^{1/2}$ is unnecessary when this reference material is used.

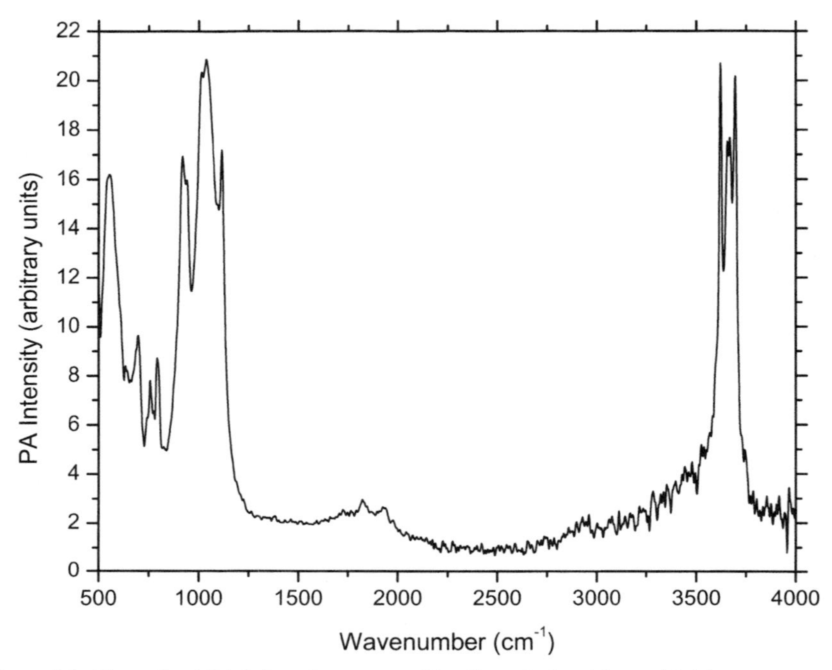


**Figure 5.4.** Normalized PA infrared spectrum of kaolin, calculated by multiplying spectrum in Fig. 5.2 by $\nu^{1/2}$.

It should also be mentioned that some early investigations of PA infrared spectroscopy advocated the use of finely divided metals for the measurement of reference spectra. The utilization of metal powders as an alternative to carbons does not appear to have ever become very popular with most analysts. However, it is the logical strategy in one type of experiment—the spectroscopic study of adsorbed species on metal surfaces! As an example, Figure 5.6 shows the PA infrared spectrum obtained for a 1% mixture of oleic acid and nickel powder. This spectrum has been corrected by ratioing the single-beam spectrum of the mixture to an appropriate reference spectrum obtained for pure nickel powder. The result is a realistic spectrum of oleic acid at this rather low concentration: The C=O stretching band near 1700 cm$^{-1}$ is clearly visible, whereas the C—H bands near 2900 cm$^{-1}$ are near the limit of detection. This experiment is yet another demonstration of the wide range of sample types that are amenable to investigation by PA infrared spectroscopy.

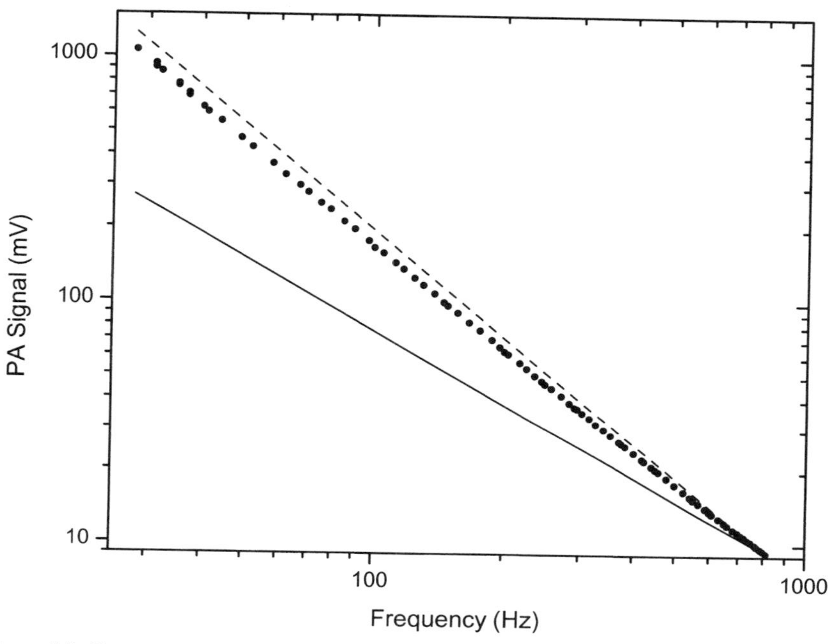


**Figure 5.5.** Frequency dependence of PA signal observed for glassy carbon using MTEC 300 cell, HeNe laser, mechanical chopper, and lock-in amplifier. Solid line has slope of −1, while dashed line has slope of −1.5.

### 5.1.4. Measurement of Spectra at Two Mirror Velocities

As a check on the reliability of the PA data obtained with an FTIR spectrometer, spectra of the sample under analysis are often obtained at two different mirror velocities. These spectra are corrected using reference spectra recorded under similar conditions. Artifacts and questionable features can generally be identified through a comparison of the results and eliminated where necessary by merging the ratioed spectra.

The availability of spectra that were measured at two velocities also makes it possible to use the normalization procedure developed by Teng and Royce (1982). According to these authors, the PA amplitude for samples with a thickness much greater than the thermal diffusion length is given by

$$q = B\{\beta\mu/[(\beta\mu)^2 + 2\beta\mu + 2]^{1/2}\} \tag{5.1}$$

where $\beta$ is the optical absorption coefficient, $\mu$ denotes thermal diffusion

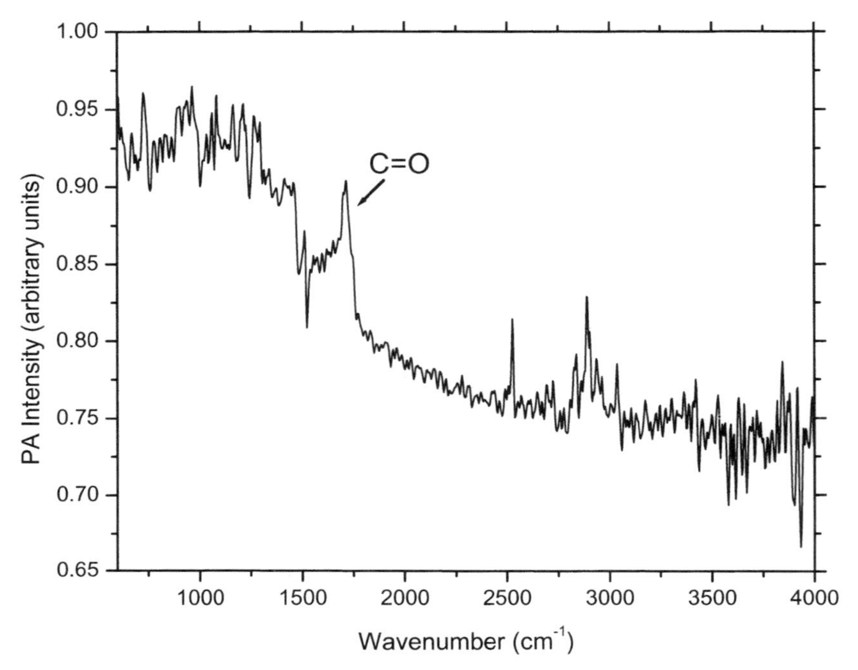


**Figure 5.6.** PA spectrum of 1% oleic acid on nickel powder. This spectrum was normalized through division by spectrum of pure nickel powder.

length, and $B$ is a constant. Labeling the spectra obtained at the two different velocities with the subscripts 1 and 2 allows definition of the ratios

$$\mu_1/\mu_2 = k \tag{5.2}$$

and

$$q_1/q_2 = r \tag{5.3}$$

Rearrangement of these equations yields

$$\beta\mu_1 = \{(k - r^2) + [(k - r^2)^2 + 2(r^2 - 1)(k^2 - r^2)]^{1/2}\}/(r^2 - 1) \tag{5.4}$$

Because $\mu_1$ is inversely proportional to the square root of modulation frequency, multiplication of this expression by $\nu^{1/2}$ yields a normalized absorption spectrum that is proportional to $\beta$.

Teng and Royce (1982) demonstrated this calculation using spectra for thick films of polymethyl methacrylate. Their results showed that the normalized PA spectrum agreed very well with an absorbance spectrum measured for a thin film of this polymer; on the other hand, the ratioed PA spectrum obtained at one particular mirror velocity exhibited relative intensities in disaccord with both normalized PA and transmission spectra.

More recently, Dubois et al. (1994) used this approach to calculate the optical penetration depth as a function of infrared wavelength for graphite-epoxy laminates. The results suggested slightly smaller penetration depths than those calculated using either transmission measurements or the phase analysis method of Bertrand and his collaborators (see below). Dubois et al. interpreted this discrepancy as an indication that the assumptions inherent in Teng and Royce's model are not valid for laminates.

## 5.2. LINEARIZATION OF SPECTRA

In principle, either the magnitude or the phase of the PA signal can be used to quantitatively characterize sample absorption. However, the accuracies with which both quantities are known suffer from limitations: The magnitude is linear over a small range at lower absorptivities, whereas the phase is subject to relatively large errors when absorption is weak. The effects of these restrictions may be minimized by combining magnitude and phase information in a single calculation. In particular, measurement of the phase can be used to extend the region in which the magnitude varies linearly with sample absorbance; this is possible because the upper limit of absorptivity in the PA phase spectrum is much higher than that in the magnitude spectrum (Burggraf and Leyden, 1981).

McClelland (1983) demonstrated the proportionality between PA intensities and absorption coefficients for weak to moderate absorption and showed that deviations from this relationship tend to occur when absorption is strong. The following equation for the linearized PA signal $q_l$ was obtained by combining expressions for the PA signal magnitude and phase:

$$q_l = q[(\cot^2(\psi - \pi/4) + 1)/2]^{1/2} \tag{5.5}$$

where $q$ is the observed signal and $\psi$ is the phase lag introduced in the PA experiment. This expression was subsequently simplified (Carter, 1992) to

$$q_l = q\left[(\cot(\psi - \pi/4) + \tfrac{1}{2}\right]^{1/2} \tag{5.6}$$

where

$$\psi = \psi_s - \psi_r + \pi/4 \tag{5.7}$$

in which the subscripts $s$ and $r$ refer to sample and reference, respectively. Substituting, the second equation becomes

$$q_l = q\left[\cot(\psi_s - \psi_r) + \tfrac{1}{2}\right] \tag{5.8}$$

$\psi_s$ and $\psi_r$ are obtained from

$$\begin{aligned} \sin \psi_j &= I_j/(R_j^2 + I_j^2)^{1/2} \\ \cos \psi_j &= R_j/(R_j^2 + I_j^2)^{1/2} \end{aligned} \tag{5.9}$$

where $I_j$ and $R_j$ are the imaginary and real intensities calculated at each wavenumber $j$.

Carter (1992) described a modification of this linearization procedure that is suitable for application in PA FTIR spectroscopy. In this more recent approach, one sample interferogram and one reference interferogram are measured. The phase and intensity reference are obtained from the same data, in contrast with the approach suggested by McClelland (1983). With this simplification, the linearized sample spectrum is given by

$$q_l = q_s^2/(R_s I_r - I_s R_r) \tag{5.10}$$

where $q_s$ is the power spectrum of the sample. The use of the reference interferogram in this procedure ensures that the linearized spectrum is also normalized with respect to intensity. This approach is valid as long as sample and reference data are acquired under the same conditions so that the instrumental phase is the same. Hence the sample interferogram should be measured immediately before or after the reference interferogram if at all possible.

Carter (1992) compared amplitude and linearized PA spectra of polycarbonate and found that linearization greatly reduced saturation effects, yielding a result that is similar to an ordinary absorption spectrum. On the other hand, the modulation frequency dependence of the sampling depth in rapid-scan PA spectroscopy persists in linearized PA spectra, although its effects are not as dramatic as in amplitude spectra. It should also be noted that bands arising from surface species, such as thin films or oxide layers, are not affected by linearization in the same way as those from components that lie deeper within a sample. This is a consequence of the fact that the phase for surface species is close to $\pi/4$ (45°).

The linearization method has been used by some groups to facilitate analysis of PA infrared spectra. For example, Sowa and Mantsch (1994) compared the linearized PA spectrum of tooth enamel with the corresponding Mertz and modulus spectra. The linearized spectrum displayed somewhat better contrast between absorption maxima and background intensity than that observed in the other two spectra, which incidentally were superimposable.

Implementation of the linearization method is illustrated in the following figures. Double-sided PA interferograms were recorded at a resolution of just under 10 $cm^{-1}$, using a Bruker IFS 113v spectrometer, for kaolin and carbon black; this is the highest obtainable resolution for a double-sided interferogram without modification of this instrument. Amplitude [using multiplicative (Mertz) phase correction] and power spectra were calculated from these interferograms with the standard software that controls the spectrometer. The arithmetical manipulations required for linearization were also performed with this software.

Figures 5.7 and 5.8 depict the square of the power spectrum and the two products in the denominator of Eq. (5.10), respectively. It can be observed that $q_s^2$ and $R_s I_r$ are positive at all frequencies, whereas the product $I_s R_r$ is

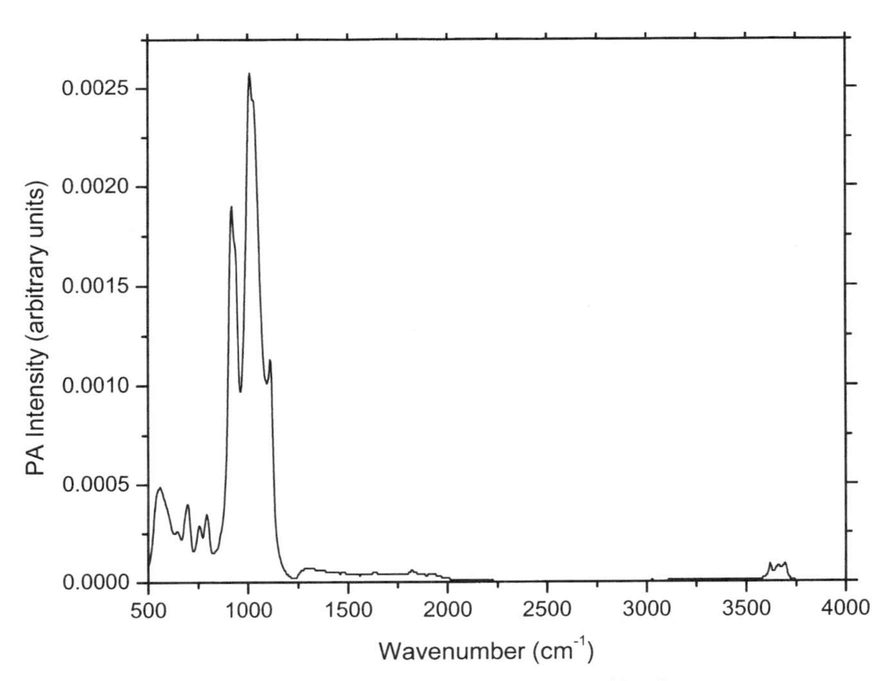


**Figure 5.7.** Square of PA power spectrum of kaolin.

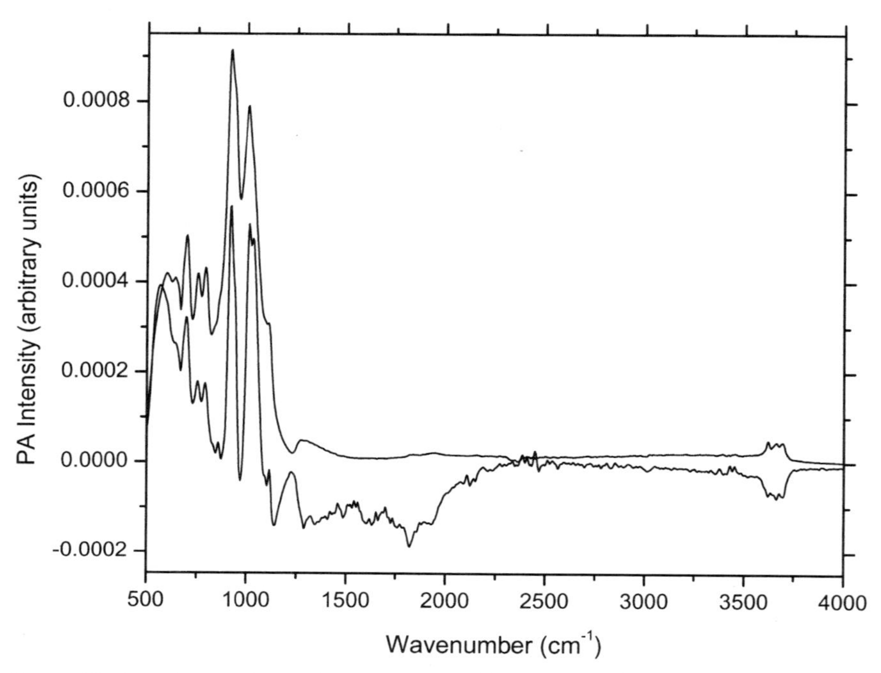


**Figure 5.8.** Quantities used to calculate linearized PA spectrum of kaolin. Upper curve, $R_s I_r$; lower curve, $I_s R_r$.

positive at frequencies below about 1000 cm$^{-1}$ and negative going at higher frequencies.

The linearized PA spectrum of kaolin is compared with the amplitude spectrum, which closely resembles the power spectrum $q_s$, in Figures 5.9 and 5.10. The amplitude spectrum has been divided by an analogous spectrum of carbon black and rescaled to facilitate comparison with the linearized spectrum in these two figures; the scaling factors in these figures are not the same, that is, the intensities are plotted on arbitrary scales. The improved band definition in the linearized spectrum is evident in both the low-frequency (Fig. 5.9) and OH-stretching (Fig. 5.10) regions. The linearized spectrum differs significantly from the PA amplitude spectrum and the published transmission spectrum of kaolinite (Farmer and Russell, 1964) insofar as the intensities of the Si–O peaks at 1010 and 1033 cm$^{-1}$ are quite unequal in the linearized spectrum. On the other hand, the four high-frequency maxima at 3620, 3652, 3669, and 3697 cm$^{-1}$ in the linearized spectrum display the expected relative intensities.

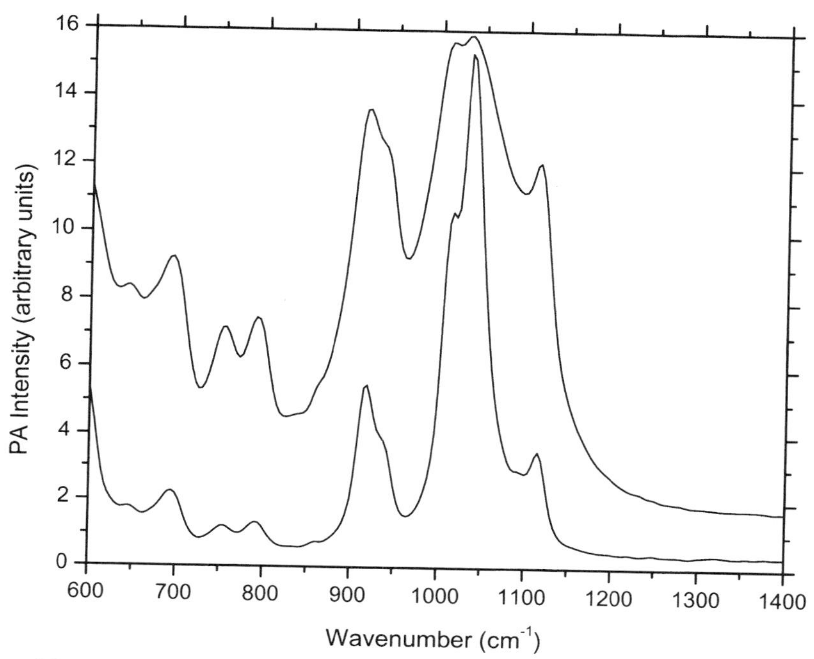


**Figure 5.9.** Low-frequency PA infrared spectra of kaolin. Upper curve, ratioed amplitude spectrum (Mertz phase correction); lower curve, linearized spectrum. Spectra have been rescaled to facilitate comparison of results.

## 5.3. PHASE ANALYSIS

Another important numerical procedure that pertains to PA infrared spectroscopy was developed a number of years ago by L. Bertrand and his colleagues at École Polytechnique de Montréal, as well as several other institutions. The objective of this method is to determine the actual absorption coefficient ($\beta$) of a sample as a function of wavenumber—in other words, to obtain quantitative PA spectra instead of the qualitative data commonly reported by many authors. The elimination of saturation effects is another goal of these calculations. Moreover, because the phase analysis method is capable of distinguishing between surface and bulk absorption in some circumstances, it provides a depth profiling capability that augments those discussed above.

Phase information can be acquired by two different approaches in PA FTIR spectroscopy. First, in what might be described as a conceptually

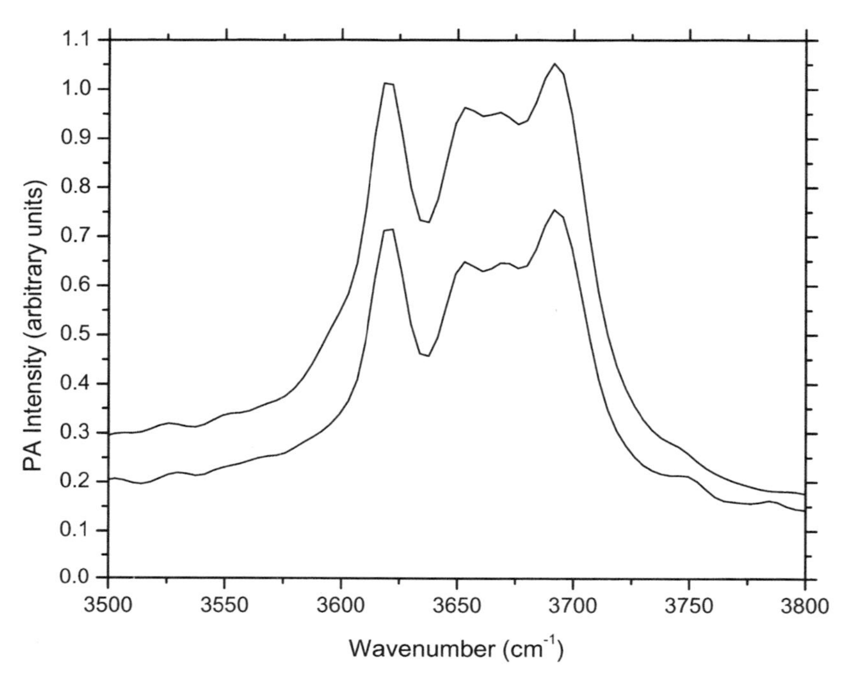


**Figure 5.10.** Hydroxyl stretching region in PA infrared spectra of kaolin. Upper curve, ratioed amplitude spectrum (Mertz phase correction); lower curve, linearized spectrum. Spectra have been rescaled to facilitate comparison of results; intensity scale is not same as in Fig. 5.9.

simple experiment, the in-phase (0°) and in-quadrature (90°) components of the PA signal are directly observed using a single- or two-phase lock-in amplifier. These measurements are most straightforward with a step-scan spectrometer but can also be carried out with a continuous (slow) scan and a relatively high external modulation frequency. In the second method, the real and imaginary spectra calculated from a double-sided interferogram recorded in either rapid- or step-scan mode can be utilized to obtain the phase. These spectra are made available to the analyst prior to the implementation of the conventional phase correction calculation in some commercial FTIR software programs. However, their resolution is significantly restricted by the limited number of data points that occur before the center burst of the interferogram in many instruments. This situation exists because many FTIR spectrometers have been designed and constructed under the realistic assumption that low-resolution phase information is adequate in most (non-PA) applications.

The availability of the in-phase and in-quadrature, or real and imaginary,

PA spectra is a prerequisite that enables the implementation of the phase analysis approach. The basic theory of this technique was outlined by Bordeleau et al. (1986), Choquet et al. (1986), and Bertrand (1988). These authors were particularly interested in the case of a thermally thin film adsorbed onto a thermally thick substrate and showed that these two components of the PA signal are given by

$$\text{INPHAS} = C(1 - \beta^s)\beta\mu/[(\beta\mu + 1)^2 + 1] \tag{5.11}$$

and

$$\text{INQUAD} = C\{\beta^s + (1 - \beta^s)(\beta\mu + 1)\beta\mu/[(\beta\mu + 1)^2 + 1] + D\} \tag{5.12}$$

where $\beta^s$ is the (dimensionless) fraction of light absorbed by the surface layer, $C$ is a constant that depends on the thermal properties of the sample, and $D$ is a thermal expansion term that is often negligible for solid samples; $\beta$ and $\mu$ have their usual meanings.

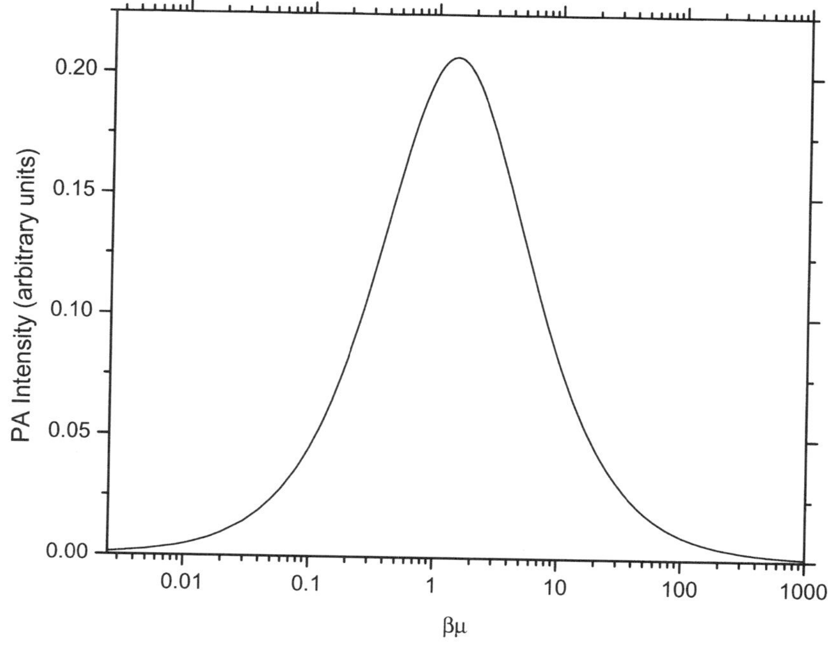


**Figure 5.11.** Variation of in-phase (real) component of PA intensity with product $\beta\mu$.

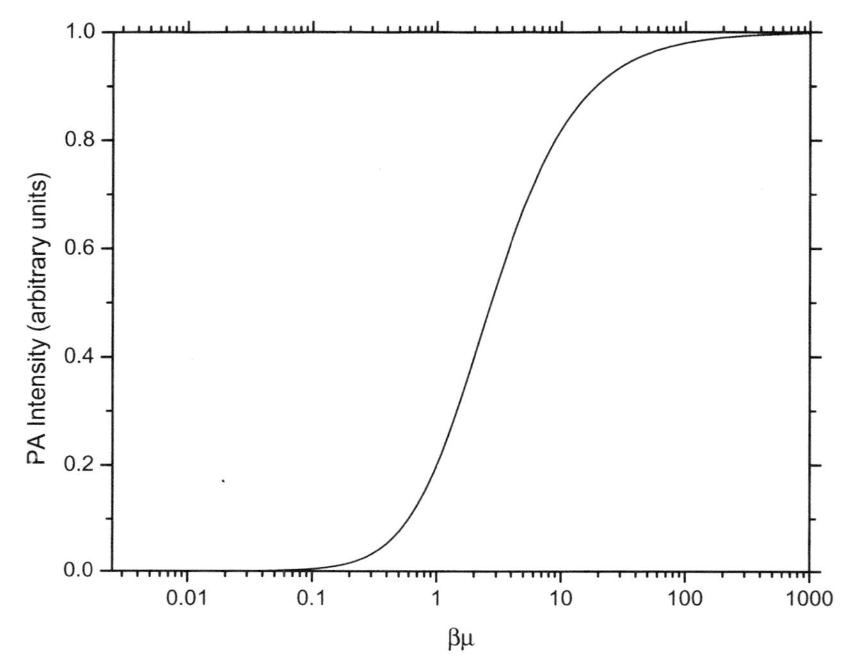


**Figure 5.12.** Variation of in-quadrature (imaginary) component of PA intensity with product $\beta\mu$.

In the absence of surface absorption (i.e., when $\beta^s = 0$) the bulk absorption $\beta\mu$ is readily obtained by rearranging these expressions to yield

$$\beta\mu = (\text{INQUAD} - \text{INPHAS})/\text{INPHAS} \tag{5.13}$$

The variations of the in-phase and in-quadrature components of the PA intensity with $\beta\mu$ are easily calculated for the case where surface absorption can be neglected. These quantities are plotted in Figures 5.11 and 5.12, respectively.

On the other hand, when $\beta^s$ is significant and $\beta\mu$ is much smaller than 1, these equations reduce to

$$\text{INQUAD} - \text{INPHAS} = C\beta^s/2 + D \tag{5.14}$$

from which the surface absorption coefficient $\beta^s$ can be easily determined. The frequency dependence of this quantity was shown to be small in PA data obtained for two polymers (Bordeleau et al., 1986).

The feasibility of phase analysis has been illustrated in several publications. Choquet et al. (1986) used the technique to correct spectra of asbestos (chrysotile) and demonstrated a significant reduction in saturation effects for this important mineral. The redistribution of intensity in the corrected PA spectrum with respect to an ordinary magnitude spectrum described by these authors qualitatively resembles that depicted in Figure 5.4. Bertrand (1988) next applied the method to data for both asbestos and ethylene-propylene rubber (EPR); the results for the latter sample included a rather impressive diminution of saturation effects, as well as the elimination of the nonzero baseline that frequently appears in PA infrared spectra (Fig. 5.13). More recently, Marchand et al. (1997) reviewed all aspects of the phase analysis method in detail and stressed the importance of correctly identifying the onset of saturation (the point where the phase angle is equal to 67.5°). These authors pointed out that further improvements to the method and the eventual realization of quantitative PA infrared spectroscopy will require that several experimental parameters be taken into account: These include sample quantity, the position of the sample with respect to the window of the PA cell, and carrier gas pressure.

The phase analysis method has also been successfully applied to samples where surface absorption is important. For example, Bordeleau et al. (1986)

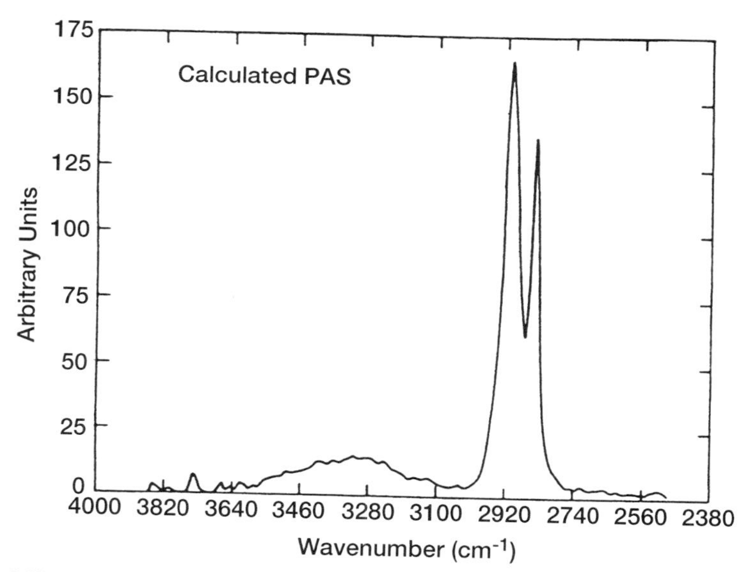


**Figure 5.13.** Corrected PA infrared spectrum of ethylene-propylene rubber, plotted as dependence of $\beta\mu$ on wavenumber. (Reproduced from Bertrand, L., *Appl. Spectrosc.* **42**: 134–138, by permission of the Society for Applied Spectroscopy; copyright © 1988.)

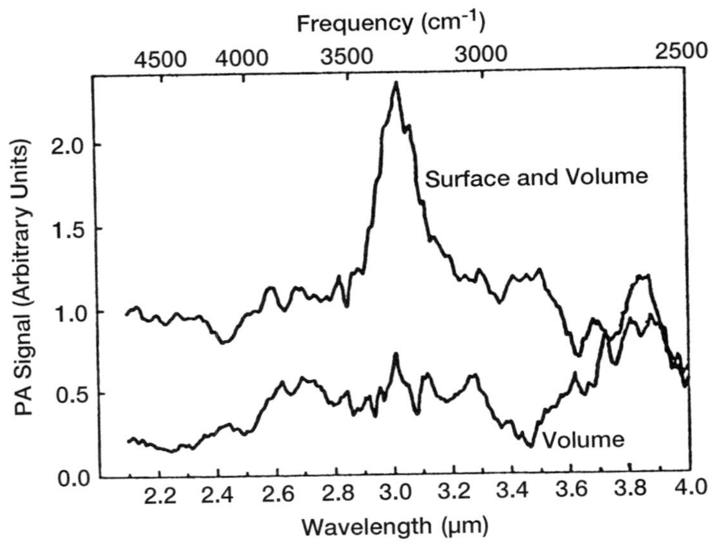


**Figure 5.14.** PA infrared spectra of hydrogenated amorphous silicon nitride in the N—$H_n$ stretching region. Upper curve, quadrature spectrum; lower curve, in-phase spectrum. (Reprinted from *Spectrochim. Acta A* **43**, Bordeleau et al., Photoacoustic infrared identification of N—$H_n$ vibrations in hydrogenated amorphous silicon nitride, 1189–1190, copyright © 1987, with permission from Elsevier Science.)

calculated $\beta^s$ and $\beta\mu$ spectra for EPR and crosslinked polyethylene in which sorbed water affects dielectric properties. $\beta^s$ increased concomitantly with relative humidity in these spectra. Mongeau et al. (1986) separated surface and bulk absorption at $CO_2$ laser wavelengths for samples of SiO deposited on $CaF_2$, and showed that the resolved bulk PA signals were similar to those obtained for uncoated $CaF_2$. Mirage detection was used in this (PBDS) experiment.

A very good example of the distinction between the surface and volume (bulk) of a sample was given in the PA infrared spectra reported by Bordeleau et al. (1987). As shown in Figure 5.14, the N—$H_n$ band at 3350 $cm^{-1}$ is prominent in the surface-sensitive (quadrature) spectrum of plasma-deposited hydrogenated amorphous silicon nitride ($a$-$SiN_x$:H) but virtually undetectable in the volume (bulk) spectrum. This clear distinction between the composition of the surface and bulk of the sample is consistent with the conclusion that the surface contaminants are localized in a layer having a thickness of about 50–200 Å. Similarly, Raveh et al. (1992) observed features due to OH and C=O groups in the quadrature spectrum of an amorphous carbon film grown in a microwave plasma. These oxygen-containing species were known to occur mostly near the film surface. The same bands were not detectable in the in-phase spectrum.

Finally, the phase analysis method was used in conjunction with several other techniques to study graphite-epoxy laminates by Dubois et al. (1994). The spectra in this study differ in a fundamental way from those discussed in the previous paragraph: Optical penetration depth is plotted as the dependent variable by Dubois et al. These authors compared results from phase analysis with those obtained by two other techniques, specifically the use of two mirror velocities (see above) and transmission. Good agreement was observed between the phase analysis results and transmission measurements. This is another clear indication of the reliability of the phase analysis method developed by Bertrand and his collaborators.

## REFERENCES

Bertrand, L. (1988). Advantages of phase analysis in Fourier transform infrared photoacoustic spectroscopy. *Appl. Spectrosc.* **42**: 134–138.

Bordeleau, A., Rousset, G., Bertrand, L., and Crine, J. P. (1986). Water detection in polymer dielectrics using photoacoustic spectroscopy. *Can. J. Phys.* **64**: 1093–1097.

Bordeleau, A., Bertrand, L., and Sacher, E. (1987). Photoacoustic infrared identification of N—$H_n$ vibrations in hydrogenated amorphous silicon nitride. *Spectrochim. Acta A* **43**: 1189–1190.

Burggraf, L. W., and Leyden, D. E. (1981). Quantitative photoacoustic spectroscopy of intensely light-scattering thermally thick samples. *Anal. Chem.* **53**: 759–764.

Carter, R. O. (1992). The application of linear PA/FT-IR to polymer-related problems. *Appl. Spectrosc.* **46**: 219–224.

Chalmers, J. M., Stay, B. J., Kirkbright, G. F., Spillane, D. E. M., and Beadle, B. C. (1981). Some observations on the capabilities of photoacoustic Fourier transform infrared (PAFTIR). *Analyst* **106**: 1179–1186.

Choquet, M., Rousset, G., and Bertrand, L. (1986). Fourier-transform photoacoustic spectroscopy: A more complete method for quantitative analysis. *Can. J. Phys.* **64**: 1081–1085.

Dubois, M., Enguehard, F., Bertrand, L., Choquet, M., and Monchalin, J.-P. (1994). Optical penetration depth determination in a graphite-epoxy laminate by photoacoustic F.T.I.R. spectroscopy. *J. Phys. IV*, **C7**: 377–380.

Farmer, V. C., and Russell, J. D. (1964). The infra-red spectra of layer silicates. *Spectrochim. Acta* **20**: 1149–1173.

Low, M. J. D. (1983). Normalizing infrared photoacoustic spectra of solids. *Spectrosc. Lett.* **16**: 913–922.

Low, M. J. D. (1985). Hexane soot as standard for normalizing infrared photothermal spectra of solids. *Spectrosc. Lett.* **18**: 619–625.

Low, M. J. D., and Parodi, G. A. (1980). Carbon as reference for normalizing infrared photoacoustic spectra. *Spectrosc. Lett.* **13**: 663–669.

Marchand, H., Cournoyer, A., Enguehard, F., and Bertrand, L. (1997). Phase optimization for quantitative analysis using phase Fourier transform photoacoustic spectroscopy. *Opt. Eng.* **36**: 312–320.

McClelland, J. F. (1983). Photoacoustic spectroscopy. *Anal. Chem.* **55**: 89A–105A.

Mongeau, B., Rousset, G., and Bertrand, L. (1986). Separation of surface and volume absorption in photothermal spectroscopy. *Can. J. Phys.* **64**: 1056–1058.

Raveh, A., Martinu, L., Domingue, A., Wertheimer, M. R., and Bertrand, L. (1992). Fourier transform infrared photoacoustic spectroscopy of amorphous carbon films. In: *Photoacoustic and Photothermal Phenomena III.* D. Bićanić (ed.). Springer, Berlin, pp. 151–154.

Riseman, S. M., and Eyring, E. M. (1981). Normalizing infrared FT photoacoustic spectra of solids. *Spectrosc. Lett.* **14**: 163–185.

Rockley, M. G., Ratcliffe, A. E., Davis, D. M., and Woodard, M. K. (1984). Examination of a C-black by various FT-IR spectroscopic methods. *Appl. Spectrosc.* **38**: 553–556.

Sowa, M. G., and Mantsch, H. H. (1994). FT-IR step-scan photoacoustic phase analysis and depth profiling of calcified tissue. *Appl. Spectrosc.* **48**: 316–319.

Teng, Y. C., and Royce, B. S. H. (1982). Quantitative Fourier tranform IR photoacoustic spectroscopy of condensed phases. *Appl. Opt.* **21**: 77–80.

Teramae, N., Hiroguchi, M., and Tanaka, S. (1982). Fourier transform infrared photoacoustic spectroscopy of polymers. *Bull. Chem. Soc. Jpn.* **55**: 2097–2100.

## RECOMMENDED READING

Choquet, M., Rousset, G., and Bertrand, L. (1985). Phase analysis of infrared Fourier transform photoacoustic spectra. *SPIE* **553**: 224–225.

Monchalin, J.-P., Bertrand, L., Rousset, G., and Lepoutre, F. (1984). Photoacoustic spectroscopy of thick powdered or porous samples at low frequency. *J. Appl. Phys.* **56**: 190–210.

# CHAPTER 6

# APPLICATIONS OF PA INFRARED SPECTROSCOPY

*"The [PA infrared] method should also be useful for the investigation of substances with virtually zero transmission."*

G. Busse and B. Bullemer, *Infrared Phys.* **18:** 631–634 (1978).

## 6.1. CARBONS

The measurement of infrared spectra of carbons obviously presents a nontrivial challenge to the spectroscopist. The dispersion of a finely divided carbon (or, more generally, a carbonaceous solid) in an infrared-transparent diluent so as to record its transmission or diffuse reflectance spectrum may be successful; however, the preparation of a well-mixed sample is sometimes difficult or even impossible. PA infrared spectroscopy allows the analyst to obtain useful data for neat carbons while eliminating the need for this sample preparation step. Therefore, it is not surprising to find that a significant number of research teams have successfully used this technique to study carbons during the last two decades. The results of these investigations are summarized in this section. In addition, previously unpublished PA spectra of hydrocarbon cokes obtained recently in the author's laboratory are presented.

A general definition of carbons is adopted here. This term is taken to include—in addition to elemental carbons—samples with very high (~80–90%) carbon contents such as chars, soots, and cokes. Since several types of functional groups exist in these substances, viable PA infrared spectra of these carbons are readily obtainable. The work discussed below generally confirms this statement. It should also be noted that coals are not included in this list: Because of the substantial literature on this subject, PA infrared spectra of coals are discussed separately in the next section of this chapter.

### 6.1.1. Research of Low's Group

Among the researchers who have investigated the PA infrared spectra of carbons during the last two decades, M. J. D. Low and his collaborators at

New York University were the most prolific, with over 35 publications to their credit. This significant body of literature is reviewed in the first part of this section. After this, the findings of another eight research groups are summarized.

Three studies in the extensive series published by Low's team mentioned the PA infrared spectra of carbons in the context of the normalization of PA spectra (Low and Parodi, 1980b; Low, 1983, 1985), a subject discussed in more detail in Chapter 5. Several chars and soots were examined in this work. A particularly salient point (Low and Parodi, 1980b) is that the results obtained for a given carbon may be qualitatively different according to the wavelength region under consideration. For example, charring sucrose for 1 and 12 h produced carbons that yielded featureless visible and near-infrared PA spectra. However, mid-infrared PA spectra of the same samples exhibited a number of well-known bands. Specifically, features due to C—O, C═C, C═O, $CH_2$, $CH_3$, and OH groups appeared in the spectrum of the 1-h char. The result for the 12-h char exhibited typical changes that accompany aromatization; only the C═C, $CH_2$, and $CH_3$ bands, plus strong aromatic C—H absorption, were observed. The identification of all of these bands shows that sucrose chars actually contain a variety of functional groups and are therefore not very well suited for normalizing PA mid-infrared spectra. In contrast with these results, soots obtained from benzene, propane, and an ordinary candle gave essentially featureless spectra from 2000 to 4000 $cm^{-1}$.

The possible use of the PA infrared spectrum of hexane soot as a reference for normalizing spectra of various solids was also considered by Low (1985). This work contains a rebuttal of the arguments of Rockely et al. (1984), which are discussed further below. The conclusion reached by Low is that the PA spectrum of hexane soot also contains some structure, which may be the source of the anomalies observed by Rockley et al. when they compared PA and other types of mid-infrared spectra.

PA infrared spectra of carbons were mentioned in a more incidental fashion in another series of studies that dealt primarily with instrumental questions and the initial implementation of the PA infrared technique by Low's group. The earlier publications in this series were discussed in Chapter 2 (Low and Parodi, 1978, 1980a,b,c). In contrast with the perspective described in the preceding paragraphs, the possible presence of particular functional groups in the carbons was not considered explicitly in this second group of papers. Instead, charcoal was used simply to obtain a background spectrum that could be utilized to correct the PA spectra of other samples. Low and Lacroix (1982) also utilized charcoal as a reference in photothermal beam deflection spectroscopy (PBDS) experiments, again assuming that the resulting spectrum was a faithful representation of the frequency dependence of the exciting radiation. Similarly, charcoal was mixed with KBr and pressed into a pellet for use as a reference in PBDS experiments on solids submerged in liquids (Varlashkin and Low, 1986). The PA spectrum of the layer of black paint on an infrared detector was measured (Low, 1984) and found to differ from the usual "empty instrument" spectrum obtained with the detector itself. The dependence of the PA spectra on mirror velocity in this investigation was attributed to differences in the sampling depth. A similar effect was thought to have influenced the PA data of Riseman and Eyring (1981) on carbons, which are discussed below.

In addition to the work described in the preceding paragraphs, Low's group published a series of 22 related works, variously entitled "IR Studies of Carbons," "Infrared Studies of Carbons," and "Spectroscopic Studies of Carbons" between 1983 and 1991. The bibliographic data for these publications, and the topics discussed in each, are briefly listed in Table 6.1. These articles are also included in the list of references at the end of this chapter. Unlike many of the other works on the PA spectra of carbons published by these researchers, these publications mostly involved the more traditional use of infrared spectroscopy, specifically for the characterization of functional groups; the issue of normalizing PA spectra was hardly discussed. The PBDS technique was again used to obtain the results in these 22 studies.

The first study on "IR Studies of Carbons" (Low and Morterra, 1983) contains a useful short literature review on the absorption of infrared radiation by carbons, covering about 50 publications. Samples studied in this work included cellulose pyrolyzed in vacuum at temperatures up to 600°C, as well as a series of chromatographic carbons. The latter are noteworthy because of their high aromaticity, indicated by the occurrence of relatively strong absorption bands near 3050 $cm^{-1}$ due to aromatic C—H groups. In fact, both the pyrolyzed cellulose samples and the chromatographic carbons yielded spectra displaying a significant number of bands. This is consistent with the general statement that the PA infrared spectra of carbons often contain a great deal of specific information regarding particular chemical groups.

The next four works in this series by Low and his colleagues discussed experiments that began with the pyrolysis of cellulose and continued with oxidation studies of the produced chars. In the first article, PBDS was used to follow the vacuum pyrolysis of cellulose at temperatures that ranged above 700°C (Morterra and Low, 1983). The infrared spectra of the pyrolysis products showed that aliphaticity is maintained up to approximately 500°C, whereas aromatization is observed initially at about 300°C and then increases rapidly at higher temperatures. The increase in aromaticity is indicated by a reduction in the intensities of the aliphatic C—H stretching

**Table 6.1. Topics Discussed in *Infrared Studies of Carbons* and *Spectroscopic Studies of Carbons*[a]**

| Paper Number | Publication | Topic |
|---|---|---|
| 1 | *Carbon* **21**: 275–281 | Cellulose, chromatographic carbon |
| 2 | *Carbon* **21**: 283–288 | Pyrolysis of cellulose |
| 3 | *Carbon* **22**: 5–12 | Oxidation of cellulose chars |
| 4 | *Carbon* **23**: 301–310 | Pyrolysis of oxidized cellulose |
| 5 | *Carbon* **23**: 311–316 | Effect of NaCl on cellulose pyrolysis |
| 6 | *Carbon* **23**: 335–341 | Effect of $KHCO_3$ on cellulose pyrolysis |
| 7 | *Carbon* **23**: 525–530 | Pyrolysis of phenol-formaldehyde resin |
| 8 | *Langmuir* **1**: 320–326 | Oxidation of phenol-formaldehyde chars |
| 9 | *Mater. Chem. Phys.* **20**: 123–144 | Pyrolysis of polyvinyl chloride |
| 10 | *Structure and Reactivity of Surfaces*, 601–609 | Medium-temperature chars |
| 11 | *Mater. Chem. Phys.* **23**: 499–516 | Pyrolysis of polyvinyl bromide |
| 12 | *Carbon* **28**: 529–538 | Polycarbonate resin chars |
| 13 | *Carbon* **28**: 855–865 | Oxidation of polycarbonate resin chars |
| 14 | *Mater. Chem. Phys.* **25**: 501–521 | Pyrolysis of polyvinyl fluoride |
| 15 | *DOE/PC/79920—10*, 1–14 | Pyrolysis of lignin |
| 16 | *Polym. Degrad. Stab.* **32**: 331–356 | Carbonization of diallyl-diglycol-polycarbonate |
| 17 | *Mater. Chem. Phys.* **26**: 193–209 | Pyrolysis of polyvinylidene fluoride |
| 18 | *Mater. Chem. Phys.* **26**: 117–130 | Rice hull char |
| 19 | *Mater. Chem. Phys.* **26**: 465–481 | Sucrose char |
| 20 | *Mater. Chem. Phys.* **27**: 155–179 | Pyrolysis of polyvinylidene chloride |
| 21 | *Mater. Chem. Phys.* **27**: 359–374 | Coconut shell char |
| 22 | *Mater. Chem. Phys.* **28**: 9–31 | Oxidation of polyvinyl halide chars |

[a] A series of PBDS investigations published by Low and co-workers between 1983 and 1991.

bands and the concomitant appearance of bands due to aromatic C—H stretching and bending modes in the spectra. The thermal decomposition of cellulose is depicted schematically in Figure 6.1.

The cellulose chars prepared by pyrolysis were subsequently oxidized under conditions that led to the formation of either acidic or basic carbons (Morterra et al., 1984). Acidic carbons were prepared through oxidation at temperatures up to about 530°C, while basic carbons required higher temperatures. Interestingly, infrared bands due to oxygen-containing functional groups did not occur in the spectra of the products of the latter process.

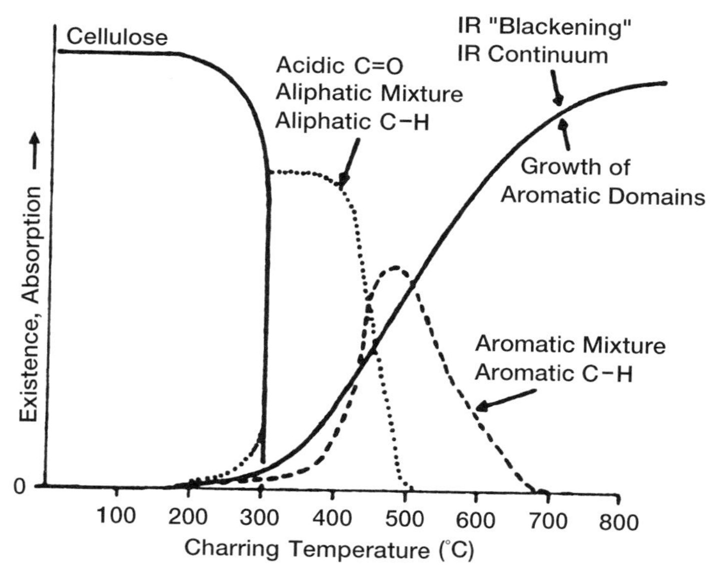


**Figure 6.1.** Decomposition of cellulose. (Reprinted from *Carbon* **21**, Morterra, C. And Low, M. J. D., IR studies of carbons–II. The vacuum pyrolysis of cellulose, 283–288, copyright © 1983, with permission from Elsevier Science.)

The next publication in this series (Morterra and Low, 1985a) described the pyrolysis of $NO_2$-oxidized cellulose (NOC). The PA spectra dramatically showed the successive changes that occurred at each stage of the pyrolysis (Fig. 6.2); this includes aromatization at temperatures above 300°C, as well as the eventual disappearance of all functionality at 700°C. The spectra obtained in this work showed that NOC degrades more easily than cellulose, yielding chars that contain different functional groups at temperatures up to 500°C.

Alkali metal compounds are known to facilitate the carbonization of cellulose and subsequent oxidation of the chars. Accordingly, Low and Morterra (1985) studied the pyrolysis of cellulose that contained NaCl. The resulting chars were oxidized and compared to those obtained from pure cellulose. NaCl was found to hasten cellulose decomposition and to change the relative amounts of aliphatic and aromatic residues. Pyrolysis at low to moderate temperatures (up to 450°C) led first to oxidation and then to aromatization; oxygen was eventually eliminated at higher temperatures. At 550°C most of the infrared bands diminished in intensity, and at 650°C only an absorption continuum remained. By contrast, aromatic groups persisted up to 600°C in the absence of NaCl.

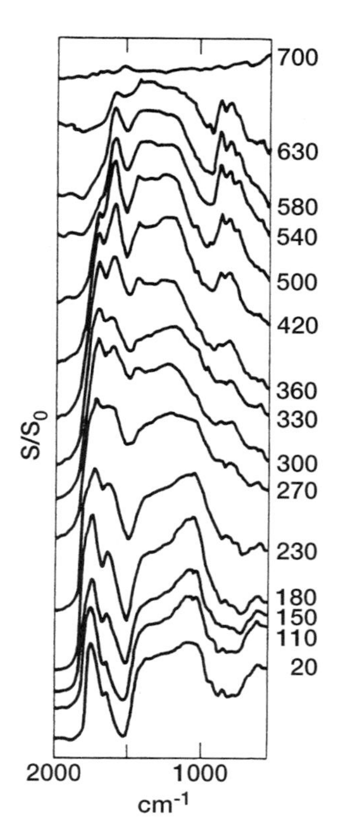


**Figure 6.2.** PBD spectra depicting pyrolysis of cellulose. Numbers on right-hand side of figure give pyrolysis temperatures in degrees Celsius. (Reprinted from *Carbon* **23**, Morterra, C. and Low, M. J. D., IR studies of carbons–IV. The vacuum pyrolysis of oxidized cellulose and the characterization of the chars, 301–310, copyright © 1985, with permission from Elsevier Science.)

The presence of $KHCO_3$ during pyrolysis and oxidation also caused significant changes with respect to the reactions in pure cellulose; the temperature at which cellulose structure disappeared was again reduced, and a higher degree of carbonization was achieved. No aromatic C—H groups were detected in the pyrolysis products of cellulose obtained with $KHCO_3$ (Morterra and Low, 1985b).

Two works in this series dealt with phenol-formaldehyde resins. Morterra and Low (1985c) found that the pyrolysis of resins produced in the presence of Novolac (an acid catalyst) in either vacuum or a nitrogen atmosphere led to the same results; branching and crosslinking occurred near 350°C, where diphenyl ethers were formed. Aryl-aryl ethers were detected at higher temperatures. The aliphatic bridges between the aromatic rings started to break down at temperatures above 500°C. Polyaromatic domains were formed, which subsequently exhibited oxidation behavior similar to that of chars derived from other precursors (Morterra and Low, 1985d). The oxidation of the low-temperature chars led to the formation of benzophenones and carboxylic acids.

The pyrolysis of several polymers that do not contain oxygen was also investigated by this group. First, Morterra et al. (1988) showed that the pyrolysis of polyvinyl chloride (PVC) exhibited many similarities with the processes described in their preceding studies. The fingerprint region of the spectra was used to show that the pyrolysis of PVC consists of the following steps: (a) the elimination of Cl, (b) the formation of alkenic structures, (c) the formation of alkanes, and (d) aromatization and polyaromatization. The appearance of the seemingly ubiquitous "1600-$cm^{-1}$ band" in the spectra of samples that lack oxygen apparently confirms that this feature is due to C=C, rather than C=O, groups. This point is taken up again later in this section.

O'Shea et al. (1989) next extended this work to include polyvinyl bromide (PVBr). The mechanism of pyrolysis was similar to that in PVC, although the formation of polyenic structures occurred at much lower temperatures for PVBr. Next, a similar study showed that polyvinyl fluoride (PVF) is more stable than the other two polymers, with the first evidence of degradation occurring above 300°C (O'Shea et al., 1990a). The oxidation of all three of these three polyvinyl halide chars was also subsequently investigated (O'Shea et al., 1991a). Low-temperature chars—in which the extent of polyaromatization was quite limited—were more susceptible to oxidation, yielding carboxylic acids, anhydrides, and lactones.

In addition to the polyvinyl halides, these authors studied the behavior of two polyvinylidene halides. The onset of degradation in polyvinylidene fluoride (PVDF) was near 300°C. This process eventually yielded a pyrolytic residue consisting of both aliphatic and fluoroaromatic structures (O'Shea et al., 1990b). Polyvinylidene chloride was investigated both as a homopolymer and in the form of a copolymer with PVC (O'Shea et al., 1991b). The spectra for the copolymer contained discrete bands even after high-temperature pyrolysis, which was not the case for the homopolymer. Because the polyvinylidene halides contain much less hydrogen than the corresponding polyvinyl halides, C—H bands disappear sooner during the pyrolysis of the former compounds.

The 1600-$cm^{-1}$ band that occurs in the PA infrared spectra of many different carbons was mentioned above. Low and Glass (1989) discussed this band in detail, emphasizing the need for caution when comparing spectra of different carbons. Despite the appearance of this band in the infrared spectra of many different types of carbons, one should not assume that a common origin applies to all of these cases. Indeed, both the frequency and the profile of the ~1600-$cm^{-1}$ band have been found to differ in the infrared spectra of various carbons.

In PA spectra of high-temperature carbons, the 1600-$cm^{-1}$ band is sometimes described as the infrared analog of the G (graphite) band in Raman spectra of carbons. Many spectroscopists may already be aware that the G band is accompanied by a D (defect) band located near 1350 $cm^{-1}$ in the Raman spectra of most carbons. In their analysis of the infrared spectra of medium-temperature chars, Low and Morterra (1989) noticed that a dip (valley) near the latter frequency tended to fill in as the pyrolysis temperature was progressively raised. These authors showed that the numerical addition of a broad Gaussian band centered at 1350 $cm^{-1}$ to the PBDS spectrum of the 480°C char yielded synthetic spectra similar to those measured for samples heated to higher temperatures. This proposal was reiterated by Wang and Low during their investigations of lignin (1989) and sucrose (1990b). It should be noted that this model suggests a possible link between the Raman and PA infrared spectra of carbons; this may serve to remind the reader of the complementarity of the many techniques available to the vibrational spectroscopist.

Two studies in this series describe the investigation of the vacuum pyrolysis of bisphenol-A polycarbonate resins and the subsequent oxidation of the chars formed in this process (Politou et al., 1990a,b). These thermoplastic resins are of particular interest because of their electrical and mechanical properties, as well as their chemical inertness. Pyrolysis at temperatures in the 420–490°C region produced the most dramatic changes, such as the formation of ester, diaryl ether, and unsaturated hydrocarbon bridges. Aromatic ring structures formed near the upper end of this temperature range and persisted up to at least 700°C. At still higher temperatures, the familiar continuum due to electronic absorption was observed. The oxidation of the low-temperature chars yielded products similar to those described above for the phenol-formaldehyde resins. In related work, Politou et al. (1991) described the spectra of pyrolyzates produced from diallyl-diglycol-polycarbonate. Low-temperature oxidation of this aliphatic thermosetting resin was found to be necessary before it would char. Pyrolysis then yielded products resembling intermediate-temperature carbons.

A number of earlier works in this series described the pyrolysis of cellulose (see Table 6.1). Lignin, which is another important component of wood, was also studied by this group of researchers. Wang and Low (1989) showed that the pyrolysis of lignin occurs in three stages, which can be described as depolymerization, consolidation of aromatic rings, and the creation of polyaromatic domains. These chemical changes take place in a more gradual manner during the pyrolysis of cellulose, where larger amounts of carbonyls are also formed.

Three of the later works in this series describe the charring of food products. Wang and Low (1990a) discussed studies of rice hulls, where the products reflect the presence of both aromatics (derived from lignins) and aliphatics (from cellulose). Silica is the other major constituent of the pyrolysis products. In one of the companion investigations, Wang and Low (1990b) described the charring of sucrose. Although the original spectrum of sucrose contains a significant amount of detail, even the mild heating employed in the preparation of confectionary products leads to a noteworthy simplification of the infrared spectrum. The charring of sucrose proceeds in several steps: (a) conversion from aliphatic to aromatic material, (b) dehydroxylation, (c) elimination of aliphatic groups, (d) loss of carbonylic species, and (e) formation of polyaromatic structures. Finally, the third study in this group (Wang and Low, 1991) described the charring of coconut shell. At temperatures up to 300°C, the observed changes were mainly textural. By contrast, in the 300–500°C region the mixture was converted from aliphatic to aromatic character. Above 550°C, the material became predominantly polyaromatic. The original coconut shell yielded an infrared spectrum similar to that of a hardwood, exhibiting bands due to cellulose, hemicellulose, and lignin.

Several other important studies on the PA infrared spectra of carbons were published by Low and his colleagues at about the same time as the series of 22 works that has just been summarized. For example, Morterra and Low (1982) studied the ~1600-cm$^{-1}$ band, observed by numerous researchers in the infrared spectra of many different types of carbons and discussed above. In general, a band at this position might be thought to arise from a highly conjugated carbonyl group or, alternatively, from a carbon–carbon double bond or an aromatic ring structure. To evaluate these possibilities, the authors obtained spectra of a commercial carbon and a cellulose that had been charred in vacuum. The cellulose char was then oxidized at 450 and 600°C in $^{16}O_2$ and $^{18}O_2$. The results showed that the occurrence and intensity of the 1600-cm$^{-1}$ band depended on the presence and concentration of surface oxidic species but that the frequency of this feature did not depend on the choice of isotope. On the other hand, a neighboring C=O band exhibited an isotope shift from 1760 to 1725 cm$^{-1}$, as would be expected from the difference in the oxygen masses. These results led to the conclusion that the 1600-cm$^{-1}$ band arises from aromatic C=C groups and that an oxidized layer crosslinking aromatic chains causes the asymmetry necessary for the appearance of this Raman-active vibration in the infrared.

In contradistinction with these results, it is shown below that the 1600-cm$^{-1}$ band is also observed in hydrocarbon cokes, which contain virtually no oxygen. This band was also observed by Low and his collaborators in spectra of polyvinyl halides, which also contain no oxygen (see above). In these cases, Morterra and Low's reasoning with regard to an oxidized layer cannot be invoked, and another cause for the occurrence of the 1600-cm$^{-1}$ band in these PA infrared spectra must be identified.

As this research progressed, PBDS was used by this group to study several other carbonaceous materials. Morterra and Low (1985e) reported spectra of wood charcoal, coals, and chromatographic carbons; the results for the coals are discussed in the next section of this chapter. Low and Morterra (1986) discussed charcoal, chromatographic carbon, and a phenol-formaldehyde resin. These samples were heated in vacuum and then oxidized. For charcoal and carbons, trends similar to those already described were observed; carbonization caused a progressive reduction in the aromatic C—H bands between 700 and 900 cm$^{-1}$ and the appearance of both C═O and C—O bands. At higher temperatures, the absorption continuum intensified and all of the bands due to specific groups eventually disappeared. This continuum may obscure the weaker absorption bands of some of the functional groups, possibly by limiting the penetration of the infrared beam into the sample under study.

The charring of sucrose and coconut shell were described above. In related work, Low and Wang (1990) studied these processes at temperatures as high as 800°C. This investigation showed that chars heated to 450°C exhibited similar spectra, displaying aromatic C—H bands between 700 and 900 cm$^{-1}$, a minimum near 1350 cm$^{-1}$, and bands at 1600 and 1700 cm$^{-1}$. Further heating to 600–650°C caused the 1350-cm$^{-1}$ dip to fill in. This phenomenon was described above. Although Low and Wang did not obtain Raman spectra of the pyrolyzates—which might have confirmed their suggestion that the latter effect was due to the infrared activation of the 1350-cm$^{-1}$ Raman band—their proposal remains an interesting possibility that could pertain to infrared spectra of many carbons.

The final work from this rather impressive body of research that will be mentioned here describes "unusual" bands between about 2050 and 2250 cm$^{-1}$ in the infrared spectra of a number of chars (Low et al., 1990). Most vibrational spectroscopists will quickly agree that only a few types of functional groups yield bands in this region. Low et al. suggested that the bands in their spectra were due to nitrogen-containing species, such as nitriles or isonitriles. The possible role of these compounds in charring is apparently an unexplored question but could be of considerable significance.

### 6.1.2. Other Groups

Research on PA infrared spectra of carbons carried out by a number of other groups will now be summarized. This work began at about the same time that the first investigations in the previous section were carried out and spans the entire period up until the publication of this book.

Riseman and Eyring (1981) obtained infrared spectra of a series of eight commercially available carbons during their investigations on the normalization of PA spectra of solids. This research demonstrated the dependence of the spectrum of a given carbon on the mirror velocity in rapid-scan PA infrared spectroscopy. Because the spectra were saturated, bands due to specific functional groups were of course not identifiable in these spectra. Moreover, the different carbons tended to produce very similar spectra at each velocity. Even though these spectra were essentially featureless, Riseman and Eyring emphasized the important fact that the PA signal passes through successive stages of saturation as modulation frequency changes. Because this frequency varies by a factor of 10 (from 400 to 4000 $cm^{-1}$) across the mid-infrared region in a rapid-scan PA experiment, the low-frequency region of the PA spectrum of a carbon may be fully saturated at the same time that the high-frequency region is only partially saturated.

Carbonaceous refractory gold ore contains organic carbon constituents that interact with the gold and make the ore nonamenable to conventional cyanidation. Nelson et al. (1982) studied oxidized and carbonaceous Carlin gold ores by means of PA infrared spectroscopy and electron paramagnetic resonance (EPR) spectroscopy. The infrared spectra of the two ores were similar to each other and, moreover, bore no resemblance to the spectrum of humic acid that was also obtained in this work. In fact, no hydrocarbons were detected in the ores. The organic carbon in the ore was therefore concluded to be an unspecified type of activated carbon.

Rockley et al. (1984) studied a carbon black (soot) prepared by combustion of hexane. These authors compared PA, attenuated total reflectance (ATR), diffuse reflectance, and transmission infrared spectra of this sample. Elemental analysis yielded a composition of 94.8% carbon, 1.2% hydrogen, and 2.4% oxygen by weight; hence, spectroscopic evidence of the existence of several functional groups was anticipated. Indeed, between 600 and 2000 $cm^{-1}$ the transmission and ATR spectra showed bands due to aromatic C—H and C=C groups. However, the diffuse reflectance and PA infrared spectra did not contain any reliable features; in fact, the PA spectrum of soot deposited on an NaCl window differed from that of soot alone. A definitive explanation for the PA results was not given, and it was concluded that the PA spectrum of carbon black should be used only as a reference spectrum. This reasoning was questioned by Low (1985), whose arguments were summarized in the previous section.

Papendorf and Riepe (1989) carried out detailed surface studies of activated carbons, using PA infrared spectroscopy to characterize adsorbed chlorinated hydrocarbons. Because carbon black is a standard material for recording background (reference) PA spectra, these authors ratioed spectra of adsorbates on carbons against a similar spectrum obtained for unloaded activated carbon. Spectra were obtained for carbons coated with tri/perchloroethylene, 1,1,1-trichloroethane, 1,1,2,2-tetrachloroethane, and pentachloroethane. The concentrations of these substances on the carbons varied between 15 and 40%. As many as five bands were observed in the 700–1000 $cm^{-1}$ region; the results resembled appropriate spectra from a reference library. The characteristic C—Cl bands disappeared at concentrations below about 7%, which establishes a rather high detection limit for these compounds using PA infrared spectroscopy.

As discussed in Chapter 5 and elsewhere in this book, finely divided carbons such as carbon black are commonly used to normalize PA infrared spectra. This use of these carbons is, of course, partly based on their very strong absorption of infrared radiation. A related question concerns the maximum concentration of carbon that a sample may contain before it gives rise to a saturated PA spectrum.

This issue was investigated by Carter et al. (1989b). These authors found that six carbon blacks with different particle sizes and reactivities gave essentially the same (featureless) PA spectra. These carbons were then incorporated into vulcanized natural rubber at a concentration of 8.5% by weight. PA spectra of these samples exhibited a series of bands due to the rubber, showing a few minor differences from one sample to the next. One of the rubbers was also studied at progressively higher carbon concentrations. A significant number of bands were clearly visible in the PA spectra of samples with carbon compositions up to 8%. On the other hand, only a few features remained at 15%, and a carbon concentration of 25% produced a PA spectrum that resembled a blackbody curve. The latter result can therefore be used to normalize PA infrared spectra (Carter and Wright, 1991); in fact, some PA infrared spectroscopists currently advocate the use of carbon-filled rubber for the measurement of reference spectra. The results of Carter et al. (1989b) provide an estimate of the maximum amount of carbon that is acceptable in many samples that are to be analyzed by PA infrared spectroscopy. However, as is shown below, far higher percentages of carbon can be tolerated in the study of cokes.

PA infrared spectra of amorphous hydrogenated carbon films (a:C—H) were reported by Raveh et al. (1992). These films were grown from a methane plasma under microwave (MW) or radio frequency (RF) discharge. The spectra were used to elucidate bonding, which is correlated with physical properties such as density and microhardness. IN-PHAS and IN-QUAD PA spectra (Chapter 5) showed that the bonded oxygen occurs near the film surface. PA spectra were generally superior to transmission spectra, exhibiting more intense and well-resolved bands that were due to C=O, $CH_3$, $CH_2$, C—C, and CH groups. The MW films are polymerlike, with lower hardness and higher hydrogen content.

Yang and Simms (1993, 1995) utilized PA infrared spectroscopy to study carbon fibers derived from petroleum pitch. The production of these fibers requires several steps: (a) the formation of a precursor (green) fiber; (b) stabilization of the precursor fiber; and (c) carbonization. PA spectroscopy offers considerable advantages with regard to the characterization of these fibers, mainly because it eliminates the need for traditional infrared sample preparation techniques. For example, transmission spectra of carbon fibers had previously been obtained after they were ground and then incorporated into KBr pellets. As an alternative, a fiber was dissolved in an organic solvent, after which the solution was deposited onto an infrared-transparent window and evaporated to dryness! It need hardly be stated that the elimination of such invasive procedures is highly desirable.

The PA spectra of these fibers resemble the corresponding spectra of some high-rank coals. Specifically, features from both aliphatic and aromatic C—H groups, as well as C=O and O—H bonds are observed. The stabilization process occurs at high temperatures and causes oxidation: This is confirmed by the occurrence of bands due to ketones, aldehydes, acids, esters, and anhydrides. Carbonization at progressively higher temperatures (up to 900°C) eventually led to the elimination of all functional groups and the observation of an essentially featureless spectrum.

Figure 6.3 shows typical data obtained by Yang and Simms (1995). Curve A is the PA infrared spectrum obtained for a green fiber and clearly displays a number of bands that are attributable to the functional groups mentioned above. After stabilization, a much simpler spectrum (curve B) was obtained. Grinding of the stabilized powder gave rise to the upper spectrum (curve C), which dramatically reveals the difference between the surface and the bulk of the fiber.

Bouzerar et al. (2001) used PBDS to study the electronic density of states in the near-infrared and visible regions for amorphous hydrogenated carbon thin films. Although no spectra were displayed in this work, variations in the absorption coefficient with photon energy were reported. The data were interpreted with regard to two density-of-state models, and a connection was established between the bandgap and the disorder in the carbon films.

### 6.1.3. Hydrocarbon Cokes

This discussion of PA infrared spectra of carbons ends with results for hydrocarbon cokes recently obtained in the author's laboratory. The cokes were provided by Syncrude Canada Ltd. and are a by-product of the process used to create Syncrude Sweet Blend (a mixture of hydrotreated components) from the vast oil sands deposits in northeastern Alberta. Typically,

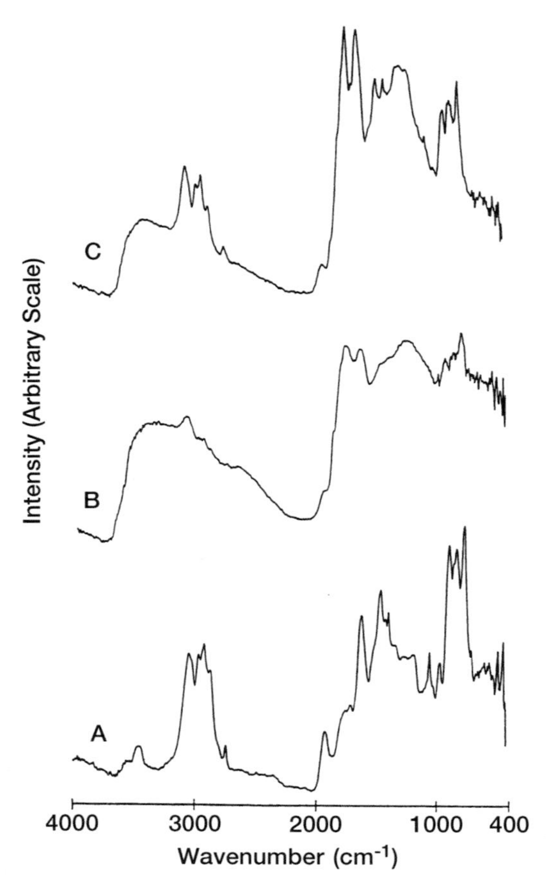


**Figure 6.3.** PA infrared spectra of carbon fibers. Curve A, green fiber; curve B, stabilized fiber; curve C, stabilized fiber (ground sample). (Reprinted from *Fuel* **74**, Yang, C. Q. and Simms, J. R., Comparison of photoacoustic, diffuse reflectance and transmission infrared spectroscopy for the study of carbon fibres, 543–548, copyright © 1995, with permission from Elsevier Science.)

these cokes consist of approximately 90% carbon, 4% sulfur, and residual aromatic hydrocarbons at concentrations up to approximately 1%.

Figure 6.4 shows representative PA spectra recorded for three of these cokes. It is obvious that aromatic hydrocarbons are detectable in some cokes but not in others. For example, prominent aromatic C—H bands occur at about 750, 810, and 870 $cm^{-1}$ in the upper spectrum. The 1600-$cm^{-1}$ peak is due to aromatic C=C stretching and was discussed above in the context of work by Morterra and Low (1982) as well as Low and Glass (1989). The third indication of aromaticity in these spectra is the absorption con-

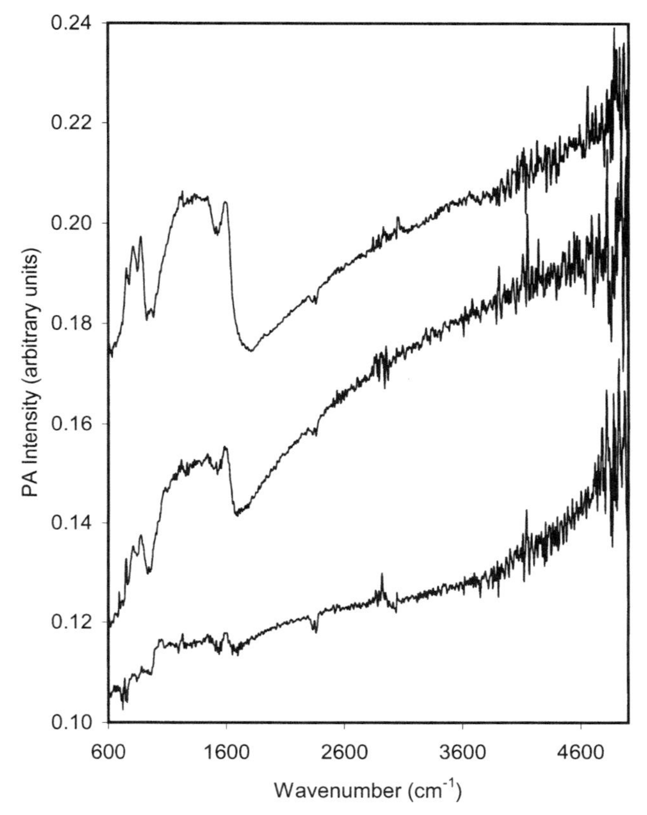


**Figure 6.4.** PA infrared spectra of three wall cokes. Coke spectra were ratioed against carbon black spectra obtained under similar conditions and have been slightly offset along the vertical axis for clarity. It should be noted that background intensity may vary due to differences in particle sizes of coke samples.

tinuum that intensifies with increasing wavenumber and extends into the near-infrared: This band is due to electronic transitions. All of these features are less intense in the middle curve and virtually absent in the lower spectrum. Thus these cokes can be classified with regard to their relative hydrocarbon contents by means of PA infrared spectroscopy; a large number of different cokes have been successfully studied in this way. In a recent study (Michaelian et al., 2002) it was shown that increased aromatic hydrocarbon contents are correlated with reduced values of the thermal conductivities, thermal diffusivities, and thermal effusivities of the cokes. This result was interpreted in terms of a model in which the aromatic hydrocarbons occupy the pore structures in the cokes, reducing the efficacy of heat propagation.

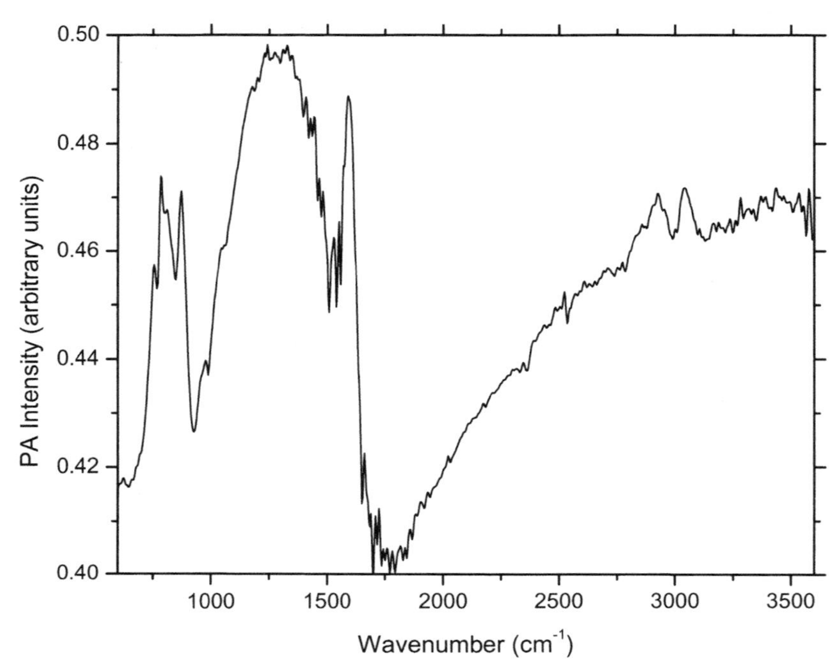


**Figure 6.5.** PA infrared spectrum of different wall coke.

A more impressive PA spectrum, obtained for a different Syncrude coke, is shown in Figure 6.5. In addition to all of the features described in the previous paragraph, both aliphatic and aromatic C—H stretching bands are clearly visible in this spectrum. Moreover, the aromatic C—H region between approximately 700 and 900 cm$^{-1}$ is observed to comprise at least four bands. These features arise from differently substituted aromatic ring structures. In general, the infrared absorption frequency in this region decreases as the number of contiguous hydrogens on an aromatic ring increases.

## 6.2. COALS

The preceding discussion of the PA infrared spectra of carbons leads naturally to the subject of coals. This progression would seem logical from the perspective of many spectroscopists, particularly those who might choose to categorize coals simply as naturally abundant carbonaceous solids. Indeed, the close relationship between carbons and coals is reflected by the fact that the same word can be used to refer to both substances in some languages: Typical examples occur in Spanish (*carbón*) and French (*charbon*). This commonality calls to mind the many similarities that occur in the spectra of carbons (as defined in the previous section) and coals. The PA infrared spectra of coals and carbons are discussed separately in this book primarily because of the considerable amount of published literature on both topics.

It is also worth noting that an even larger body of literature exists on the more general subject of the infrared spectroscopy of coals. These publications span the entire period of modern infrared spectroscopy, that is, they cover approximately six decades of research in coal science. No attempt has been made to review this literature here. The interested reader may wish to consult the many conference proceedings, review articles, and texts on coal characterization for more information. Moreover, while the present discussion is restricted to the PA infrared spectroscopy of coals, it should also be mentioned that many present-day workers continue to utilize alternative infrared techniques such as transmission and diffuse reflectance for the analysis of coals with good success.

### 6.2.1. Early PA Infrared Spectra of Coals

As discussed in Chapter 2, many of the initial successful demonstrations of PA infrared spectroscopy emphasized its capabilities with regard to the analysis of problematic solid samples. Coals were included among these solids because they generally require extensive grinding before they can be properly dispersed in alkali halides and studied by the classical infrared transmission technique. This grinding of coals is undesirable for several reasons: It is time consuming, leads to heating and oxidation, and tends to liberate entrained clays and minerals. Hence it is clearly preferable to eliminate (or at least minimize) the use of this sample preparation step when possible. The early accomplishment of this objective and the accompanying measurement of research-quality PA spectra are thus particularly relevant. The investigations of Mead et al. (1979), Vidrine (1980), Low and Parodi (1980), Herres and Zachmann (1984), Larsen (1988), as well as Low (1986) should all be mentioned in this context. Although the infrared spectra of coals in these articles were generally not interpreted in detail, the data were certainly of sufficient quality to justify the subsequent application of the technique by other researchers.

Another group of early publications can be identified in which the PA infrared spectra obtained for coals are discussed more thoroughly. For example, a pioneering contribution by Rockley and Devlin (1980) compared spectra of aged and freshly cleaved surfaces of sub-bituminous, bituminous, and anthracitic coals. (These terms are arranged in order of increasing rank, or fixed carbon content, of the samples.) C—H stretching, $CH_2$ deformation, and C=C stretching bands due to the coals, as well as several features arising from clay components, were identifiable in the PA spectra. The results suggested that volatilization and oxidation occur during the ageing process. Similarly, Royce et al. (1980) reported a PA study of coal pyrolysis and showed that the bands due to aliphatic groups diminished in intensity during this process. This behavior resembles that described for carbons in the previous section.

Krishnan (1981) described the close similarity between the PA and diffuse reflectance spectra of a finely powdered coal sample of unspecified origin. The results in this study also serve to demonstrate the enhanced sensitivity of PA spectroscopy to the clays that are frequently present in coals. This effect can be readily put down to the intrinsically small particle sizes of clays and the fact that PA intensities are generally greater for samples comprised of smaller particles. In another comparison of infrared techniques, Solomon and Carangelo (1982) examined PA, diffuse reflectance, and transmission spectra of a Pittsburgh seam coal with regard to their abilities to quantify hydroxyl absorbance; the region above 3500 $cm^{-1}$ in the PA spectrum was partly obscured by bands due to water vapor, hindering the study of the much broader hydroxyl band. A greatly improved spectrum of this sample was subsequently obtained by McClelland (1983).

A more detailed discussion of the PA infrared spectra of coals was presented by Zerlia (1985). Semianthracite, medium- and high-volatile bituminous, and sub-bituminous coals were examined using standard (nonspectroscopic) analytical techniques in addition to PA infrared spectroscopy. As shown in Figure 6.6, detailed band assignments were given for most of the features in the infrared spectra. The ability of PA infrared spectroscopy to provide relevant technological data on coals was also established in this work. Specifically, the volatile matter contents of the coals were shown to vary linearly with the integrated area of the 2800–3800 $cm^{-1}$ region in the PA spectra. This section of the spectra includes contributions from bands due to the OH, NH, aromatic CH, and aliphatic $CH_n$ groups in the coals. The correlation between the spectroscopic data and the results from traditional coal analysis is particularly significant in view of the considerable time and effort required for the latter measurements.

The aromatics contents of coals can be compared using the three low-frequency aromatic C—H bands at approximately 750, 810, and 865 $cm^{-1}$ or, alternatively, the broad band between about 3000 and 3100 $cm^{-1}$. Gerson et al. (1984) obtained PA spectra of a series of eight coals that represented a wide range of ranks and showed that higher volatile matter contents corresponded to lower aromatic/aliphatic ratios in the C—H stretching region. The same trend was reported in the above-mentioned work of Zerlia

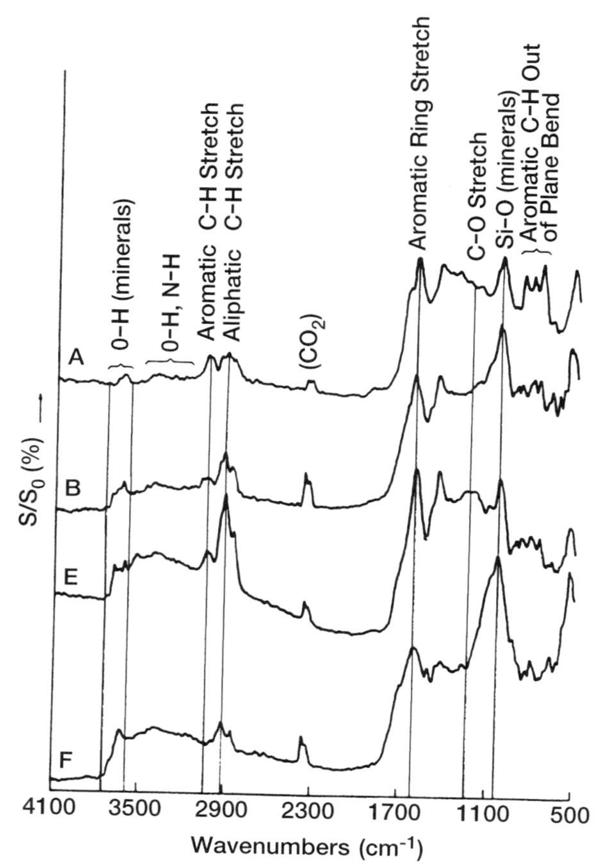


**Figure 6.6.** PA infrared spectra of four coals. A, semianthracite; B, medium-volatile bituminous; E, high-volatile bituminous; F, sub-bituminous. (Reprinted from *Fuel* **64**, Zerlia, T., Fourier transform infrared photoacoustic spectroscopy of raw coal, 1310–1312, copyright © 1985, with permission from Elsevier Science.)

(1985). Both findings were consistent with rank determinations based on reflectivity, a widely accepted parameter in coal characterization. To summarize, increased aromatics contents in coals are correlated with higher reflectivity and rank; greater aliphatics contents imply the presence of more volatile matter and the opposite trends in the first three parameters.

PA infrared spectroscopy can also be used to analyze fly ash, the fine solid waste material produced during the combustion of coal. Seaverson et al. (1985) obtained spectra of four different coal fly ashes that were known to contain several minerals as well as water. These species were eliminated by thermal treatment at temperatures of about 400°C and higher. In three of the fly ashes, the eliminated inorganic component was confirmed to be $Ca(OH)_2$.

### 6.2.2. PA Near-Infrared Spectra of Coals

The near-infrared spectra of many coals exhibit a featureless absorption continuum that spans the entire region and usually intensifies with increasing wavenumber. This band arises from electronic transitions in aromatic structures and is analogous to that observed in the PA spectra of many high-temperature carbons. In addition, electronic absorption is thought to be responsible for part of the underlying intensity that is visible between the more well-defined bands in the mid-infrared spectra of coals. The very broad electronic absorption bands of coals thus extend from the mid-infrared, throughout the near-infrared, and into the visible wavelength regions.

The near-infrared absorption continuum in the spectra of coals can, in principle, be observed by various means. On the other hand, its identification is sometimes complicated by the effects of scattering when techniques such as transmission and diffuse reflectance are employed. PA spectroscopy offers a distinct advantage in this situation because results can be obtained for various particle sizes and the influence of scattering can be isolated. The elimination of the diluent in PA near-infrared spectroscopy also simplifies this experiment.

The assignment of the absorption continuum to electronic transitions in aromatic species is consistent with the results reported for separated macerals of a sub-bituminous Alberta coal (Donini and Michaelian, 1986). As shown in Figure 6.7, the PA spectrum of fusinite displays the most intense continuum; the relative intensities of the aromatic and aliphatic bands in the C—H stretching region confirm that this maceral is the most highly aromatic of the three studied. By contrast, the continuum is significantly weaker in the vitrinite spectrum and practically unidentifiable in resinite. The C—H stretching bands show that the latter maceral consists primarily of aliphatic material.

Two other publications also described the near-infrared PA spectra of coals. McAskill (1987) found that dispersive PA spectra of oil shales and coals of several different ranks (brown, sub-bituminous, and bituminous) all displayed the near-infrared continuum mentioned above. In fact, this broad band was considered to be somewhat detrimental in this investigation since it tended to obscure several interesting vibrational (combination and overtone) bands. This interference was less problematic for brown coals and more significant for coals of higher rank. Several years later, Michaelian et al. (1995a) confirmed that the near-infrared continuum intensified as the

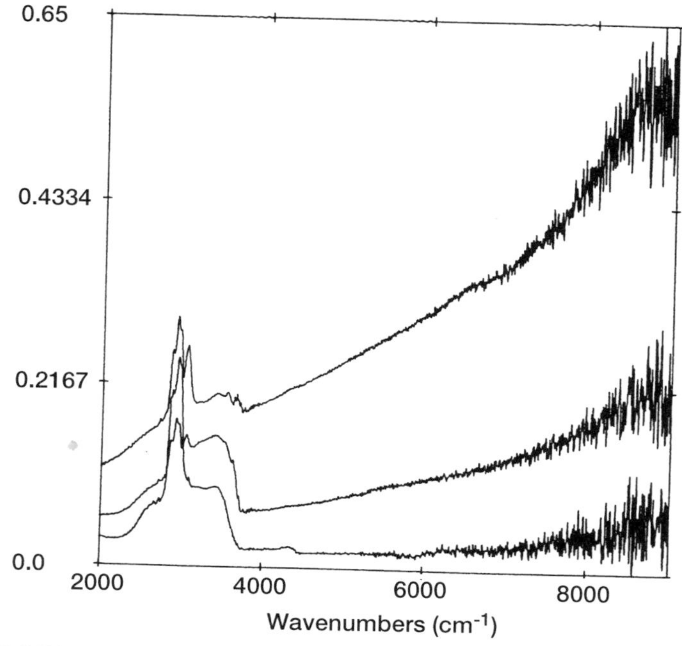


**Figure 6.7.** Mid- and near-infrared PA spectra of separated macerals from sub-bituminous Alberta coal. Upper curve, fusinite; middle curve, vitrinite; lower curve, resinite. (Reprinted from *Infrared Phys.* **26**, Donini, J. C. and Michaelian, K. H., Near infrared photoacoustic FTIR spectroscopy of clay minerals and coal, 135–140, copyright © 1986, with permission from Elsevier Science.)

ranks of bituminous, sub-bituminous, and lignitic coals were artificially increased by thermal treatment in both oxidizing and inert atmospheres.

### 6.2.3. Study of Coal Oxidation by PA Infrared Spectroscopy

The oxidation of coals results in a number of important changes to their physical and chemical properties. For example, the propensity of a coal for coking, its calorific value, and its floatability are all adversely affected by oxidation. A number of analytical methods—including infrared spectroscopy—have been applied to this problem, with varying degrees of success. The use of PA infrared spectroscopy in the study of coal oxidation, which is naturally motivated by both the minimal sample preparation and the possibility of depth profiling, is reviewed in the following paragraphs.

The capability of infrared spectroscopy with regard to the detection of oxygen-containing functional groups in coals was established decades ago. Thus it is logical that several groups investigated the feasibility of using PA infrared spectroscopy for this particular application when the technique was first gaining acceptance. The PA spectra of coals obtained in several early studies readily displayed a number of changes due to oxidation: These include the elimination of aliphatic and aromatic $CH_n$ groups, as well as the production of anhydrides, esters, carboxylic acids, and other carbonylic species (Hamza et al., 1983; Chien et al., 1985a,b; Angle et al., 1988). It should also be mentioned that oxidation under laboratory conditions is not necessarily representative of the processes that occur during stockpiling (Mikula et al., 1985) or natural weathering.

The research of M. J. D. Low's group on the PA infrared spectra of carbons was discussed in detail in the preceding section of this chapter. Several works published by this group also mention coals or coal chars. For example, Morterra and Low (1985e) proposed a model in which an anthracitic coal is comprised of a carbonaceous material and a second, independent hydrocarbon phase. Oxidation of the hydrocarbon component was thought to yield compounds such as anhydrides. This investigation showed that C=O groups are formed initially in oxidation and that C—O functionality appears later. The charring of coal yielded a product with a spectrum similar to those for charred wood, oxidized cellulose, and several other similar substances (Low and Morterra, 1989).

Infrared and Raman spectroscopists are well aware that the region between about 1900 and 2600 $cm^{-1}$ exhibits bands from only a few types of functional groups. Low (1993a) observed bands in the 2150–2250 $cm^{-1}$ region in PA spectra of several oxidized coals and coal chars. These bands were attributed to nitrogen-containing species, thought to be present in pyridinic or pyrrolic structures. This proposal is consistent with the fact that the concentration of nitrogen in the coals was on the order of 2%. Part of the significance of this work stems from the fact that nitrogen-containing heterocycles may also occur in other hydrocarbon fuels, such as those derived from bitumen and heavy oil.

A series of studies on the oxidation and derivatization of coals was published in the 1980s by B. M. Lynch and his colleagues at St. Francis Xavier University and several other institutions. After initial investigations of both PA and diffuse reflectance spectroscopy (Lynch et al., 1983), the former technique was chosen for further work. Both naturally weathered and artificially oxidized coals were studied in this research. Bands in the carbonyl region were examined thoroughly, with four features at 1650, 1690, 1720, and 1750 $cm^{-1}$ being of particular interest. The intensities of these bands were determined by subtracting the PA spectra of the fresh coals from those of the corresponding oxidized coals.

One of the major conclusions of this work refers to peroxide species, which were proposed as precursors of the oxidation products. The authors concluded that peroxides are present on the surfaces of virtually all coals, except for those that are freshly prepared. Carbonyl functionalities were produced by base-promoted and thermal decompositions of the peroxides (Lynch et al., 1986, 1987a) to yield products thought to be cyclic or open-chain esters. Importantly, total carbonyl intensity varied linearly with the bulk oxygen content of the coal (Lynch et al., 1987b). Moreover, carbonyl intensities were shown to be correlated with the so-called plastic properties (Gieseler fluidity, dilatation, and melting range) of the coals (Lynch et al., 1988). The findings of this research were subsequently reviewed by Lynch and MacPhee (1989). One important conclusion from this work is that laboratory coal samples should be stored in sealed ampoules to prevent unwanted oxidation.

The oxidation of hand-picked resinites from western and Arctic Canada and Australia was studied by Goodarzi and McFarlane (1991). The authors found that PA infrared spectroscopy was particularly well suited for the analysis of naturally exposed and weathered resinite surfaces, irrespective of their morphology. More than 20 bands were observed in the infrared spectra and assigned to various functional groups in the coal or the clay that was also present in the samples. Resinites are predominantly aliphatic, although oxidation at temperatures above 250°C leads to partial aromatization (McFarlane et al., 1993). The natural oxidation products depended primarily on the rank of the coal that was the source of the resinite and the degree of weathering, which was influenced by the depositional environment. The surface sensitivity of PA infrared spectroscopy played a key role in this research.

### 6.2.4. Depth Profiling of Coals

Oxidized coals present an excellent opportunity for PA infrared depth profiling studies since the oxidation process is generally expected to begin at the surface of the coal particle and progress toward its interior. Moreover, differences in composition are expected to occur over dimensions on the order of a few micrometers, which are comparable to sampling depths in PA infrared spectroscopy. Thus depth profiling experiments that discriminate between the surface and the bulk of a coal sample are pertinent to the study of oxidized coals.

Zerlia (1986) carried out rapid-scan depth profiling experiments on a laboratory-oxidized coal sample. Bands due to $CH_n$ and hydroxyl groups exhibited lower intensities at higher mirror velocities, implying that these species were less abundant in the surface layer of the coal. Because $CH_n$ groups are known to be consumed in oxidation, the results confirmed that the surface of the coal was oxidized. Different oxidation mechanisms were identified for the interior and surface regions of the relatively coarse (4-mm) coal particles in this study. There was no spectroscopic evidence for the production of several oxygen-containing functional groups (ether, carbonyl, or carboxyl) in the interior of the particles to compensate for the loss of $CH_n$. This is expected if oxidation occurs by oxygen uptake and peroxide formation.

The most prominent change in the infrared spectrum of a coal that accompanies oxidation is often the intensification of a C=O band near 1750 $cm^{-1}$. As mentioned above, this peak may contain contributions from several carbonylic species, depending on the composition of the coal and the oxidation mechanism. The 1750-$cm^{-1}$ band was investigated in amplitude modulation step-scan depth profiling experiments on fresh and oxidized coals (Michaelian, 1989a, 1991). These studies confirmed that the oxidation of the coal was more extensive in the surface layer than in the interior of the coal particles. Furthermore, variation of the sampling depth indicated that the first 12 μm of the coal particles was oxidized uniformly. These results were discussed in more detail in Chapter 4.

Yamada et al. (1996) performed phase modulation depth profiling experiments on oxidized coals. At a modulation frequency of 400 Hz, the quadrature (surface-sensitive) spectrum showed the intensification of the carbonyl band and the diminution of C—H stretching intensity that accompany oxidation. These results are consistent with those in the earlier experiments described above. As the phase angle was reduced and greater depths were examined, the change in the C=O band persisted, while that for the C—H bands vanished. On the other hand, the change in C—H intensity persisted to greater depths at a frequency of 100 Hz. These results suggested an oxidation mechanism in which dehydrogenation occurs mainly in the surface layer of the coal. Alternatively, aliphatic compounds could migrate deeper inside the coal during oxidation.

### 6.2.5. Numerical Analysis of PA Spectra of Coals

PA infrared spectra of coals have also been used to demonstrate numerical methods such as phase correction, signal averaging, curve-fitting, and deconvolution. Some of the publications on these subjects emphasize the relevant numerical methods rather than the chemical analysis of coals, these samples being used primarily for illustrative purposes. While these articles contribute relatively little information on coal structure, their emphasis on the accurate calculation of the PA spectra of coals tends to make them relevant in the present context.

Most commercial FTIR software utilizes a procedure known as multiplicative (Mertz) phase correction to minimize the effects of phase errors on the spectra. An alternative technique—interferogram symmetrization—

is thought to yield more accurate spectra by some researchers, who often develop the necessary software in their own laboratories. This method was applied to PA data for separated coal macerals and other types of samples several years ago by the present author (Michaelian, 1989b, 1990). In fact, this research showed that the PA spectra of a given sample calculated using either method of phase correction were generally in good agreement. It should be noted that these calculations implicitly assume that all phase information should be removed from the computed spectra and do not take into account the role played by the phase of the PA signal (see Chapters 4 and 5).

Coals sometimes yield weak PA spectra, which can make signal averaging strategies particularly important (Michaelian, 1987, 1990). In this regard, it should be mentioned that it is normally preferable to average a smaller number of "scans" (interferograms) in FTIR spectroscopy so as to ensure short-term instrument stability. When a large number of scans is required to achieve an adequate signal-to-noise ratio, it is better to acquire a series of spectra, each corresponding to a relatively short measurement time. These spectra can then be averaged to produce the final result. It is less advisable to accumulate a large number of scans directly to produce the final spectrum, even though the time for data acquisition is the same in both cases. This situation exists because the phase correction calculation is intrinsically more accurate for shorter experiments.

The complexity of coals is manifested in the C—H stretching region of their PA infrared spectra. Observed spectra generally exhibit three broad, overlapping bands at about 2870, 2930, and 2960 $cm^{-1}$ that are due to aliphatic carbon–hydrogen groups. In addition, many coals display an aromatic CH band centered near 3050 $cm^{-1}$. To increase the amount of information discernible in the C—H region of the PA spectra, numerical techniques such as second derivative computation, curve-fitting, and deconvolution have been used (Friesen and Michaelian, 1986, 1991; Michaelian and Friesen, 1990). These methods show that the part of the spectrum between about 2800 and 3000 $cm^{-1}$ is actually comprised of about eight to nine bands—roughly the same number used to fit the aliphatic region in the PA infrared spectra of hydrocarbon fuels (see below). These bands can be readily assigned to $CH_n$ groups in several environments that are known to exist in coals. The consistency of the results and the plausibility of these band assignments both suggest the credibility of the findings of these investigations.

An example of the results obtained by these methods is depicted in Figure 6.8. Curve-fitting and deconvolution were both employed in this study (Friesen and Michaelian, 1991). The agreement between the deconvolved experimental spectrum (upper curve) and the analogous result for the fitted spectrum (lower curve) indicate that the number of bands used, as well as their parameters, are realistic.

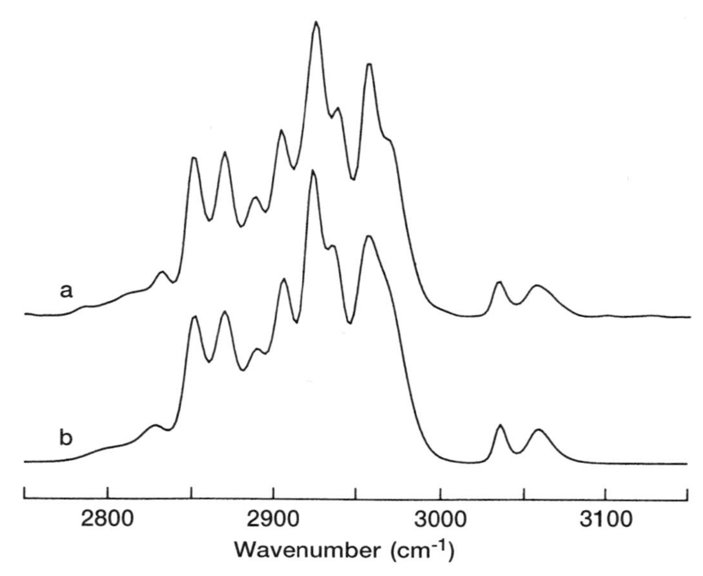


**Figure 6.8.** Deconvolution of measured (curve a) and fitted (curve b) PA infrared spectra of lignite. Gaussian bands were used in the curve-fitting calculations. (Reproduced from Friesen, W. I. and Michaelian, K. H., *Appl. Spectrosc.* **45**: 50–56, by permission of the Society for Applied Spectroscopy; copyright © 1991.)

The availability of PA infrared spectra of coal macerals enables their correlation with a number of important technological properties. For example, an inverse procedure in which the petrographic composition of a coal is determined from the resolved PA infrared spectra of separated macerals has been described (Gagarin et al., 1993). Similarly, PA intensities at 2867 and 3019 $cm^{-1}$ have been used to predict heats of combustion (Gagarin et al., 1995a), as well as the pseudo-first-order rate constant for liquefaction in tetralin (Gagarin et al., 1994). The Roga index, a measure of the caking propensity of coals, can also be predicted from these data (Gagarin et al., 1995b). It should also be mentioned that the region between about 1000 and 1800 $cm^{-1}$ in the PA spectra has also been analyzed by inverse methods (Michaelian et al., 1995b).

## 6.3. HYDROCARBONS

A familiarity with the PA infrared spectra of aliphatic and aromatic hydrocarbon compounds is essential for a thorough interpretation of spectra of hydrocarbon fuels such as coals and middle distillates. Accordingly, published near- and mid-infrared PA spectra of hydrocarbons are discussed in this section. This literature summary is followed by some typical (previously unpublished) examples of PA infrared spectra of aromatic hydrocarbons, including polycyclic aromatics.

Not surprisingly, an examination of the literature shows that many of the early publications on PA spectra of hydrocarbons emphasized the near-infrared region. This situation reflects the general status of PA infrared spectroscopy during the development of the technique, which began more than 20 years ago. As discussed elsewhere in this book, the accessibility of the near-infrared region to PA researchers led to the early primacy of work in this region; however, because of the widespread availability of FTIR spectrometers and commercial PA cells, the majority of current applications of PA spectroscopy to the study of hydrocarbons and hydrocarbon fuels are based on the mid-infrared region.

The work of Adams et al. (1978) is among the first near-infrared PA investigations of hydrocarbons. These authors obtained dispersive spectra between about 0.8 and 2.7 μm (12,500–3700 $cm^{-1}$) for *n*-hexane, benzene, anthracene, and benzanthracene. Between three and six prominent bands were observed for each compound. A peak at about 2.2 μm (~4545 $cm^{-1}$) occurred in the spectra of all three aromatics but was absent from the spectrum of hexane. Hence this band can be used for quantitation of aromatics in aromatic-aliphatic mixtures. Indeed, Adams et al. examined mixtures of benzene in hexane and found that the intensity of the 2.2-μm band varied in direct proportion to the benzene concentration. A similar result was observed in the near-infrared PA spectra recorded for mixtures of benzene with an Iranian crude oil. These results nicely illustrate the suitability of PA near-infrared spectroscopy for the analysis of hydrocarbons and hydrocarbon fuels. They also demonstrate the suitability of PA spectroscopy for quantitative analysis, a subject that is discussed more thoroughly in Chapter 7. These conclusions were reiterated in a separate review by Kirkbright (1978).

Several years later, Lewis (1982) reported dispersive near-infrared (1.0–2.6 μm) PA spectra of both organic and organometallic compounds. The goal of this work was to provide band assignments and promote the use of PA near-infrared spectroscopy to "fingerprint" compounds—the latter objective usually being addressed by means of mid-infrared spectroscopy. The hydrocarbons studied in this investigation included benzene, benzene-$d_6$, toluene, and cyclohexane. Like most dispersive near-infrared spectra, the PA spectra in this publication are displayed on a linear wavelength scale; nevertheless, band positions are reported in wavenumbers. For example, the three bands observed in the PA spectrum of benzene by Adams et al. (1978) were stated by Lewis to occur at 4100, 4675, and 6025 $cm^{-1}$. The peak at

4675 $cm^{-1}$ is the same one used by Adams et al. for quantitation of benzene in binary mixtures. Lewis (1982) assigned the near-infrared bands in the PA spectra of the organic compounds to combinations and overtones of mid-infrared frequencies, which were observed in separate (transmission) experiments.

Broadening the scope of this discussion slightly to include compounds that contain elements other than carbon and hydrogen leads to the next relevant PA near-infrared study of substituted benzenes. Sarma et al. (1987) obtained dispersive spectra of three benzonitriles, five acetophenones and three benzylbromides. About 20 bands were identified in the near-infrared spectra of these substances. These features were readily assigned using mid- and far-infrared band frequencies observed by these authors in a related investigation. Because of the structural similarity of the samples, the same set of band assignments was applicable for all 11 compounds. Many of the bands were identified in the spectra of most or all of the substituted benzenes.

The final near-infrared PA study to be discussed here is that of Manzanares et al. (1993). These authors investigated fundamental and overtone spectra in the C—H stretching region for *cis*- and *trans*-3-hexene. To observe the fourth and fifth overtones, which occur near the high-frequency limit of the near-infrared and in the long-wavelength part of the visible region, respectively, the authors employed PA detection. Experimental apparatus included an acousto-optically modulated argon ion laser, a dye laser, and piezoelectric detection. The $Ar^+$ laser pumped the dye laser, which supplied the near-infrared and visible radiation used to obtain the spectra. The use of PA detection is rather incidental to this work, which is primarily concerned with curve-fitting calculations for the C—H bands and assignment of the spectra according to the local mode model. This implies that piezoelectric PA detection of near-infrared spectra of alkenes is sufficiently reliable to allow researchers to emphasize the interpretation of data, rather than the technique itself.

Two studies that discuss mid-infrared PA spectra of hydrocarbons adsorbed on various substrates will now be mentioned. In the first, Saucy et al. (1985b) reported spectra of polynuclear aromatic compounds adsorbed on alumina and silica. The impressive detection sensitivity of PA spectroscopy, which sometimes extends to submonolayer coverage, was demonstrated in this work. The absence of sample preparation and the capability for analysis of milligram quantities of sample were recognized by the authors as the most desirable attributes of PA infrared spectroscopy.

This work showed that a monolayer of phenanthrene on silica was sufficient for the detection of several well-defined PA bands due to phenanthrene. Two additional analyte bands were identified after subtraction of the

PA spectrum of silica. All of these features were also observed in the PA spectrum of pure phenanthrene. Moreover, the authors used PA spectroscopy to show that ultraviolet (UV) irradiation of 9-nitroanthracene adsorbed on silica led to the production of anthraquinone. Finally, through a comparison of spectra of intact and ground pellets, Saucy et al. (1985b) demonstrated the very significant increase in signal that accompanies an increase in surface area in PA spectroscopy.

A more recent example of the use of PA spectroscopy for the characterization of organic compounds in the solid state has been described by Gosselin et al. (1996). These authors utilized PA spectroscopy to monitor the synthesis of resino compounds, whose structures are much more complicated than those of the simple hydrocarbons referred to earlier in this section. Transmission spectra obtained for KBr pellets of resins are complicated by light scattering and reflection. On the other hand, the PA gas-microphone technique removes the need for sample grinding and preparation of pellets and yields superior spectra that are free from these troublesome effects. Hence Gosselin et al. concluded that PA infrared spectroscopy can serve as an effective analytical method in solid-phase organic chemistry.

The PA infrared spectra of cokes and coals described in the first two sections of this chapter are generally interpreted under the logical assumption that these materials contain a large number of different aromatic hydrocarbons. Aromatics are also an integral component of hydrocarbon fuels, which are discussed in the next section. Hence it is appropriate to examine the mid-infrared PA spectra of pure aromatic hydrocarbons, which can be taken as models for all three types of samples. PA infrared spectra of most aromatic hydrocarbons are not available in the published literature; consequently, spectra were recently obtained for a series of 12 readily available compounds in the author's laboratory and are presented here.

The results are organized according to the number of aromatic rings in each hydrocarbon that was studied and to the degree of substitution of the parent compound with alkyl side chains. Thus Figure 6.9 displays PA infrared spectra of naphthalene and three alkylated derivatives, 1-methyl-, 2-ethyl-, and 2,3,5-trimethylnaphthalene. It should be noted that 2-ethylnaphthalene is a liquid at ambient temperatures. The figure shows that more than 20 bands occur in the fingerprint region below 2000 $cm^{-1}$ in each spectrum, with another series of peaks appearing in the C—H stretching region. The prominent bands just below 3000 $cm^{-1}$ in the top three curves arise from the alkyl substituents, while the bands above this frequency are due to aromatic CH groups. Importantly, all of the spectra display a large number of well-defined, relatively narrow bands—in contrast with the general appearance of the PA spectra of the carbonaceous materials mentioned above. Moreover, it should be recognized that the PA infrared spectra of

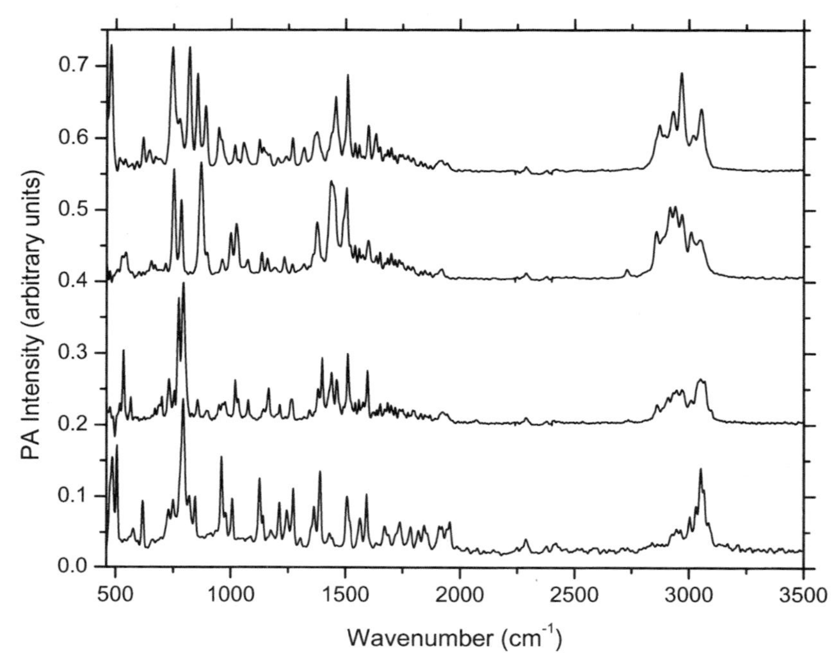


**Figure 6.9.** PA infrared spectra of substituted naphthalenes. Spectra have been rescaled and offset along vertical axis for clarity. Bottom curve, naphthalene; second curve, 1-methylnaphthalene; third curve, 2,3,5-trimethylnaphthalene; top curve, 2-ethylnaphthalene. Bands near 2350 $cm^{-1}$ due to atmospheric $CO_2$ have been subtracted from some of spectra.

pure aromatic hydrocarbons are qualitatively similar to infrared spectra obtained using traditional infrared sampling methods.

The spectra of several tricyclic aromatics (phenanthrene, anthracene, and two substituted anthracenes) are depicted in Figure 6.10. The results resemble those for the less complex compounds discussed in the previous paragraph, with about 30 or more bands appearing in the spectrum of each tricyclic compound.

Four unsubstituted polycyclic aromatics (pyrene, chrysene, benzo[*a*]-pyrene, and 1,2:3,4-dibenzanthracene) are characterized in Figure 6.11. The PA spectra are similar to those in the previous two figures, with the obvious exception that bands due to alkyl groups are not present in the third group.

The literature reviewed in this section, and the new results obtained for a series of aromatic compounds, clearly demonstrate the suitability of PA infrared spectroscopy for the characterization of hydrocarbons. The PA spectra of these hydrocarbons are generally similar to those obtained with

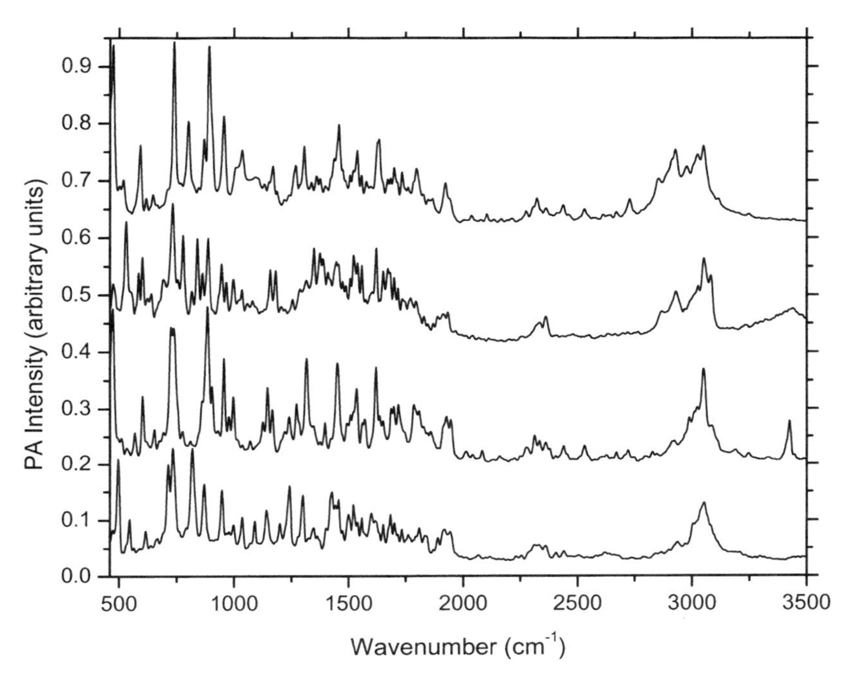


**Figure 6.10.** PA infrared spectra of tricyclic aromatic hydrocarbons. Spectra have been rescaled and offset along vertical axis for clarity. Bottom curve, phenanthrene; second curve, anthracene; third curve, 9-methylanthracene; top curve, 2-methylanthracene. Some of features near 2350 $cm^{-1}$ are due to atmospheric $CO_2$.

other infrared techniques but do not resemble those of natural hydrocarbons discussed elsewhere in this book very closely.

## 6.4. HYDROCARBON FUELS

The published results and other data presented in the first three sections of this chapter clearly illustrate the suitability of PA infrared spectroscopy for the characterization of carbonaceous solids (carbons, coals, cokes) and several different types of aromatic hydrocarbons. The many successful analyses of these substances carried out during the last two decades suggest that PA spectroscopy should also be used to augment other techniques, such as near-infrared and FT-Raman spectroscopies, in the characterization of typical hydrocarbon fuels. The capabilities of PA infrared spectroscopy with regard to the study of these fuels, which were established quite recently, are briefly summarized in this section.

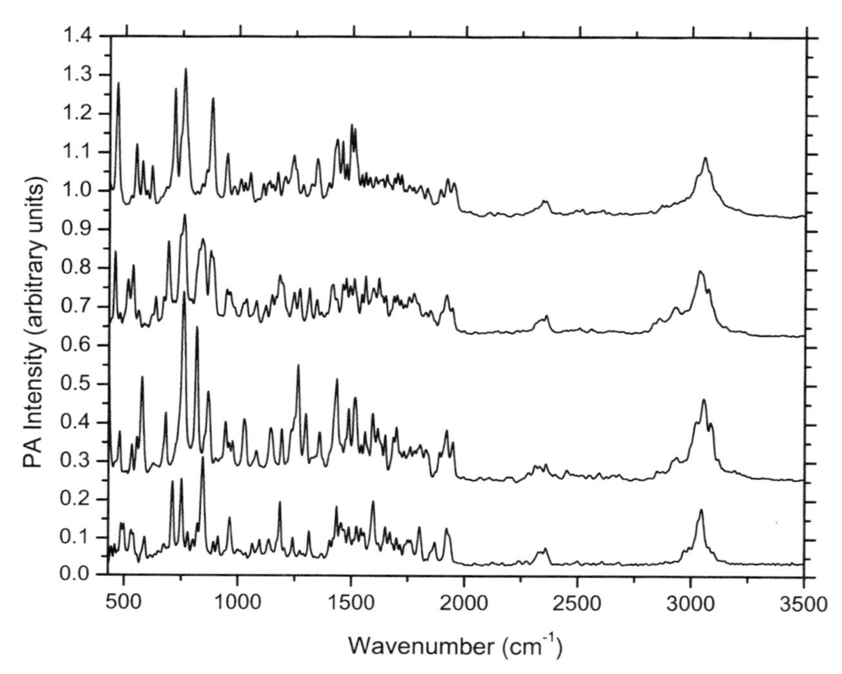


**Figure 6.11.** PA infrared spectra of polycyclic aromatic hydrocarbons. Bottom spectrum has been rescaled, and other three spectra have been offset along vertical axis for clarity. Bottom curve, pyrene; second curve, chrysene; third curve, benzo[*a*]pyrene; top curve, 1,2:3,4-dibenzanthracene. Some of the features near 2350 $cm^{-1}$ are due to atmospheric $CO_2$.

Most of the literature discussed so far in this book refers to the investigation of samples that are solid at ordinary temperatures. Only a few examples of spectra of liquids have been given; some of these are found in the immediately preceding description of near- and mid-infrared PA spectra of hydrocarbons. While it is clear that there is no substantive reason why PA infrared spectroscopy cannot be successfully utilized in the study of liquid samples, some might argue that the choice of the technique in this context should still be justified. As is shown below, the measurement of infrared spectra of hydrocarbon fuels—particularly highly aromatic, viscous liquids—provides this justification.

Recently, Michaelian et al. (2001) presented a detailed PA infrared study of a series of distillation fractions derived from Syncrude Sweet Blend (SSB). This darkly colored liquid (sometimes alternatively referred to as synthetic crude oil or bitumen-derived crude) is obtained by extraction and upgrading of the bitumen in the vast oil sands deposits in northeastern Alberta. SSB

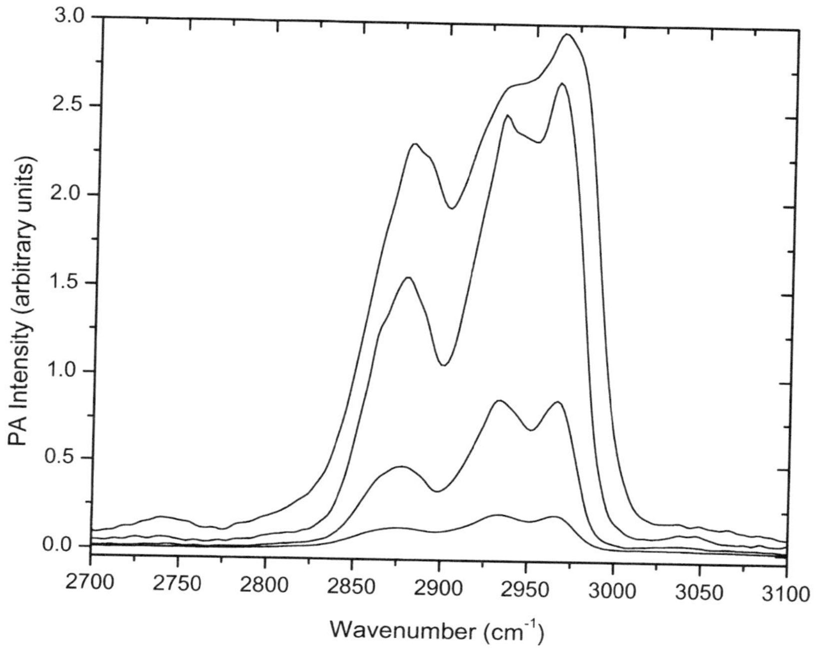


**Figure 6.12.** C—H stretching region of PA infrared spectra for four distillation fractions from SSB. Boiling points: top curve, 30–71°C; second curve, 71–100°C; third curve, 100–166°C; bottom curve, 166–177°C.

was separated by distillation into 12 fractions, with boiling point intervals that span the range from 30 to 524°C, in this work. Insofar as some of these fractions resemble commercial middle distillate fuels (e.g., gasoline, diesel, and jet fuel), this study demonstrates the applicability of PA infrared spectroscopy to the analysis of hydrocarbon fuels.

This investigation showed that the majority of the information in the PA spectra of these fractions is conveyed by the C—H stretching region. As depicted in Figure 6.12, the intensity and shape of the characteristic profile between about 2800 and 3000 $cm^{-1}$ change significantly for the first four distillation fractions; the boiling point dependence of the spectra continues, in a less dramatic fashion, for the higher fractions. The C—H stretching region of these spectra was also analyzed by curve fitting and integration (Fig. 6.13). The eight bands due to aliphatic ($CH_2$ and $CH_3$) groups systematically shift to lower frequencies as successively higher boiling point ranges are analyzed. Moreover, the percentage of total C—H intensity attributable to $CH_2$ groups increases with boiling point, while that due to

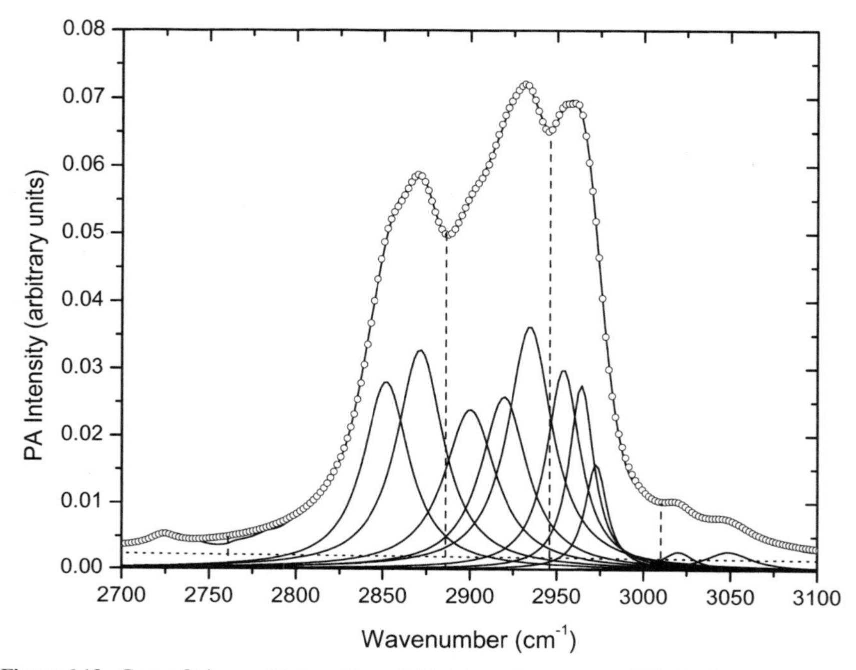


**Figure 6.13.** Curve fitting and integration of PA infrared spectrum of distillation fraction with boiling point range 232–288°C. Circles represent ordinates in sum of plotted component bands; dashed vertical lines show integration limits used to calculate areas of asymmetric $CH_2$ and $CH_3$ stretching bands and total area due to aliphatic C—H stretching. Dotted straight line at about 0.03 intensity units is fitted baseline.

$CH_3$ groups concomitantly decreases. Although the origin of the trend in band positions is still unclear, the growth in the $CH_2/CH_3$ intensity ratio is readily attributed to an increase in the average length of the alkyl side chains attached to the aromatic ring structures in these fractions. These results confirm that the PA infrared spectra of these hydrocarbon distillation fractions are of research quality, enabling detailed analysis and interpretation.

As might be expected, the physical characteristics of the hydrocarbon distillation fractions differ considerably. For example, the fractions with boiling points below 100°C are colorless liquids that are easily transferred into various sample accessories. Their PA infrared spectra can be obtained with a single scan of the moving mirror in an FTIR, that is, in about one second. In these cases, traditional methods (e.g., ATR and transmission) and PA detection of infrared spectra are equally convenient.

The fractions with higher boiling points (and higher aromatics contents) are progressively darker and more viscous. These liquids are not well suited for more conventional infrared sampling techniques but are still quite amenable to PA infrared spectroscopy. This is an example of a situation where the elimination of sample preparation in PA spectroscopy enables the analysis of samples that would otherwise probably be considered as nearly intractable.

The distillation fraction in the wide boiling range between 343 and 524°C is referred to as heavy gas oil (HGO) and is particularly important with regard to the refining of SSB into commercial fuels. PA infrared and FT-Raman spectra of a second series of distillation fractions, which comprise the entire HGO region, were obtained in a subsequent investigation (Michaelian et al., 2003). Curve-fitting and integration calculations were again used to analyze the C—H stretching region of the PA spectra. In contrast with the results obtained in the first study, the positions and relative areas of the individual bands were observed to be approximately constant within the HGO region. This shows that the changes in the infrared spectra of the fractions occur mainly at lower boiling points. Although the HGO distillation fractions are quite viscous, PA infrared spectroscopy enables their analysis in a relatively straightforward manner.

The residue that remains after removal of the distillation fractions, aptly referred to as "resid," is a brittle solid that tends to shatter when attempts are made to grind it with a mortar and pestle. A PA spectrum obtained for several millimeter-size pieces of a typical resid is shown in Figure 6.14. The high aromatics content of this resid is indicated by the prominence of the three aryl C—H bands between 700 and 900 $cm^{-1}$, as well as the broad aromatic C=C stretching band at approximately 1600 $cm^{-1}$. These features were discussed earlier with regard to the PA spectra of both cokes and coals. In addition, aliphatic groups in the resid give rise to the bands at 1377 and 1463 $cm^{-1}$, while the C—H stretching region of this spectrum resembles that of the higher HGO distillation fractions.

The investigation of resid demonstrated another noteworthy phenomenon that is characteristic of PA infrared spectroscopy. As the measurement of successive spectra was continued for more than one hour, a progressive diminution in their intensities was observed. After this experiment was completed, the original small particles were found to have coalesced in the sample cup, forming a disk about 5 mm in diameter and 1 mm thick. The decrease in surface area thus accounted for the loss of PA signal. The softening of the resid was obviously a consequence of sample heating by the infrared beam, mentioned briefly in Chapter 3. This observation—which might be considered surprising in light of the high boiling point of the sample—illustrates what might be regarded as a minor disadvantage of PA infrared spectroscopy.

The spectroscopic characterization of the bitumen from which all of the

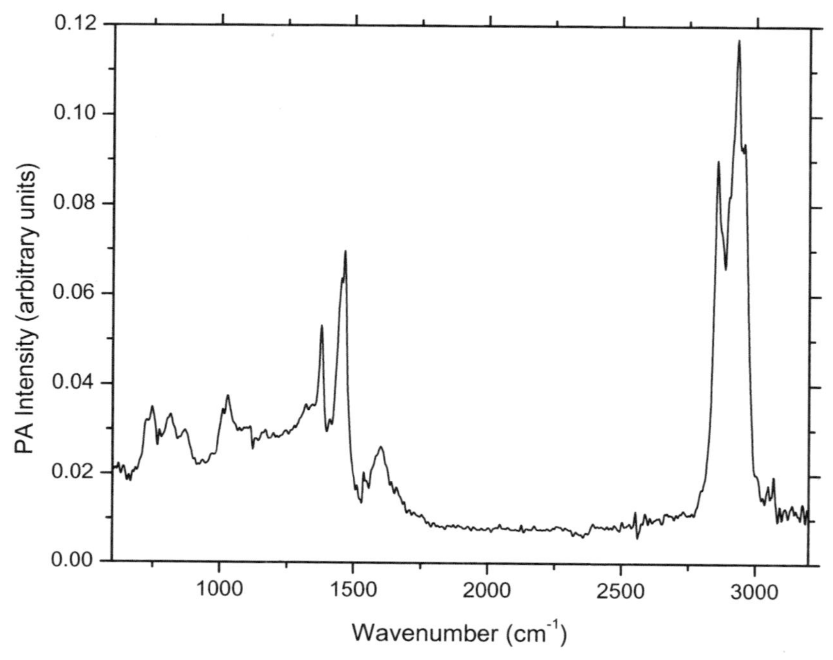


**Figure 6.14.** PA infrared spectrum of "resid" from distillation; boiling point greater than 524°C.

above samples originate is clearly a nontrivial problem for the analyst. At room temperature, bitumen has a consistency similar to that of molasses. Not surprisingly, this extremely thick black liquid has rarely been investigated by infrared spectroscopy; its problematic physical character is obviously part of the reason for this situation.

Figure 6.15 illustrates a typical PA infrared spectrum of bitumen obtained in the author's laboratory with a Bruker IFS 88 FTIR spectrometer and an MTEC 200 PA cell. As noted in the figure, bands due to aliphatic and aromatic CH groups are clearly identifiable. An additional peak near 1030 $cm^{-1}$ is probably due to carbon–carbon stretching involving aliphatic groups bonded to aromatic rings. This spectrum was acquired in about 15 min.

The sample characterized in Figure 6.15 was stored in the laboratory for about 4 months. Because the bitumen was not removed from the sample cup, its surface was continuously exposed to the ambient air during this period. After this delay, another PA spectrum was recorded. As shown in Figure 6.16, prominent new bands appeared at about 1030 $cm^{-1}$ (due

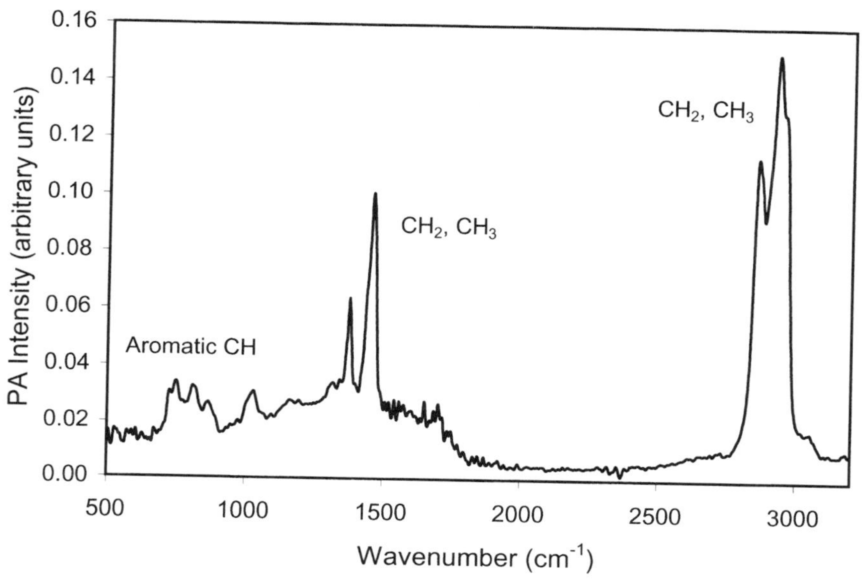


**Figure 6.15.** PA infrared spectrum of fresh Syncrude bitumen.

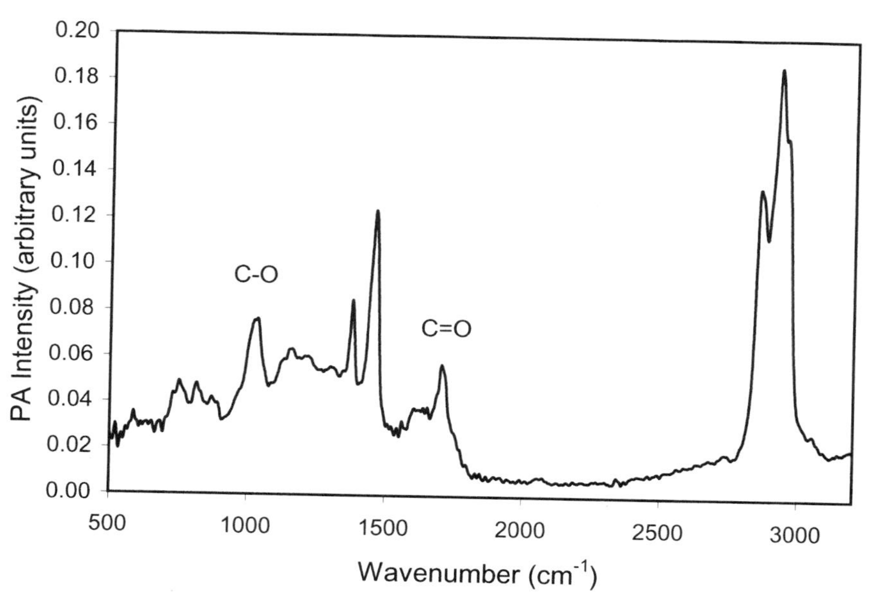


**Figure 6.16.** PA infrared spectrum of aged Syncrude bitumen.

to C—O or S=O stretching) and 1700 $cm^{-1}$ (C=O stretching). Several weaker bands between 1100 and 1200 $cm^{-1}$ are also attributable to C—O groups. Thus the aging of bitumen led to significant oxidation. However, it should be kept in mind that the PA infrared spectrum corresponds to a surface layer about 10–20 μm in thickness: This experiment provides no information about the possible bulk oxidation of the sample.

The extraction of bitumen from the oil sands produces vast quantities of tailings, thereby raising a number of environmental issues. Moreover, this waste material contains residual bitumen, which amounts to an unused resource. One aspect of the latter problem was recently studied by Bensebaa et al. (2001), who used PA infrared and X-ray photoelectron spectroscopies to develop a better understanding of the structure of fine tailings. The solids were separated into fractions referred to as fine solids (FS), bi-wetted solids, and hydrophobic solids, respectively. Step-scan PA infrared spectra were obtained for these fractions using phase modulation. These spectra contained bands arising from clays (primarily kaolinite) and the unextracted bitumen. Depth profiling experiments yielded important information regarding the arrangement of these two components in the solids. For example, the spectra of the FS platelets exhibited stronger inorganic (clay) bands at higher modulation frequencies: This suggested that the organic component was concentrated in the interior of the particles. The ordering of the layers was greatly diminished when the platelets were ground to a powder. This allowed the authors to develop a conceptual model in which the residual bitumen is trapped between the particulate layers. It can be noted that both depth profiling and the minimal sample preparation in PA infrared spectroscopy were key components of this work.

The examples in this section were selected to illustrate the suitability of PA infrared spectroscopy for the analysis of liquid and solid hydrocarbon fuels, as well as waste materials that result from their production. Although this research topic is still relatively new, the available results demonstrate that the technique can be used for virtually any sample relevant to this area. On the other hand, the possibility of significant sample heating should always be kept in mind during the measurement of PA spectra of strongly absorbing hydrocarbons.

## 6.5. CORROSION

The suitability of PA infrared spectroscopy for the analysis of corrosion products was first demonstrated by Eyring's research group (Gardella et al., 1982, 1983a). These authors used both PA infrared spectroscopy and X-ray photoelectron spectroscopy (XPS) to analyze a powdery blue coating formed on large iron gratings during the production of HCN. The atmosphere above the gratings also contained $NH_3$, water vapor, and a number of trace gases. The PA infrared spectrum of the deposit displayed bands characteristic of the $NH_4^+$ and $CN^-$ groups and also indicated the presence of a metal–CN bond. These results, together with those from XPS, permitted identification of the primary corrosion product as ammonium ferro-(ferrocyanide), or $(NH_4)_xFe_y[Fe(CN)_6]$, where $x + 2y = 4$. Minor products included ammonium ferrocyanide, ammonium cyanide, and ferrous ammonium cyanide. Interestingly, the results from XPS were consistent with those from PA infrared spectroscopy, even though the sampling depth in the former experiment was on the order of 2–5 nm—three to four orders of magnitude smaller than that in the PA experiment.

A second, more recent, investigation in which PA infrared spectroscopy was used to study corrosion occurs in the work of Salnick and Faubel (1995, 1996). These authors studied the corrosion of copper used as a roofing material at Stockholm City Hall. Prolonged exposure of this metal to the atmosphere results in the formation of a green/green-blue corrosion layer, or patina, that comprises a number of copper-containing minerals. PA spectroscopy was used to analyze an 8- × 8- × 600-μm section cut from the roof, without any further sample preparation. This method was chosen mainly because of the opaque and highly scattering nature of patinas; traditional infrared sample preparation methods would, of course, disrupt the morphology of the patina layers.

The PA infrared spectra obtained for copper patina implied the existence of about 10 different components. For example, comparison with published spectra of pure compounds allowed the authors to identify brochantite, $Cu_4(OH)_6SO_4$; antlerite, $Cu_3(OH)_4SO_4$; copper sulfate, $Cu_4(OH)_4OSO_4$; copper (II) nitrite, $Cu_4(OH)_6(NO_3)_2$; copper oxysulfate, $Cu_3O_2SO_4$; and copper carbonate, $Cu_2CO_3(OH)_2H_2O$. Some compounds that were not previously known to occur in copper patinas were also observed: A noteworthy example was copper (II) fluoride, $CuF_2 \cdot xH_2O$.

In addition to the identification of the minerals present in the patina, Salnick and Faubel (1995) presented a description of several photothermal properties of the samples. Their methods included photothermal beam deflection (PBD), which was used to calculate thermal diffusivity, calculation of the thermal diffusion length of the patina, estimation of the patina layer thickness, and determination of the optical absorption length. The thermal diffusivity for the outdoor patina layer was 0.022 $cm^2/s$, which is close to the value of 0.025 $cm^2/s$ calculated for brochantite but much lower than that for pure copper, 1.1 $cm^2/s$. Knowledge of the thermal diffusivity allowed calculation of thermal diffusion lengths at the seven FTIR mirror velocities utilized in this study. For the patina, this dimension ranged from

15.9 to 39.7 μm at 4000 $cm^{-1}$ and from 50.1 to 125.5 μm at 400 $cm^{-1}$. The optical absorption length, which was calculated from absorption data obtained for KBr pellets, ranged from 250 μm to 1 cm. Because these dimensions were much greater than the 25-μm thickness of the patina surface layer, the samples were optically transparent in the PA experiments.

Salnick and Faubel (1995, 1996) described an additional numerical technique in their study of the corrosion of copper, which could be applied much more widely in PA infrared spectroscopy. In the "power-index" method, the dependence of the PA intensity on the modulation frequency $f$ is used to determine the degree of PA saturation. As discussed in Chapter 5, the PA signal $S$ varies as $S \sim f^{-1}$ for optically transparent, thermally thin solids, and as $S \sim f^{-1.5}$ for optically opaque, thermally thick samples. The authors suggested that the power-index spectrum corresponding to the PA signal frequency dependence $S \sim f^n$ be calculated as an alternative to the ordinary PA infrared spectrum. The value of $n$ is obtained from

$$n(v) = \log(S_1/S_2)/\log(V_1/V_2) \tag{6.1}$$

where $V_1$ and $V_2$ are two different mirror velocities (cm/s) and $S_1$ and $S_2$ are the corresponding PA intensities at waveumber $v$. For the patina, $n$ ranged from $-1.5$ in regions of low absorption to $-1.1$ for the intense peak at 3401 $cm^{-1}$ and the region from 400 to 1200 $cm^{-1}$. Thus, even in regions of strong absorption, the experimental conditions are far from PA saturation ($n = -1.5$); the thermal properties of the Cu backing play only a limited role in the generation of the PA signal. In contrast, from 2000 to 2800 $cm^{-1}$ where absorption is weak, the optically opaque and thermally thick copper substrate plays a more dominant role with regard to the observed PA signal. It is important to recognize that $n$ is not a constant in a particular experiment and varies throughout the infrared spectrum; a value for $n$ has significance only with regard to a particular infrared frequency.

A previously unpublished result for a different type of corrosion product is shown in Figure 6.17. This PA infrared spectrum was obtained for a darkly colored solid that had been scraped from a section of corroded pipe at an active wellhead site in central Alberta. The PA spectrum displays a broad strong band just above 1400 $cm^{-1}$, which is characteristic of carbonates. In fact, a comparison of the band positions in this infrared spectrum with the published literature indicates that the dominant mineral in this sample is siderite ($FeCO_3$). This result was confirmed by X-ray diffraction. The sample also contained minor amounts of hematite ($Fe_3O_4$) and other minerals. The familiar bands at 2852 and 2924 $cm^{-1}$ signify the presence of residual aliphatic hydrocarbons in the corrosion product.

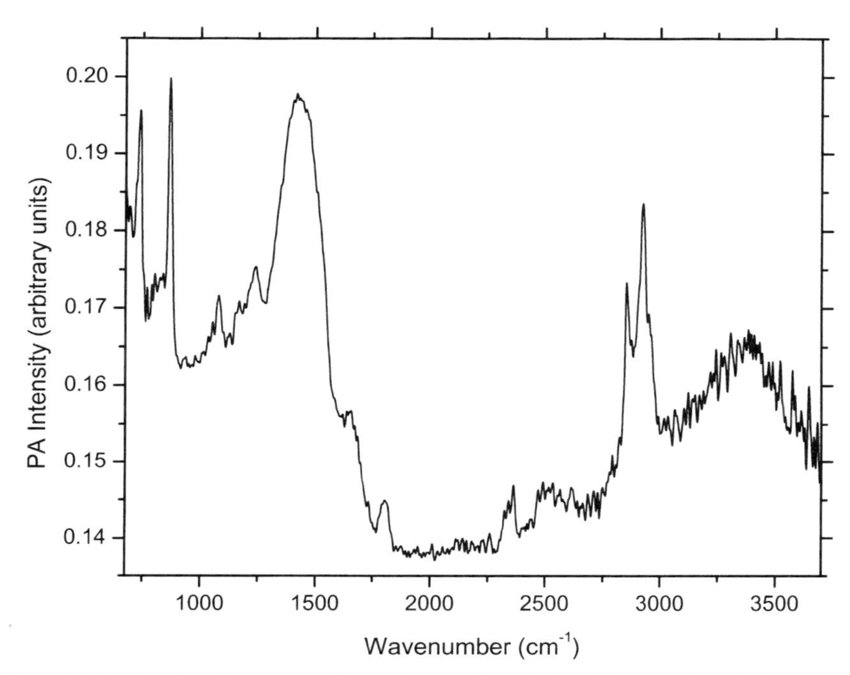


**Figure 6.17.** PA infrared spectrum of corrosion product obtained from section of pipe at wellhead site in central Alberta.

## 6.6. CLAYS AND CLAY MINERALS

The reader may already be aware of the extensive literature on the infrared spectroscopy of clays and minerals: Numerous texts and review articles have been dedicated to this subject. Indeed, some clay scientists might question whether it is even necessary to use PA detection in order to measure infrared spectra of these industrially important materials. A few exploratory studies of PA infrared spectroscopy utilized clays and minerals simply as examples of common solids; these circumstances could hardly be considered auspicious and did not provide a particularly strong justification for the use of PA spectroscopy in the characterization of these substances.

Investigations carried out by several research groups during the last 20 years have clearly shown much more compelling reasons for the acquisition of PA infrared spectra of clays and minerals. Most importantly, the minimal sample preparation in PA spectroscopy means that these solids generally require little or no grinding prior to measurement of their spectra. The elimination of grinding is highly desirable since this procedure can lead to delamination of layered clays and to other phase changes. Similarly, it is particularly important to avoid the grinding of clays in situations where observation of both longitudinal and transverse optic bands is possible; this is illustrated below.

As is discussed elsewhere in this book, the preparation of pellets is unnecessary in PA infrared spectroscopy. This fact is particularly relevant with regard to the study of clays and minerals, where the pressing of pellets containing a diluent such as KBr can sometimes result in unwanted ion exchange reactions. This undesirable effect was recently reported by Pelletier et al. (1999), who observed the transformation of Na-saponites into K-saponites during the preparation of KBr pellets. Fortunately, such an effect can be entirely eliminated through the measurement of PA infrared spectra.

As noted above, a number of studies that were published when PA infrared spectroscopy was first gaining widespread acceptance presented spectra of clays and minerals. Among these publications was the work of Kirkbright (1978), already mentioned in several other contexts. This work includes a dispersive near-infrared PA spectrum of kaolinite, a common phyllosilicate with the empirical composition $Al_4Si_4O_{10}(OH)_8$. In the region of the PA spectrum between about 1.2 and 2.6 μm (8330–3850 $cm^{-1}$), five bands were observed. Features near 1.4 and 1.9 μm (7140 and 5260 $cm^{-1}$) were attributed to free water, while the bands at slightly longer wavelengths were assigned to combinations involving hydroxyl groups. A number of years later, the near-infrared PA spectrum of kaolinite was obtained in the laboratory of the present author using an FTIR spectrometer. The results of that experiment are described in more detail below.

Two publications relevant to this discussion arise from the research of Rockley's group, part of which was also summarized in Chapter 2. In the first study, Rockley et al. (1982) reported the PA infrared spectrum of asbestos (chrysotile). The O—H stretching doublet at 3655 and 3692 $cm^{-1}$ was well defined in the PA spectrum and showed the expected 1:3 intensity ratio. The authors also noted the strong absorption of asbestos at 950 $cm^{-1}$ and suggested that a $CO_2$ laser might be a suitable radiation source when measuring PA spectra in this region. This proposal might not seem viable to present-day researchers with access to modern FTIR instruments. However, it was entirely reasonable at the time it was made, inasmuch as the use of $CO_2$ lasers in PA infrared spectroscopy was still fairly common. The PA spectrum of asbestos has been studied by other groups (Monchalin et al., 1979) and is mentioned again later in this section; it is also described in Chapter 5.

Several years later, Rockley and Rockley (1987) compared PA and transmission infrared spectra for interlamellar cation-exchanged montmorillonites. Dimethyl methylphosphonate (DMMP), a phosphonate pesticide model compound, was adsorbed on the clays. Some DMMP absorption bands were shifted by a few reciprocal centimeters ($cm^{-1}$) as a consequence of adsorption, leading to the conclusion that the interaction between DMMP and montmorillonite occurs through the P=O bond. The PA spectra obtained in this work contain a few weak bands that were not detectable in transmission spectra obtained for KBr pellets; the increase in the relative intensities of these bands suggests that the PA spectra may have been partially saturated.

The authors reported unsuccessful attempts to measure the PA spectrum of pure DMMP. Unfortunately, the samples were heated by the infrared beam and eventually evaporated; this led to the observation of gas-phase PA spectra. Practicing PA spectroscopists may, of course, be all too familiar with this phenomenon, which is quite readily observed when attempts are made to record PA spectra of liquids with relatively high vapor pressures.

Low and his colleagues at New York University studied a number of minerals during their development of the PBDS technique, reviewed in Chapter 3. In one of their initial publications on this subject (Low et al., 1982a), a PBDS spectrum of calcite ($CaCO_3$) was shown. This spectrum contains about 10 well-defined, narrow peaks; in addition, a broad, strong band near 1400 $cm^{-1}$ arises from the carbonate group. Subsequently, a much more extensive study (Low and Tascon, 1985) reported spectra of series of calcite-type minerals (magnesite, calcite, rhodocrosite, siderite, smithsonite), aragonites (aragonite, strontianite, witherite, cerussite), and lead minerals (crocoite, endlichite, vanadinite, mimetite, wulfenite). The spectra of the aragonites are shown in Figure 6.18. The authors observed that slight grinding of the minerals increased band intensities and reduced specular reflection effects, which sometimes caused severely distorted bands to occur in the spectra of unground samples. For example, PBDS spectra of a single crystal of calcite showed strong orientation effects but were not easily interpreted because they displayed unusual band shapes caused by specular reflection. Spectra of ground calcite were free of this effect.

The near-infrared PA spectrum of kaolinite was mentioned above. The measurement of this spectrum by Kirkbright (1978) was one of the first successful applications of PA infrared spectroscopy to the study of clays. Several years later, Donini and Michaelian (1985, 1986) obtained mid- and near-infrared PA spectra (approximately 2000–10,000 $cm^{-1}$) of several clays, including kaolinite, using an FTIR spectrometer. This more recent PA spectrum of kaolinite (Figure 6.19) displayed bands at 4526 and 4620 $cm^{-1}$. These features are due to the combinations of the OH stretching bands at 3620 and 3695 $cm^{-1}$ with the Al—OH bending frequency at 915 $cm^{-1}$. At higher frequencies, the overtones of the same two stretching frequencies

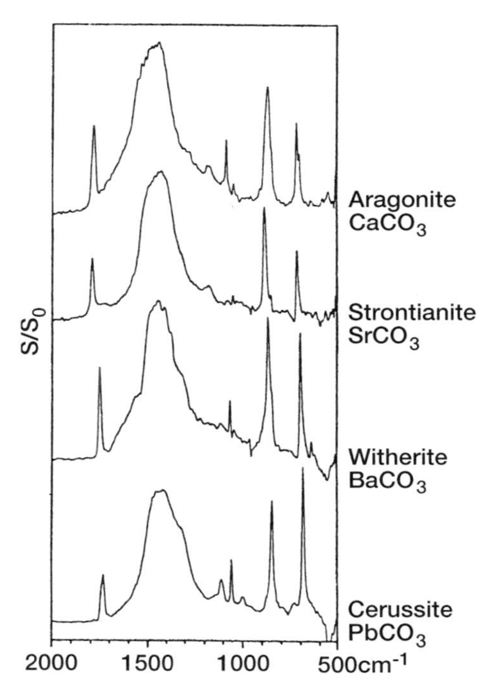


**Figure 6.18.** PBDS spectra of aragonites. (Reproduced from Low, M. J. D. and Tascon, J. M. D., *Phys. Chem. Minerals* **12**: 19–22, Fig. 7, by permission of Springer-Verlag; copyright © 1985.)

occurred at 7065 and 7160 $cm^{-1}$, respectively. It should also be noted that the latter assignment conflicts with that of Kirkbright, who had previously attributed the ~7160-$cm^{-1}$ band to adsorbed water instead of structural hydroxyl groups. Moreover, the observation of the two high-frequency near-infrared bands in the more recent work permitted the calculation of anharmonicities for the 3620-$cm^{-1}$ ("inner" hydroxyl) and 3695-$cm^{-1}$ ("inner surface" hydroxyl) stretching vibrations. The PA near-infrared spectra of some cation-exchanged bentonites were also obtained in this investigation, and several bands were assigned by analogy with the results for kaolinite.

After this work was completed, an even wider spectral region (250–10,000 $cm^{-1}$) was studied for partially deuterated kaolinite (Michaelian et al., 1987). Far-, mid-, and near-infrared PA spectra were obtained for a sample that was about 20% deuterated in this study. It was observed that deuteration shifted the five OH stretching bands at 3620, 3651, 3669, 3684, and 3695 $cm^{-1}$ to 2675, 2691, 2706, 2718, and 2725 $cm^{-1}$, respectively. The $\nu(OH)/\nu(OD)$ ratios for these bands are thus all approximately equal to 1.355. Part of the significance of these data lies in the fact that the 3684-

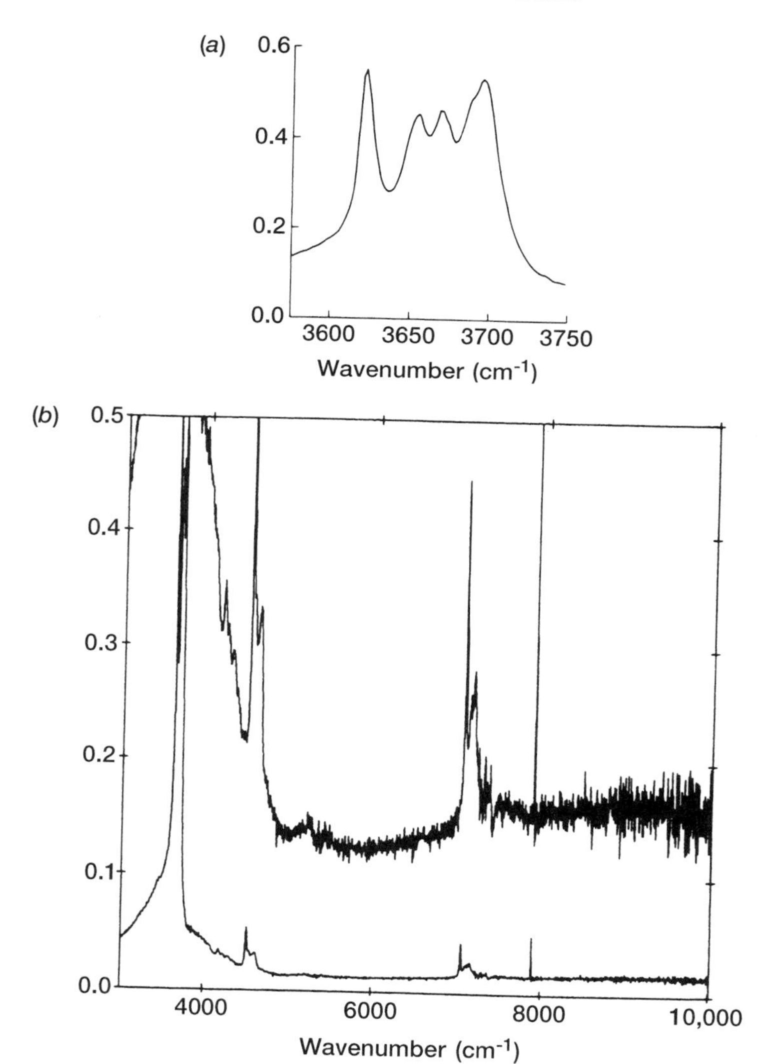


**Figure 6.19.** PA mid- and near-infrared spectra of international kaolinite. (*a*) OH-stretching region; (*b*) near-infrared region. The upper curve in (*b*) is scale-expanded by factor of 10. Spike at 7900 $cm^{-1}$ is submultiple of frequency of HeNe laser. (Reprinted from *Infrared Phys.* **26**, Donini, J. C. and Michaelian, K. H., Near infrared photoacoustic FTIR spectroscopy of clay minerals and coal, 135–140, copyright © 1986, with permission from Elsevier Science.)

$cm^{-1}$ band had been identified only in a few Raman spectra of natural kaolinite when the PA spectrum of the deuterated clay was recorded; the observation of the 2718-$cm^{-1}$ band was therefore one of the first indications that this fifth hydroxyl stretching band is actually infrared active. Subsequent investigations of infrared (both PA and transmission) and Raman spectra of kaolinites obtained from different sources confirmed this point. Those results are summarized below.

Far-infrared PA spectra of clays and minerals were discussed further in a subsequent study by Donini and Michaelian (1988). The weakness of these spectra mandated the systematic elimination of noise sources that had particularly deleterious effects on the data. In addition to some obvious sources of noise (Donini and Michaelian, 1985), the sequence of data averaging was considered in this work; it was found that the average of several individually calculated spectra exhibited significantly less noise than the result obtained when the corresponding interferograms were first averaged and then used to calculate a single spectrum. This effect is now known to arise from the fact that phase correction of the individual PA spectra is inherently more accurate than the single phase correction calculation carried out for an average interferogram. The far-infrared PA spectrum of asbestos obtained in this investigation is depicted in Figure 6.20.

The development of commercial step-scan FTIR spectrometers in the late 1980s made the investigation of signal saturation and phase in PA spectroscopy much more feasible than had been the case previously. In an early investigation of amplitude modulation step-scan PA infrared spectroscopy (Michaelian, 1990), mid- and near-infrared spectra were obtained for kaolinite. In general, the quadrature spectra were observed to be superior to the corresponding in-phase data, with less saturation, a lower background, and better band definition. One of the more important results in this study was the appearance of the ~3686-$cm^{-1}$ hydroxyl stretching band in the quadrature spectrum of kaolinite; this was among the first direct observations of this band in the infrared. This result is shown in Figure 6.21. This band was mentioned above with regard to deuterated kaolinite and will now be discussed in more detail.

The combination of results from transmission and PA infrared spectra with those from Raman spectroscopy has shown that a total of five hydroxyl stretching bands occur in infrared and Raman spectra of kaolinites. Transmission infrared spectroscopy had identified the four bands at 3620, 3651, 3669, and 3695 $cm^{-1}$; the fifth OH band was first observed at about 3684 $cm^{-1}$ in Raman spectra of several kaolinites more than 20 years ago and subsequently identified in rapid- and step-scan PA infrared spectra (Friesen and Michaelian, 1986). The interpretation of this fifth band has been the subject of recent research. For example, Shoval et al. (1999a) showed that its

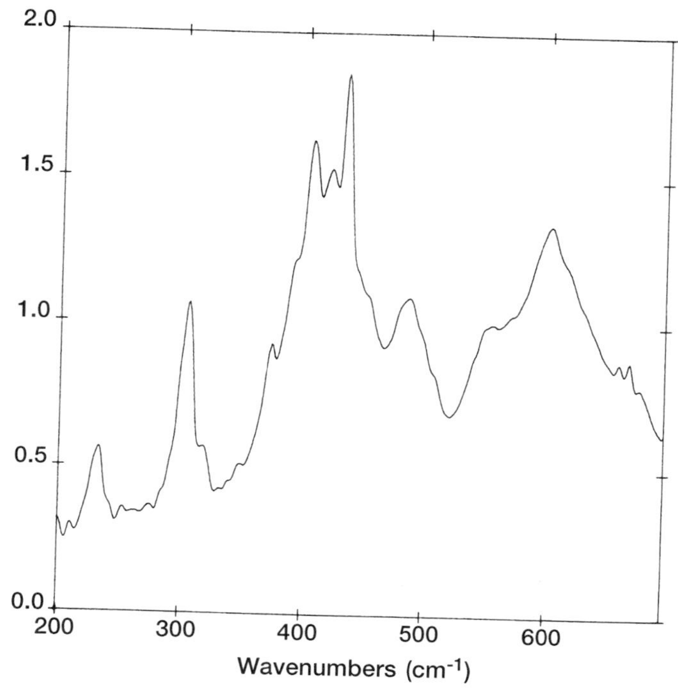


**Figure 6.20.** PA far-infrared spectrum of chrysotile asbestos. (Reproduced from Donini, J. C. and Michaelian, K. H., *Appl. Spectrosc.* **42**: 289–292, by permission of the Society for Applied Spectroscopy; copyright © 1988.)

intensity was proportionately greater in Raman and PA infrared spectra of highly crystallized kaolinites, which have large coherent domains. A comparison of spectra for kaolinites with different particle sizes led these authors to conclude that the bands at 3695 and 3686 $cm^{-1}$ are the longitudinal optic (LO) and transverse optic (TO) modes, respectively, of the high-frequency inner-surface OH stretching band.

Curve-fitting results for the PA spectrum of a highly crystallized sample are shown in Figure 6.22. It should be noted that the relative intensity of the 3686-$cm^{-1}$ band in this PA spectrum may be enhanced due to partial saturation (Michaelian, 1990). This band has now also been identified in transmission infrared spectra of hydrothermal and authigenic kaolinites, which have a high degree of crystallinity (Shoval et al., 1999b). The continued grinding of a highly crystallized kaolinite causes the destruction of these domains and the disappearance of the 3686-$cm^{-1}$ band (sometimes labeled as band Z); hence, the transmission spectrum differs from the PA spectrum by an increasing amount (Shoval et al., 2002a).

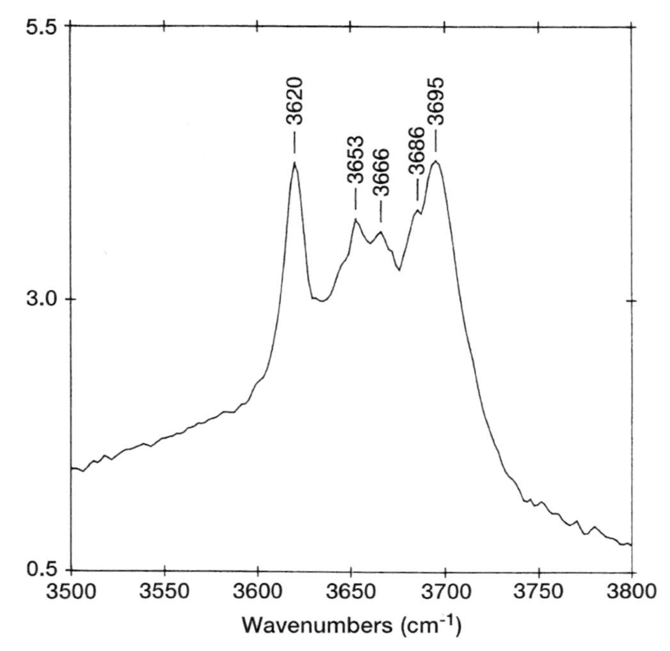


**Figure 6.21.** OH-stretching region in quadrature step-scan PA infrared spectrum of kaolinite, measured at resolution of 4 $cm^{-1}$ and modulation frequency of 40 Hz. (Reprinted from *Infrared Phys.* **30**, Step-scan photoacoustic infrared spectra of kaolinite, 181–186, copyright © 1990, with permission from Elsevier Science.)

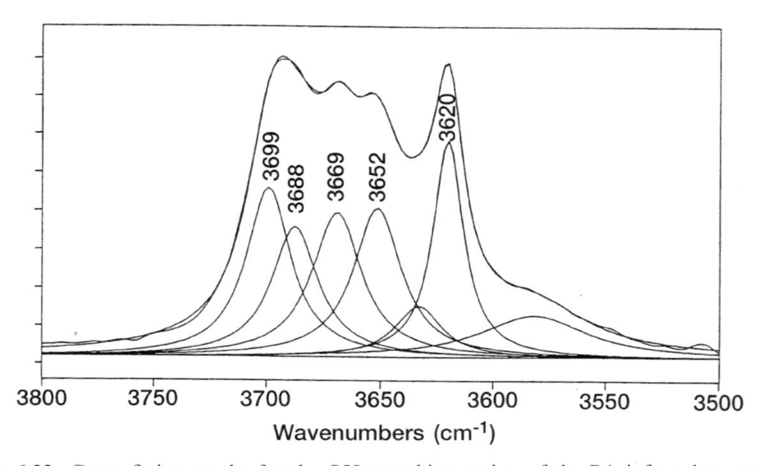


**Figure 6.22.** Curve-fitting results for the OH-stretching region of the PA infrared spectrum of highly crystallized MS-6 kaolinite, obtained from Makhtesh-Ramon, Israel. (Reproduced with the permission of the Mineralogical Society of Great Britain & Ireland from the paper by S. Shoval et al.: *Clay Minerals* **34**, p. 561; copyright © 1999.)

Kaolinite is usually referred to as a nonexpanding clay. In fact, kaolinite can be intercalated, either by small, polar organic molecules or by alkali halides, particularly in the presence of small amounts of water. The formation of mechanochemical complexes between kaolinite and alkali halides has been investigated by PA infrared spectroscopy; the motivation for this work was the critical need to avoid grinding of the complexes after they are formed. Mid- and far-infrared PA spectra of kaolinite complexes with caesium and potassium halides will now be considered.

Both PA and diffuse reflectance infrared spectroscopies were used to study the reactions between CsBr and kaolinite that occur during shaking, grinding at ambient atmosphere, and grinding in the presence of a few drops of water (Michaelian et al., 1991). An intercalation complex was formed by wet grinding, but not by the other two techniques. It is important to note that this grinding was an integral step in the intercalation process; once the CsBr–kaolinite complex was formed, it was not ground again before its PA spectrum was obtained. Figure 6.23 shows the PA spectra of untreated kao-

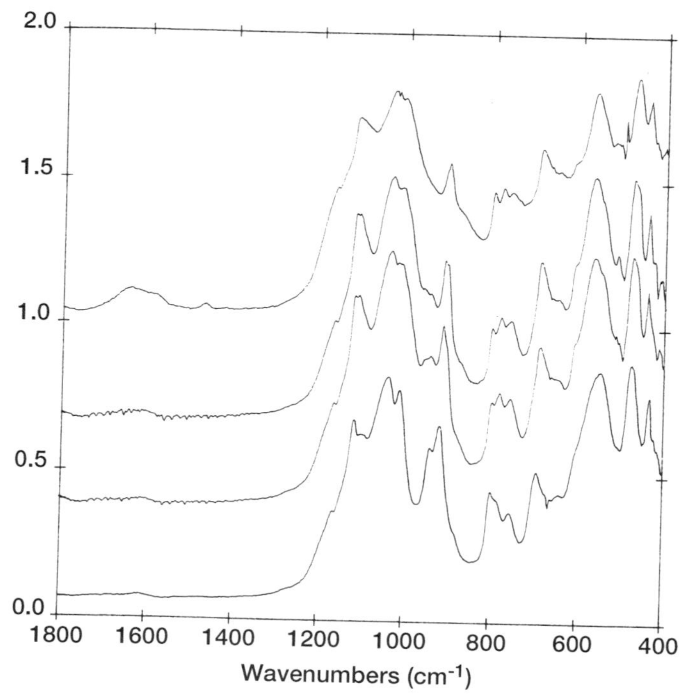


**Figure 6.23.** PA infrared spectra of kaolinite (bottom curve) and kaolinite/CsBr mixtures. Middle curve, air ground; top curve, wet ground. (Reprinted from Michaelian K. H. et al., *Can. J. Chem.* **69**: 749–754, by permission of NRC Research Press; copyright © 1991.)

linite and the CsBr-kaolinite mixtures; the spectrum of the air-ground mixture was similar to that of kaolinite, implying that no complex was formed. On the other hand, the spectrum of the wet-ground sample exhibited a number of changes that indicate the intercalation of both CsBr and $H_2O$. In particular, the Al—OH deformation bands near 900 $cm^{-1}$, the Al—O band at 552 $cm^{-1}$, and the Si—O bands in the spectrum of kaolinite were all shifted to different frequencies; moreover, the PA spectrum of the wet-ground sample also displayed the water $\nu_2$ band at 1630 $cm^{-1}$. These observations led to the development of a model in which a water molecule coordinated to the $Cs^+$ ion donates one proton to a siloxane oxygen and the other to the halide, while accepting a proton from an inner surface hydroxyl group of kaolinite. A similar structure exists for the kaolinite–CsCl complex, which has been prepared by a variety of different procedures and characterized by thermal analysis and PA infrared spectroscopy (Yariv et al., 1994). The kaolinite–CsI complex cannot be obtained mechanochemically, although it has been produced by an indirect technique (Michaelian et al., 1998).

It has already been stated that relatively few publications describing PA far-infrared spectra are known to exist. Nevertheless, far-infrared PA and diffuse reflectance spectra have been obtained for the kaolinite–alkali halide complexes mentioned in the previous paragraph, and for complexes involving the potassium halides (Michaelian et al., 1997). The PA far-infrared spectra of all of the complexes were quite weak, partly because of the presence of excess alkali halide in each sample. Another complication arose because of the low-frequency cutoff of the salts, which restricted the observable spectral region in some cases. Despite these limitations, useful PA spectra were obtained (Fig. 6.24). Intercalation caused some kaolinite bands to become narrower or to shift by as much as 13 $cm^{-1}$. These bands are due to mixed Si–O deformations and octahedral sheet vibrations.

Dickite is another, less common, member of the kaolin subgroup of dioctahedral nonexpanding clays. It has the same composition as kaolinite, $Al_4Si_4O_{10}(OH)_8$, differing only in the stacking of its layers. A recent investigation has shown that LO and TO crystal modes also occur in the hydroxyl stretching region of the Raman and infrared spectra of dickite; the assignments of the six identifiable OH bands to three LO/TO pairs were determined from polarized micro-Raman spectra of single-crystal dickite and PA infrared spectra of coarse, nonoriented crystals (Shoval et al., 2001). All six bands were used to fit the PA spectrum in which the observed bands are much broader than their counterparts in the Raman spectrum of dickite.

The thermal treatment of dickite, which is pertinent to the ceramic industry, was studied by Shoval et al. (2002b). PA and transmission infrared, and micro-Raman, spectra were used to identify the products of the treat-

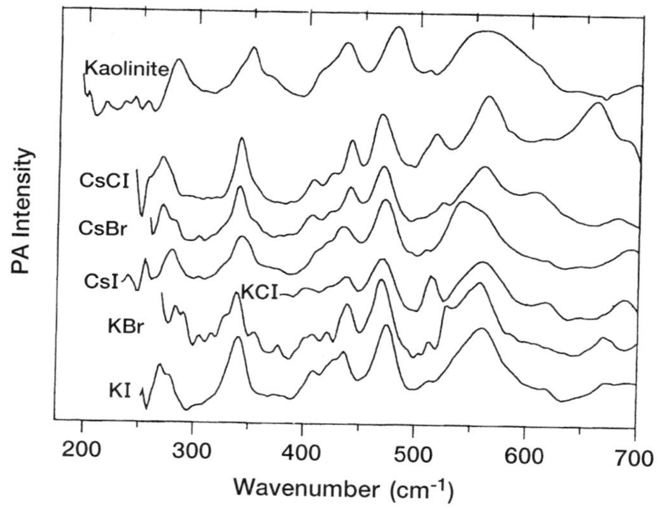


**Figure 6.24.** Far-infrared PA spectra of kaolinite/alkali halide/water complexes. (Reproduced from Michaelian, K. H. et al., *Mikrochim. Acta* **14** (Suppl.): 211–212, by permission of Springer-Verlag, copyright © 1997.)

ment. The PA spectra were curve-fitted with a series of bands—many of which are quite broad—that prove the existence of materials such as mullite, Al-spinel, corundum, and amorphous silica. Treatment at temperatures between 700 and 1000°C results in the production of amorphous meta-dickite, whereas still higher temperatures (1000–1300°C) cause the minerals to recrystallize.

Mid- and far-infrared PA spectra of asbestos were described earlier in this section. Returning briefly to this topic, it is important to mention a series of publications by Bertrand and his colleagues at École Polytechnique de Montréal. Bertrand et al. (1982) compared magnitude and phase PA infrared spectra of asbestos obtained with a dispersive spectrometer and found the phase spectrum to be the more accurate representation of the absorption spectrum (obtained in a transmission experiment). Specifically, the intensity of the broad band near 3400 $cm^{-1}$ (Rockley et al., 1982) due to adsorbed water is reduced in the phase spectrum of asbestos. This work was continued with an FTIR spectrometer (Choquet et al., 1985), where real and imaginary spectra calculated from double-sided interferograms were used to calculate a spectrum that was more accurate than that obtained using the conventional FTIR phase correction algorithm. In addition to the diminution of the broad band arising from adsorbed water, the high-frequency part of the PA spectrum is intensified with regard to the low-frequency region in these calculations. This work was put on a more theoretical basis in two subsequent studies (Choquet et al., 1986; Bertrand, 1988). The numerical methods used by these authors were described in detail in Chapter 5.

As a final comment to this discussion, it can be noted that the minimal sample preparation afforded by PA infrared spectroscopy has provided the motivation for much of the published work on clays and minerals. The advantages offered by the other major capability of PA spectroscopy—depth profiling—do not appear to have been recognized in this context. The study of organo-clay complexes, which are important in a variety of industrial applications, will certainly benefit from both attributes of PA infrared spectroscopy.

## 6.7. WOOD AND PAPER

The measurement of infrared spectra of wood, paper, and wood products presents some rather obvious challenges to the analyst, particularly with regard to sample preparation. While many of these substances are insoluble, or in general are not very amenable to traditional infrared measurements, PA infrared spectroscopy proves to be an entirely viable means for their characterization. Some representative published results, augmented by previously unpublished data, for wood-based samples are summarized in this section.

A group based at the Pulp and Paper Research Institute of Canada and McGill University was among the first to utilize PA spectroscopy in research on wood products. St.-Germain and Gray (1987) chose the technique in a study of mechanical pulp brightening, primarily because of the minimal sample preparation required. PA infrared spectra were obtained for both pulp sheets and ground pulp in this work. About a dozen bands were observed, the most prominent occurring at 1056 $cm^{-1}$ and at about 2900 and 3400 $cm^{-1}$. The authors noted that intensities varied with sample packing in the PA cell, and therefore scaled the PA spectra based on the observed intensities of either the broad 3400-$cm^{-1}$ band, the 1056-$cm^{-1}$ (C—O) band, or the 1510-$cm^{-1}$ aromatic band. This use of internal standards improved reproducibility to $\pm 5\%$ or better. Brightening of pulp with peroxide led to the weakening of the bands at 1740 and 1650 $cm^{-1}$, implying deacetylation and the removal of conjugated carbonyl structures, respectively. The spectra were also consistent with the formation of quinones during brightening. An absorption spectrum of a KBr pellet of the pulp was recorded to check on the PA results; the result was similar to the corresponding PA spectrum.

At least 20 bands are known to occur in the infrared spectra of various types of wood. In an exploratory study carried out at about the same time as the above-mentioned work, Kuo et al. (1988) convincingly demonstrated the suitability of PA infrared spectroscopy for the characterization of a number of different kinds of wood. In fact, these authors found PA spectroscopy to be superior to more traditional infrared techniques (transmission, ATR, and diffuse reflectance) with regard to this analytical problem. In their PA experiments, they examined $\frac{1}{4}$-in. circular samples cut from microtomed sections that were 400 μm thick. The thermal diffusivity of wood is approximately 0.002 $cm^2/s$, implying sampling depths ranging from 18 to 56 μm at the mirror velocities used to obtain the PA FTIR spectra in this investigation.

Kuo et al. (1988) obtained spectra for transverse and oblique sections of ponderosa pine; the cellulose bands near 1050 and 3350 $cm^{-1}$ are stronger in the spectrum of the oblique section, consistent with the fact that the cellulose chains are oriented perpendicularly to the transverse section normal. This observation of differences due to microfibrillar orientation would, of course, not have been possible in a sample that had been ground for a transmission or diffuse reflectance measurement. In a related experiment, these authors examined eastern cottonwood samples with various degrees of decay by the brown-rot fungus *Gleoephyllum trabeum* and found that several bands due to carbohydrates decreased in intensity as decay progressed; on the other hand, features arising from lignin intensified. Using only small sample quantities, the authors also demonstrated very significant differences between the PA infrared spectra of eastern cottonwood kraft and groundwood fibers or groundwood vessel elements (Fig. 6.25). The relative lignin content of these samples was also deduced from these spectra.

This successful analysis of wood by PA spectroscopy eventually led to further work, some of it involving other infrared methods. For example, Bajic et al. (1998) investigated PA and transient infrared spectroscopy of wood in light of the need for rapid analysis of wood-based materials. The latter technique, which records emission spectra of moving samples, is mentioned again in the section on PA spectroscopy of food products.

In this work, Bajic et al. (1998) found that PA infrared spectra of milled wood specimens displayed bands that characterized the samples as hardwoods or softwoods and could be correlated with particular functional groups. Principal component analysis was then employed to determine whether wood species could be distinguished on the basis of their PA spectra. Indeed, the hardwoods and softwoods clustered separately, showing that this approach is feasible. ASTM methods were also used to analyze the wood samples; the PA results agreed with the American Society for Testing and Materials (ASTM) data within 1% with correlation coefficients ranging from 0.85 to 0.94. These encouraging results show that PA infrared spectra can be used for feedstock identification and analysis of the chemical com-

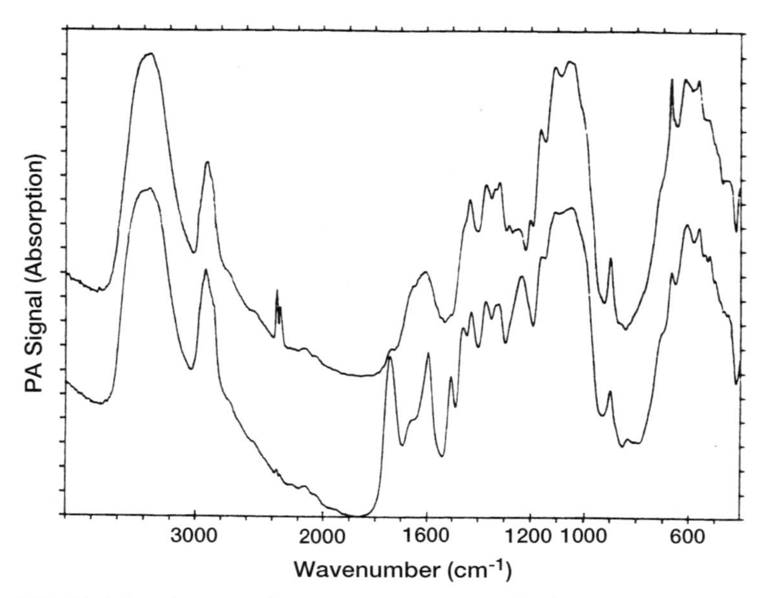


**Figure 6.25.** PA infrared spectra of eastern cottonwood kraft (KAPPA No. 25, top curve) and groundwood fibers (bottom curve). (Reproduced from Kuo, M.-L. et al., *Wood Fiber Sci.* **20**: 132–145, by permission of the Society of Wood Science and Technology; copyright © 1988.)

position of wood before it is processed. Moreover, quantitative analysis of wood extracts, lignin, and carbohydrates by PA infrared spectroscopy is certainly feasible. Similar work in the present author's laboratory has shown that preservatives added to wood are readily detected using PA infrared spectroscopy.

Sawdust is, of course, another familiar product derived from wood. Because it is so finely divided, sawdust can be analyzed by dispersing it in an alkali halide and acquiring transmission or diffuse reflectance spectra; however, undiluted sawdust can be studied even more readily using PA infrared spectroscopy. As an example, Figure 6.26 shows a previously unpublished PA infrared spectrum of sawdust derived from hemlock; approximately 20 bands are identifiable. Sawdust obtained from fir yielded a very similar result. Not surprisingly, these PA spectra closely resemble those measured for samples of the original woods.

The results discussed so far in this section clearly show that PA infrared spectroscopy is very well suited for the analysis of wood, pulp, and sawdust. Paper is, of course, another important wood product still to be included in this discussion. It is relevant to mention that paper may contain—in addition to cellulose—lignin, hemicellulose, sizing components such as starch

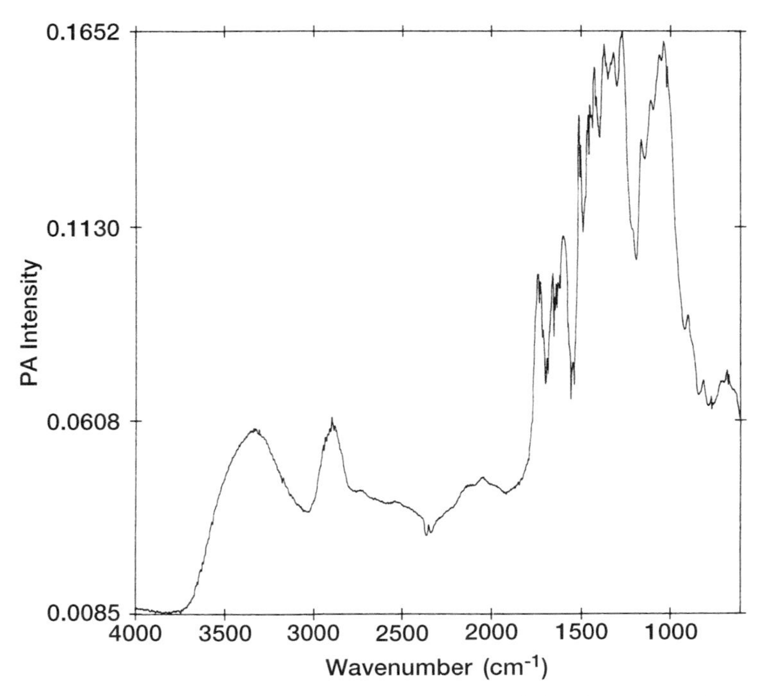


**Figure 6.26.** PA infrared spectrum of hemlock sawdust. Sample fraction with particle sizes less than 250 μm was analyzed.

and clay, and polymer coatings. The heterogeneity of some papers thus provides another justification for their analysis with PA infrared spectroscopy: In addition to the obvious desirability of minimal sample preparation, the capability of depth profiling makes it possible to characterize the layers in coated papers. Both attributes of PA infrared spectroscopy were utilized in the research discussed in the following paragraphs.

The suitability of PA infrared spectroscopy for the analysis of paper was briefly illustrated in an early survey of applications of PA spectroscopy (Krishnan, 1981). PA spectra were recorded for both a coated paper and the corresponding uncoated paper. Subtraction of the second spectrum from the first canceled the bands due to the paper and revealed the spectrum of the coating; several important bands due to aromatic and aliphatic C—H groups, carbonyl functionality, and other species were identified in this way. The experiment was then repeated using the specular reflectance technique. Subtraction was less successful in the second case because the bands arising from the paper were too strong to permit reliable cancellation. This investigation was among the first to show that PA infrared spectroscopy is particularly well suited for the analysis of paper.

Another demonstration of the capability of PA infrared spectroscopy for the analysis of paper occurs in the work of Jin et al. (1982). These authors showed that dispersive near-infrared PA spectra of paper contained three bands in the 1.3–2.2 μm region. Features due to cellulose were observed at 1.55 and 2.05 μm (6450 and 4880 $cm^{-1}$, respectively). A third band at 1.9 μm (5265 $cm^{-1}$) was due to moisture. The intensity of the latter peak was found to vary systematically with the moisture content of the paper; interestingly, even prolonged drying at 105°C did not result in its complete disappearance.

As an alternative to the more common gas-microphone PA infrared technique, PBDS can also be utilized to study paper. For example, Low et al. (1984) obtained PBDS infrared spectra of paper and observed that treatment of the sample with ethylene oxide resulted in diminution of the broad band at 3400 $cm^{-1}$ and intensification of the C—H stretching bands near 3000 $cm^{-1}$. The open geometry and nondestructive sampling features of PBDS were put to particularly good use in this experiment in which a piece of paper manufactured in 1577 was also analyzed; a blue discoloration on one side of this paper was concluded to be a type of mold and found to yield a spectrum distinct from that of the paper itself.

Chemists may anticipate using PA infrared spectroscopy to analyze precipitates or other solid samples deposited on filter paper. To be successful, this experiment requires that the spectrum of filter paper itself be obtained, so that its contribution to the spectrum of the filtered solid can be removed numerically. In this context, it is relevant to note that PA spectra of filter paper were described in the literature a number of years ago. Harbour et al. (1985) obtained near-infrared PA spectra for Whatman filter paper and observed six well-defined bands between about 3800 and 8000 $cm^{-1}$. Exposure of the filter paper to atmospheres with varying relative humidities showed that the bands at 3816, 5155, and 6944 $cm^{-1}$ intensified as humidity increased. It was therefore proposed that the ratio of the intensity of the 5155 $cm^{-1}$ band to that of a cellulose band at 4670 $cm^{-1}$ be used as a measure of water content in paper and other related materials.

The mid-infrared PA spectrum of Whatman filter paper was also obtained by Harbour et al. (1985) and found to contain bands due to cellulose at about 700, 1100, 1350, 1450, 2900, and 3300 $cm^{-1}$. A similar PA infrared spectrum of a commercial filter paper, extending from 500 to 4500 $cm^{-1}$, is depicted in Figure 6.27. This spectrum can reliably be subtracted from the PA spectrum of a solid or oil deposited on a similar sheet of paper so as to reveal the bands of the analyte: The strategy mentioned above is, in fact, readily implemented.

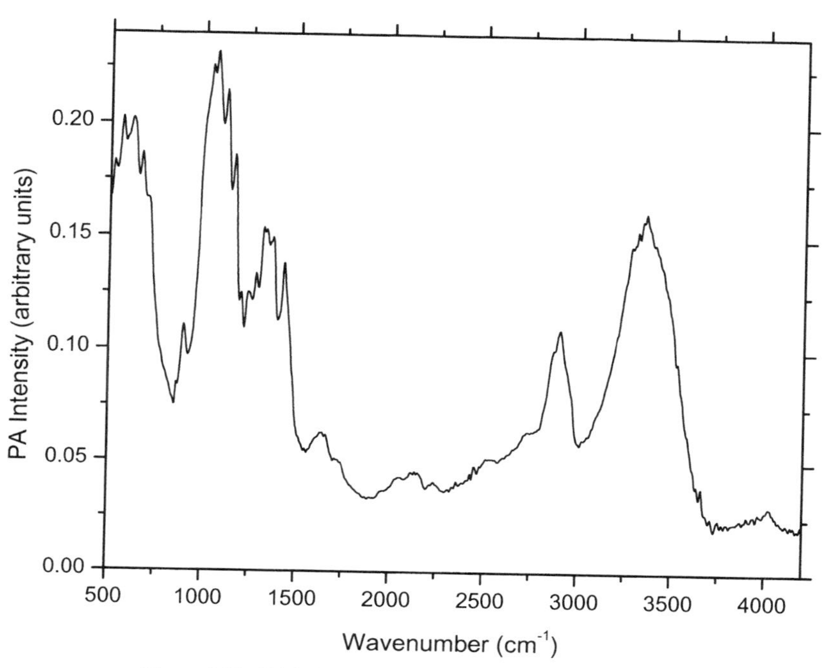


**Figure 6.27.** PA infrared spectrum of ordinary filter paper.

PA infrared spectra of bleached kraft papers were described by Gurnagul et al. (1986). This investigation was carried out at about the same time as the study of mechanical pulp brightening by this group that was mentioned above. Gurnagul et al. examined the effects of beating the pulp for various periods of time and of exposing the paper samples to a moist atmosphere. They observed that PA intensities were lower for the more highly beaten sample than for the unbeaten sample, which had a higher specific surface area and higher surface roughness. These results are somewhat reminiscent of numerous PA studies of finely divided solids, where intensity generally increases as particle size decreases. Another significant finding from this investigation was the fact that PA intensities were lower for sheets of kraft paper with high moisture contents. The authors reasoned that this might have been due to a lowered efficiency of heat transfer between the moist cellulose surface and the carrier gas; PA spectroscopists may well recall their own experiences in which moist or oily samples yielded spectra that were weaker than expected. In the case under discussion here, the moist paper samples had lower surface areas, which would have also given rise to reduced PA intensities.

Photothermal radiometry (PTR) and PA infrared spectroscopy were used in a combined study of specialty papers with various cotton contents (Garcia et al., 1998, 1999). Two important thermophysical properties (thermal diffusivity, $\alpha_s$; thermal conductivity, $\kappa_s$) were calculated from the PTR data, utilizing a one-dimensional photothermal model of a sheet of paper in air. $\alpha_s$ ranged from $1.3 \times 10^{-7}$ m$^2$/s to $1.9 \times 10^{-7}$ m$^2$/s for seven different types of paper, while $\kappa_s$ took values between 1.00 and 3.60 W m$^{-1}$ K$^{-1}$. PTR measurements were effected with both transmission and backscattering geometries; both signals depended on the infrared emission coefficient $\beta_{IR}$, which is also equal to the average absorption/extinction coefficient across the bandwidth of the infrared detector used in this experiment. $\beta_{IR}$ was calculated at $1.79–3.10 \times 10^{-4}$ m$^{-1}$ and was approximately equal to $\beta_s$, the analogous quantity in the visible region. It should be noted that $\beta_{IR}$ cannot be correlated with specific infrared absorption bands because the PTR measurements were carried out with a broadband infrared detector, that is, no specific wavelength discrimination was involved in these experiments.

The relationship between the PA infrared spectra of these papers and the PTR results is partly illustrated in Figures 6.28 and 6.29. Figure 6.28 shows

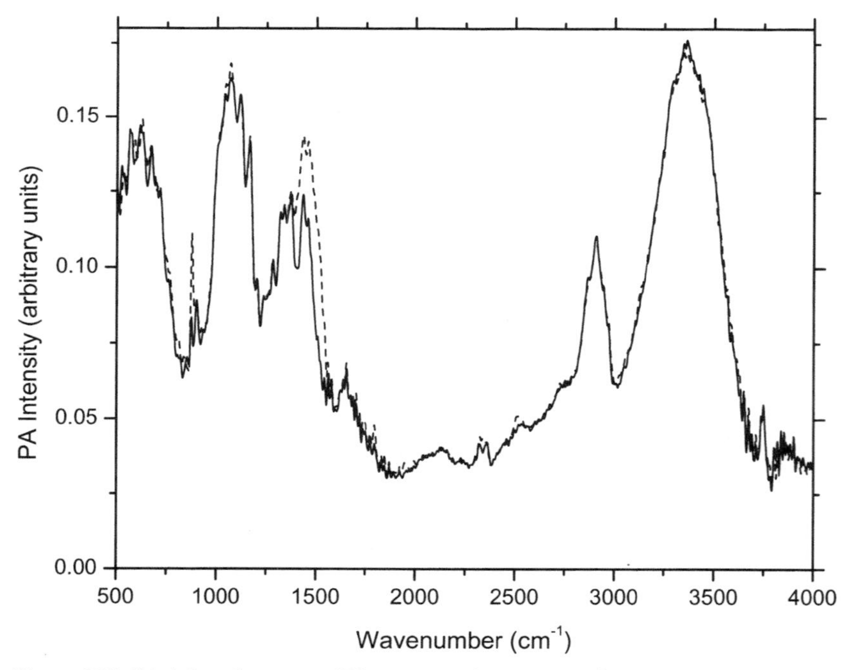


**Figure 6.28.** PA infrared spectra of Krypton parchment paper (high cotton content) provided by Centre Innovation DOMTAR. Solid and dashed curves represent spectra obtained for two sides of a single sheet.

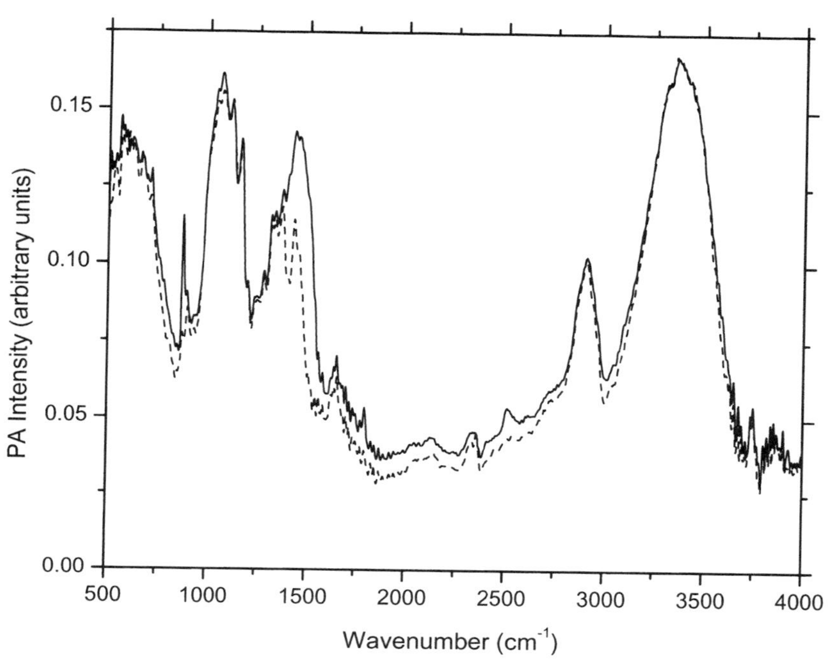


**Figure 6.29.** PA infrared spectra of Belfast bond paper (intermediate cotton content). Details are the same as in Fig. 6.28.

the PA spectra obtained for the front and back surfaces of a sheet of paper for which the corresponding $\beta_{IR}$ values were determined as $2.85 \times 10^{-4}$ $m^{-1}$ and $1.79 \times 10^{-4}$ $m^{-1}$, respectively. Indeed, the differences in the infrared spectra in the 1300–1700 $cm^{-1}$ region suggest varying depth profiles in both optical and thermophysical properties. In Figure 6.29, the difference between the PA infrared spectra of the two surfaces is more pronounced and corresponds to an even larger differential in the thermophysical properties obtained from the transmission and backscattering PTR experiments. In general, a decrease in cotton content results in higher $\alpha_s$ values for these papers.

Two publications that deal with coated papers recently appeared in the PA literature. In the first, Halttunen et al. (1999) carried out depth profiling experiments on paper coated with sodium oleate. The base paper thickness was 78 μm, with coating thicknesses of 18–58 μm. PA magnitude spectra obtained at higher modulation frequencies exhibited more intense bands due to the oleate, consistent with the fact that it was present on the surface of the paper. Concomitantly, a $CaCO_3$ band from the base paper diminished in intensity.

In the carbonyl stretching region, the frequency of the asymmetric —$COO^-$ band shifts from 1563 $cm^{-1}$ in pure sodium oleate to 1573 $cm^{-1}$

when the oleate is mixed with starch to form the coating. Attachment of the coating to paper produced bands at 1539 and 1573 $cm^{-1}$. The asynchronous two-dimensional correlation spectrum of the coated paper showed that the band due to pure oleate (shifted slightly to 1560 $cm^{-1}$) was of shallower origin, with the 1573-$cm^{-1}$ feature arising from the intermediate layer and the 1539-$cm^{-1}$ band arising from the deepest region of the layer. Hence the concentration of sodium oleate was increased on the surface of the coating layer. Phase spectra further confirmed that the 1539-$cm^{-1}$ band arose from the sodium oleate at the bottom of the coating layer. These impressive results illustrate the considerable detail that can be derived from depth profiling experiments on coated paper by PA infrared spectroscopy and suggest further research on this topic.

Wahls et al. (2000) also reported depth profiling experiments for coated paper. These authors used the difference between magnitude PA spectra of the coated paper and the uncoated (base) paper to distinguish the spectrum of the bulk from that of the coating. Step-scan magnitude spectra, which were calculated as $[I^2 + Q^2]^{1/2}$, where $I$ and $Q$ are the in-phase and quadrature spectra, respectively, were obtained using a digital signal processor (DSP) lock-in amplifier. The spectra of the coated and uncoated papers were scaled based on their intensities in the 2200–2300 $cm^{-1}$ region, where no absorption bands were observed. As shown in Figure 6.30, this strategy yielded positive and negative differences, rather than nulling the bands of one component. The differences between the PA spectra of the base paper and the coated paper arose from the fact that less base paper was sampled in the second case; the difference bands varied with modulation frequency because of the corresponding changes in $\mu_s$. This behavior was modeled with synthetic triangular bands, assuming several typical values of the absorption coefficient. Absorption by the coating yielded positive difference bands, while bands from the bulk gave negative differences; these differences were greater when $\mu_s$ was roughly equal to the thickness of the coating. As pointed out by Wahls et al. (2000), this technique could also be used to study laminates.

The results presented in this section show that PA infrared spectroscopy is particularly well suited to the study of wood, paper, and other wood products. Both major attributes of PA spectroscopy—its requirement for little sample preparation and the capability of depth profiling—provide a strong justification for the use of the technique in this context. The future analysis of coated papers by PA infrared spectroscopy appears to be particularly promising.

## 6.8. POLYMERS

The patient reader is probably already aware of the existence of a substantial body of published literature on the PA infrared spectroscopy of polymers. An inspection of Appendix 2 confirms that polymers comprise one of the

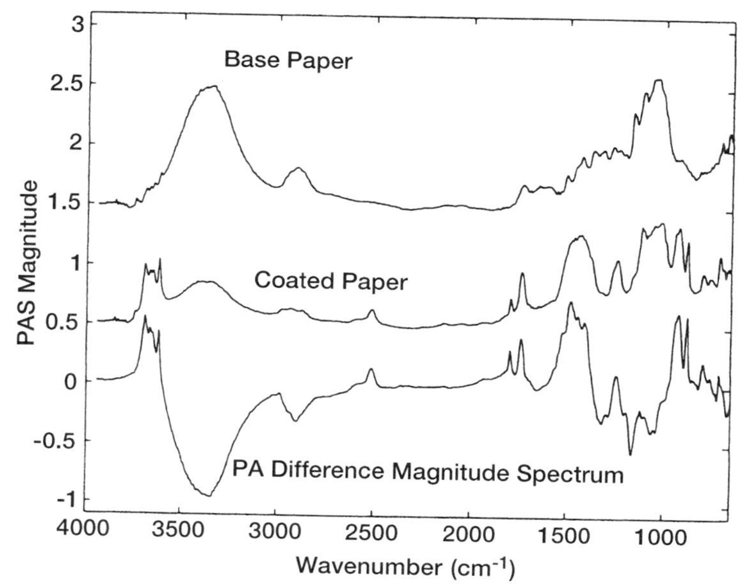


**Figure 6.30.** Normalized and scaled PA infrared magnitude spectra obtained at modulation frequency of 800 Hz for base paper (upper curve) and coated paper (middle curve). Bottom curve is the difference (coated – base paper). Spectra have been offset along $y$ axis for clarity. (Reproduced from Wahls, M. W. C. et al., *Appl. Spectrosc.* **54**: 214–220, by permission of the Society for Applied Spectroscopy; copyright © 2000.)

most popular sample classes that have been analyzed using PA detection: More than 150 articles that refer to spectra of polymers are included in this list. This popularity is at least partly due to the fact that polymer films and laminates are quite amenable to depth profiling studies. Indeed, polymer layers with thicknesses on the order of a few micrometers—dimensions that are comparable to typical thermal diffusion lengths in PA infrared spectroscopy—can be prepared by several well-known techniques. A number of studies on the subject of depth profiling in layered polymer samples are discussed in Chapter 4; most of these articles are not included in the present section.

Fortunately, the considerable extent of the scientific literature pertaining to the PA infrared spectroscopy of polymers has prompted several authors to write helpful review articles during the last two decades. In chronological order (and, therefore, in sequence of increasing literature coverage) these include Vidrine and Lowry (1983), Koenig (1985), Jasse (1989), and finally Dittmar et al. (1994). The latter review was also mentioned in Chapter 1 because of the considerable amount of useful information that it conveys in general with regard to PA infrared spectroscopy.

During the period that might be referred to as the demonstration phase of PA infrared spectroscopy (the late 1970s and early 1980s), some researchers chose polymers as typical solid samples that might be used to validate the technique. While the resulting publications presented spectra that were mostly of good quality, the data tended to be accompanied by little or no interpretation (Low and Parodi, 1980a; Royce et al., 1980; Vidrine, 1980; Chalmers et al., 1981; Zachmann, 1984). Another group of somewhat more rigorous investigations examined the differences between the PA and ATR infrared spectra of polymers, particularly with regard to sampling depth. These studies showed that PA infrared spectroscopy usually characterizes depths that are significantly greater than those analyzed in ATR spectroscopy, where the evanescent wave penetrates the sample to a distance on the order of 2 μm or less (Krishnan, 1981; Krishnan et al., 1982; Gardella et al., 1984; Vidrine, 1984; Saucy et al., 1985a). The requirement for little or no sample preparation in PA infrared spectroscopy is another feature that is relevant in this context, since many of the polymer samples studied were irregular in shape.

PA infrared spectra of both undoped and doped polyacetylene were described by two research groups at about the same time as the work just mentioned. The emerging PA technique was selected primarily because it allowed the measurement of infrared spectra of these opaque samples without the usual requirement that they be available as thin films suitable for transmission experiments.

Two studies on this topic were published by Eyring's group. In the first, Riseman et al. (1981) obtained spectra of a mixture of cis and trans isomers of undoped polyacetylene. An *n*-doped sample containing the tetrabutylammonium ion was also investigated in this study. Although the spectra reported by these authors were rather noisy, several new bands were successfully identified for the doped sample. The second study by this group (Yaniger et al., 1982) showed that acetylene doped with iodine yielded a PA spectrum containing a strong band at 1400 $cm^{-1}$ due to carbon–carbon stretching, as well as a very broad band near 950 $cm^{-1}$ and a narrower peak at 750 $cm^{-1}$ (Fig. 6.31). These results were consistent with a model involving charge transfer into soliton levels. Metallic semiconducting and insulating regions were both concluded to exist within moderately doped bulk samples.

In subsequent work, Eckhardt and Chance (1983) reported near-infrared PA spectra of thick films of cis and trans polyacetylene. The only band in these spectra was at 0.74 eV (5970 $cm^{-1}$). Doping with oxygen led to the appearance of another band at 0.67 eV (5400 $cm^{-1}$) that exhibited a shape consistent with soliton absorption. The data allowed the authors to develop a model of polyacetylene fibers in the defect structure.

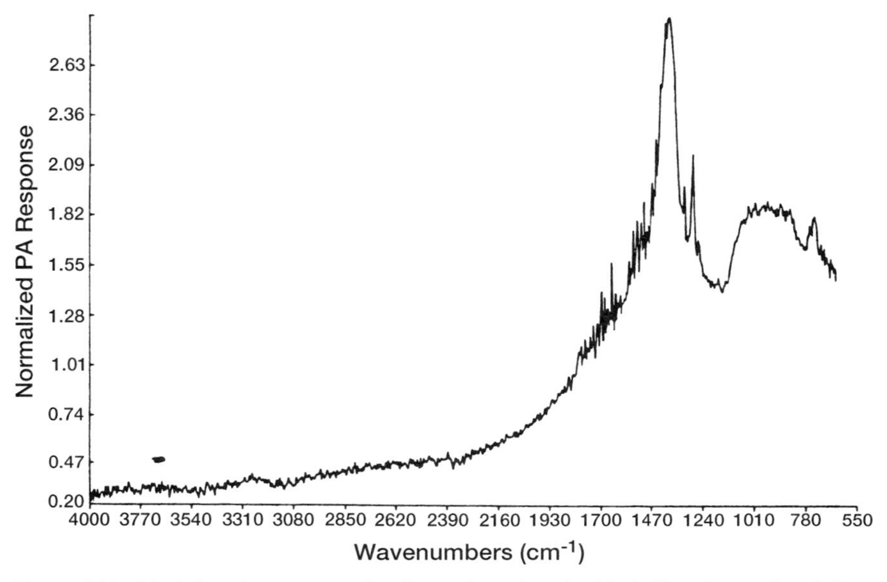


**Figure 6.31.** PA infrared spectrum of polyacetylene doped with iodine. (Reproduced from Yaniger, S. I. et al., *J. Chem. Phys.* **76**: 4298–4299; used with permission. Copyright © American Institute of Physics 1982.)

A series of studies on the PA infrared spectra of polymers was published at about the same time as the work mentioned in the previous paragraphs by Teramae and Tanaka of The University of Tokyo. This work began with two exploratory investigations of textiles and polymers that are also mentioned in the later section on PA spectra of textiles (Teramae and Tanaka, 1981; Teramae et al., 1982). The design of the PA cell constructed by the authors, and its Helmholtz resonance, were also discussed in these works. Very satisfactory PA infrared spectra of carbon-filled rubbers and phenolic resins were obtained in these early studies.

These authors then turned their attention to the characterization of polymer films. A sample that consisted of a 40-μm layer of polyethylene and a 10-μm layer of PET yielded a surprisingly complicated spectrum containing bands from both layers, even though the Rosencwaig–Gersho theory predicted that only the thicker top layer should produce a PA signal (Teramae and Tanaka, 1984, 1985). This result was shown to arise from heat generation at the rear surface of the sample (Teramae and Tanaka, 1988) and to depend on the magnitudes of the optical absorption coefficients of the bands. The positioning of the polymer film in the PA cell also had a major effect on the results; better spectra were obtained when the film was either immediately behind the entrance window or in the cell cavity.

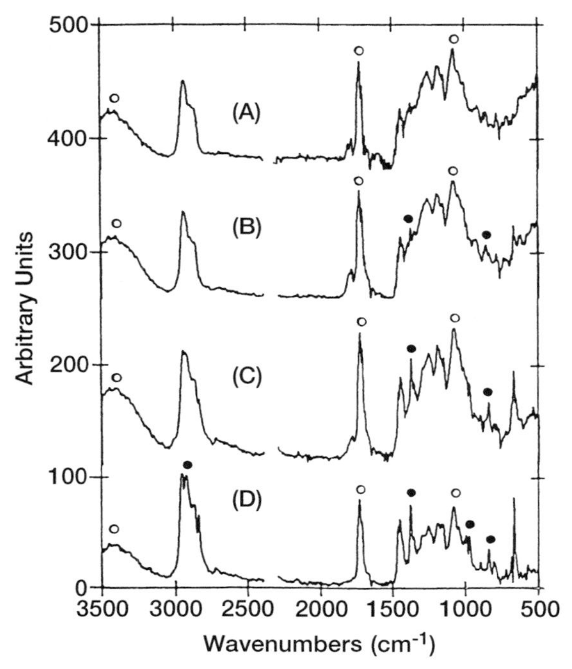


**Figure 6.32.** PA infrared spectra of a series of bilayered films of epoxy resin on 23-μm polypropylene. Thickness of top layer was (A) 23 μm, (B) 9 μm, (C) 5 μm, and (D) 3 μm. Filled circles denote bands from the lower layer, whereas the open circles signify the upper layer. (Reproduced from Teramae, N. and Tanaka, S., *Appl. Spectrosc.* **39**: 797–799, by permission of the Society for Applied Spectroscopy; copyright © 1985.)

The detection of the subsurface layer of a bilayered film can also be turned to advantage in a relatively simple depth profiling experiment. Teramae and Tanaka (1985) showed that a progressive reduction of the thickness of the top layer from about 25 to 3 μm in layered samples of epoxy resin on either polypropylene or polyethylene terephthalate led to the concomitant intensification of several bands from the lower layer (Fig. 6.32). Thus depth profiling of polymer films is possible even in situations where only a single rapid-scan mirror velocity is available to the analyst. The authors reviewed all of their results in a subsequent publication (Teramae and Tanaka, 1987).

PA and ATR infrared spectroscopies were compared with regard to their capabilities for characterizing polymeric coatings in more recent publications by several research groups. Carter et al. (1989a), who also utilized the transmission, specular reflection, and diffuse reflection techniques, investigated the weathering of paints used in the automotive industry; the greater sampling depth in PA infrared spectroscopy mentioned above was again observed in this work. Factor et al. (1991) investigated a photocurable coating for a polycarbonate substrate and obtained infrared results that agreed with those from photodifferential scanning calorimetry. A saturated band in the PA spectra was successfully used to normalize the intensities of two bands at 1620 and 1635 $cm^{-1}$ that were employed to monitor curing in this work. Finally, Wetzel and Carter (1998) utilized ATR microspectroscopy as well as PA infrared spectroscopy to study the degradation of an acrylic polymer by ultraviolet radiation. The disappearance of the acrylic bands and the corresponding appearance of new bands due to degradation products were both noted in the PA spectra.

It is relevant to mention several studies in which PA infrared spectroscopy was used to characterize polymers primarily because of the minimal sample preparation involved. The adsorption of polymers on *gamma*-iron oxides, which pertain to magnetic memory media, was investigated by Cook et al. (1991), as well as Nishikawa et al. (1992). The interactions between the polymers and the iron oxides were successfully elucidated using the PA spectra obtained in both studies. Hocking et al. (1990a,b,c) employed PA infrared spectroscopy to characterize a series of newly synthesized imide-amide copolymers; the use of PA detection was rather incidental to this work.

The preceding summary discusses some representative articles on the PA infrared spectroscopy of polymers. The considerable extent of the published literature on this subject (see Appendix 2) precludes a complete review. The interested reader may wish to study the article by Dittmar et al. (1994) or the original studies cited in Appendix 2.

## 6.9. GASES

Photoacoustic spectroscopy has achieved perhaps its greatest accomplishments with regard to the analysis of gases. The quite spectacular sensitivity of the PA effect in gases sets a standard that may never be attained by many vibrational spectroscopists who investigate solids and liquids: Trace gas analysis by PA spectroscopy routinely realizes detection limits at the parts-per-million (ppm) or even the parts-per-billion (ppb) levels. Such concentrations are, of course, several orders of magnitude lower than those accessible in infrared and Raman spectra of condensed phases in most ordinary circumstances.

The success of the PA infrared spectroscopy of gases has been established over the last three decades; indeed, initial research predates the development of PA FTIR spectroscopy by a number of years. The comparatively long lifetime of this area of specialization has naturally led to the existence of a large body of scientific literature. A few important early references were included in the historical review in Chapter 2.

Much of the relevant work on the PA spectroscopy of gases completed prior to 1994 was summarized in *Air Monitoring by Spectroscopic Techniques*, edited by M. W. Sigrist and published as Volume 127 in the Chemical Analysis series. A number of other important texts and review articles have also been published. A comprehensive review of this literature would certainly be a major undertaking and, moreover, is well beyond the scope of the present book. Instead, the 1994 publication by Sigrist will be taken as a convenient point of demarcation in this extensive literature, and the present discussion will be restricted to work completed after this date. Readers interested in a more comprehensive account may wish to consult the other books and review articles or the original literature on the subject.

It should also be appreciated that much of the research on PA infrared spectroscopy of gases has been carried out using near- and mid-infrared lasers as radiation sources; these include the well-known CO and $CO_2$ gas lasers, laser diodes, and various other devices. On the other hand, the number of references that specifically describe PA FTIR spectra of gases is relatively small. Hence—recalling the definitions introduced in Chapters 1 and 2—most of the work on this subject can be classified as either single- or multiple-wavelength (sequential) spectroscopy. This means that the perspective from which the PA infrared spectroscopy of gases must be discussed is different from that for most of the other types of (condensed-phase) samples described in this book. This use of several different infrared sources also provides the basis for the organization of the following discussion.

### 6.9.1. CO Laser Excitation

Some representative publications describing PA infrared spectra of gases, obtained with a CO laser as the infrared source, are listed in Table 6.2. In its normal mode of operation, this laser produces a series of emission lines between about 1250 and 2000 $cm^{-1}$ (5–8 μm), making it particularly well suited for the study of compounds containing the C=C and/or C=O functional groups. Ethylene and the carboxylic acids are typical examples and are included along with several other compounds in these investigations.

The research team of Persijn and co-workers at the University of Nijmegen has utilized PA infrared spectroscopy to study gases associated with the ripening of fruits. Three studies published by these investigators describe rather elegant experiments in which three intracavity PA cells were employed simultaneously in a CO laser for trace gas detection. In the first publication, Bijnen et al. (1998) cooled the laser to 77 K to achieve single-

**Table 6.2. Recent PA Studies of Gases**[a]

| Gases | Frequency ($cm^{-1}$) | Reference |
|---|---|---|
| $CH_3(CH_2)_nCOOH$, $n = 8, 10, 12, 14$ | 1730–1890 | Jalink et al. (1995) |
| $CH_3COOH$, $CD_3COOD$, $CH_3CH_2COOH$ | 1600–1950 | Kästle and Sigrist (1996a) |
| $CD_3COOD$ | 1700–1840 | Kästle and Sigrist (1996b) |
| $C_2H_4$, $CH_3CHO$, $CO_2$, $CH_3CH_2OH$, $H_2O$ | 1250–2080 | Bijnen et al. (1998) |
| $CH_3CH_2OH$, $CH_3CHO$, $C_2H_4$, $CO_2$, $H_2O$ | 1500–1700 | Persijn et al. (1999) |
| HCOOH | 1500–2000 | Merker et al. (1999) |
| $C_2H_4$, $C_2H_6$, $C_5H_{12}$ | 2630–3600 | Santosa et al. (1999) |
| $CH_3CH_2OH$, $CH_3CHO$, $C_2H_4$, $CO_2$, $H_2O$ | 1300–2000 | Persijn et al. (2000) |

[a] CO laser used as the infrared source.

line operation over the relatively wide tuning range from 1250 to 2080 $cm^{-1}$; ethylene, acetaldehyde, $CO_2$, ethanol, and water vapor were studied with regard to both aerobic and anaerobic emission of gases by cherry tomatoes. Absorption coefficients were obtained for each of these gases for a series of laser lines between approximately 1350 and 1950 $cm^{-1}$. Ethanol and acetaldehyde were successfully detected at ppb levels in the multicomponent gas mixtures.

Trace gases emitted by pears were discussed in two subsequent publications by this group. Persijn et al. (1999) established detection limits that ranged between 0.3 ppb and 10 ppm for the five gases mentioned above during a study of the effect of $CO_2$ on fermentation. The other study (Persijn et al., 2000) reported a detailed list of absorption coefficients for these gases, obtained by scaling intensity data to the Hitran database. Low-resolution PA spectra between 1300 and 2000 $cm^{-1}$, plotted as the variation of absorption coefficient vs. infrared frequency, were also reported in this work.

Long-chain saturated fatty acids, which exist as solids at ambient temperatures, were studied in the gas phase by Jalink et al. (1995). These authors utilized a heat pipe cell and CO laser excitation to measure PA spectra at elevated temperatures for the even-carbon-numbered capric, lauric, myristic, and palmitic acids. About 25 different laser lines were available in the 1730–1850 $cm^{-1}$ region, permitting detection of the ~1780-$cm^{-1}$ C=O stretching band for each compound. For capric acid, the PA signal was measured as a function of temperature, allowing the authors to confirm the validity of the Clausius–Clapeyron equation.

PA spectra of two of the lower carboxylic acids were described by Kästle and Sigrist (1996a,b). These authors obtained spectra of acetic acid -$h_4$ and

-$d_4$, as well as propionic acid, in the C═O stretching region (1700–1840 $cm^{-1}$). The PA gas cell was immersed in a temperature-controlled water bath so as to control its temperature between about 280 and 350 K in these experiments. Bands due to both the monomeric (1780 $cm^{-1}$) and dimeric (1730 $cm^{-1}$) forms of the acids were observed, the former being partially resolved into P and R branches. The thermodynamic quantities $\Delta H$ and $\Delta S$ associated with dimerization were calculated from the temperature dependence of the spectra. In addition, absolute integrated absorbances were determined from transmission measurements performed in conjunction with the PA experiments.

PA spectra of the simplest carboxylic acid were recently studied in an innovative experiment that also used a CO laser (Merker et al., 1999). High-resolution spectra of formic acid vapor were recorded by mixing CO laser lines with microwave radiation; this permitted tuning over the 1500–2000 $cm^{-1}$ region with a spectral coverage of about 50%. Despite the existence of a number of gaps among the available laser frequencies, the authors were able to obtain a result that was believed to be the first rotationally resolved spectrum of the formic acid dimer. Definitive assignments of some of the newly observed features in this PA spectrum are not yet available—a typical situation when pioneering spectroscopic data are first obtained.

In a significant departure from the research described in the preceding paragraphs, Santosa et al. (1999) reported the use of an overtone CO laser, which potentially produces hundreds of lines between 2.6 and 4.0 μm (2500–3850 $cm^{-1}$). These frequencies are, of course, well suited for the study of C—H, O—H, and N—H stretching vibrations. Ethylene, ethane, and pentane were studied with an intracavity PA cell in this work. The experimental apparatus was designed to analyze gases produced during the storage of fruit and to detect ethane and pentane in human breath or in gases produced by plant tissue. The latter two gases are generated during the course of lipid peroxidation in both experiments. Excellent detection limits at or below the ppb level were established in this investigation. These results, in conjunction with those summarized above, lead to the encouraging conclusion that a significant fraction of the mid-infrared region (1250–2000 and 2500–3850 $cm^{-1}$) can be successfully studied by gas-phase PA spectroscopy when a CO laser is employed as the source of infrared radiation.

### 6.9.2. $CO_2$ Laser Excitation

The $CO_2$ laser has also been widely used as an infrared source in the study of PA spectra of gases. A number of studies pertaining to this topic, published between 1994 and 2000, are listed in Table 6.3. The $CO_2$ laser produces tunable radiation near 10 μm (1000 $cm^{-1}$); hence, it is a logical choice for

**Table 6.3. Recent PA Studies of Gases[a]**

| Gases | Frequency ($cm^{-1}$) | Reference |
|---|---|---|
| $C_2H_4$, $CH_3OH$, $C_2H_5OH$, $C_7H_8$ | 1040–1057 | Repond and Sigrist (1994) |
| $NH_3$, $NH_2D$, $NHD_2$, $ND_3$ | 930–1085 | Petkovska and Miljanić (1997) |
| $C_2H_4$, $SO_2$ | 930–960 | Gondal (1997) |
| $C_2H_4$, $NH_3$ | 944 + others | Radak et al. (1998) |
| $O_3$ | 1064 | Zeninari et al. (1998) |
| $C_2H_4$ | 949, 953 | Schäfer et al. (1998a) |
| $C_2H_4$, $N_2$ | 1069 | Zeninari et al. (1999) |
| $C_2H_4$ | 925–1085 | Calasso and Sigrist (1999) |
| $O_3$, $O_2$, $N_2$, He, Ne, Ar, Kr, Xe | 1064 | Zeninari et al. (2000) |
| $C_2H_4$, $CH_3OH$, $C_2H_5OH$, $C_6H_6$, $CO_2$, $H_2O$ | 949–1079 | Nägele and Sigrist (2000) |

[a] $CO_2$ laser used as an infrared source.

the measurement of spectra of oxygen-containing compounds, many of which exhibit characteristic bands in this region. In addition, PA infrared spectra of several small molecules that do not contain oxygen have also been obtained between 900 and 1100 $cm^{-1}$ with this laser. These results are briefly discussed in the following paragraphs.

It is widely recognized that M. W. Sigrist and his colleagues at the Institute of Quantum Electronics at the Eidgenössische Technische Hochschule (ETH) in Zürich have made a very substantial contribution to the PA spectroscopy of gases for many years. One important area of their research involves the use of the $CO_2$ laser as an infrared source. For example, Repond and Sigrist (1994) used a high-pressure $CO_2$ laser, which was tunable from about 1039 to 1057 $cm^{-1}$, and a nonresonant cylindrical stainless steel cell to obtain PA spectra of ethylene, methanol, ethanol, and toluene. Ethylene was detected at concentrations as low as 50 ppb in this investigation. The capabilities of the technique for quantitation were clearly demonstrated by two important results: First, the PA signal varied linearly with concentration over an especially wide range (four orders of magnitude); and second, the spectrum of a multicomponent mixture was shown to be equal to the sum of the spectra of the individual gases. The later use of a resonant multipass PA cell (Nägele and Sigrist, 2000), which is capable of flow-mode operation, resulted in the reduction of the detection limit of ethylene to an even more impressive 70 parts per trillion (ppt). In a related publication, the characteristics of the two types of microphones (condenser and electret) used in nonresonant gas-phase PA experiments were discussed in detail by Calasso and Sigrist (1999).

Ammonia is another important molecule that absorbs in the region made accessible by the $CO_2$ laser. In the first of two related studies, Petkovska and Miljanić (1997) described PA infrared spectra of ammonia and its various deuterated forms. The PA signal was observed to decrease with increasing deuterium content; $NH_3$ absorbs at more of the available laser emission wavelengths than the other isotopomers, and with greater intrinsic intensity. The companion publication (Radak et al., 1998) examined the coincidences between the $CO_2$ laser lines and the absorption bands of both $NH_3$ and $C_2H_4$. As might be expected, pressure broadening had a significant effect on the overlap of the infrared bands with the narrower laser lines. The PA signal also depended on the pressure of the absorbing gas.

Gondal (1997) constructed a resonant $CO_2$ laser PA spectrometer that was designed to detect air pollutants. $C_2H_4$ and $SO_2$ were analyzed with this system; the minimum detectable concentrations were 50 ppt and 50 ppb, respectively. Cell resonance was studied in detail for 1-ppm mixtures of $C_2H_4$ in $N_2$, Ar, and He. Three or four resonances were identified at frequencies that ranged up to 8 kHz for each case. The longitudinal resonance frequency did not vary with buffer gas pressure for $N_2$ and Ar but increased with pressure when He was used. In gas-phase PA infrared spectroscopy, these acoustic resonances must be fully characterized so that the response of the apparatus [and the so-called Q (quality) factor] can be maximized.

Trace gas monitoring was also discussed by Schäfer et al. (1998b), who noted the advantages of the use of a pulsed $CO_2$ laser for PA detection of $C_2H_4$. These authors also observed a linear relationship between PA signal intensity and analyte concentration; this implies that it is possible to determine the optical absorption coefficient if the concentration is known. Optical saturation can also be studied with this technique.

A collaboration between scientists at the Université de Reims (France) and the Russian Academy of Sciences on the PA spectroscopy of ozone and other gases has yielded several recent publications that are relevant to the current discussion. Zeninari et al. (1998) utilized a $CO_2$ laser to selectively excite the $\nu_3$ vibrational level of ozone. A lock-in amplifier was used to measure the phase shift between the PA signal and the incident radiation. The observed phase lag was consistent with a model based on a two-step deexcitation process, rather than a simple single-step mechanism. A subsequent investigation (Zeninari et al., 2000) confirmed the appropriateness of this three-level model and reported rate constants for the collision of $O_3$ with various noble gases (He, Ne, Ar, Kr, and Xe). These authors also described a differential resonant Helmholtz cell in considerable detail (Zeninari et al., 1999), the low-pressure case being of greatest relevance to their research. Ethylene and nitrogen were studied in this work.

### 6.9.3. Other Lasers

As noted above, a variety of other lasers have also been utilized as mid- and near-infrared sources in PA investigations of gases. The recent literature on this subject is summarized in Table 6.4 and reviewed in the following paragraphs. The mid-infrared region is considered first.

The research of Sigrist and his colleagues has already been discussed with regard to the use of both CO and $CO_2$ lasers. In addition to those studies, this group has investigated the use of a number of other mid-infrared lasers in recent years. For example, Bohren and Sigrist (1997) employed a nanosecond pulsed optical parametric oscillator (OPO) to obtain PA spectra of gases in the 2.5–4.5 μm (4000–2200 $cm^{-1}$) region. Benzene, toluene, methanol, ethanol, and isopentane vapors were examined at concentrations of 100 ppm in synthetic air in this work. The detection limits for most of these gases with this system were on the order of a few ppm; sub-ppm sensitivity was achieved for methane and isopentane.

Seiter and Sigrist (2000) used a pulsed laser based on difference-frequency mixing of the outputs from a continuous-wave (cw) external-cavity diode laser and a Q-switched Nd:YAG laser in periodically poled $LiNbO_3$ (a nonlinear optical material) for trace gas analysis in the 3–4 μm region. This source was used with a 36-m multipass cell to detect formaldehyde at concentrations below 8 ppb. In a related investigation (Fischer et al., 2001) a single-pass resonant cell was used to obtain PA spectra of $CH_4$. A detection limit of 1.5 ppm, which is roughly equal to the typical concentration of this gas in air, was established in this work.

The quantum cascade distributed-feedback laser is a novel mid-infrared source that has also been used very recently to obtain PA spectra of gases. These lasers have been available commercially for about 3 years. Hofstetter et al. (2001) obtained spectra of $CO_2$, $CH_3OH$, and $NH_3$ using a quantum cascade laser and a Herriott multipass arrangement around a PA cell equipped with a 16-microphone array. Ammonia was detected in the 966–968 $cm^{-1}$ region at a concentration of 300 ppb with this apparatus. Earlier PA spectra of $NH_3$ were obtained with a quantum cascade laser by Paldus et al. (1999).

Two additional publications on PA mid-infrared spectroscopy of gases should be mentioned in the present review. Kühnemann et al. (1998) also utilized an optical parametric oscillator as a source of infrared radiation. An innovative aspect of this work was the use of a cw OPO, rather than a pulsed oscillator such as that described by Bohren and Sigrist (1997). Kühnemann et al. (1998) established a detection limit of less than 1 ppb for ethane with this equipment, which compares favorably with data from systems based on

**Table 6.4. Recent PA Studies of Gases**[a]

| Source | Gases | Wavelength (μm) | Frequency ($cm^{-1}$) | Reference |
|---|---|---|---|---|
| DFB[b] diode laser | $NH_3$ | 1.53 | 6,529 | Fehér et al. (1994) |
| Color center laser | $C_2H_2$ | 1.56–1.59 | 6,400–6,300 | Hornberger et al. (1995) |
| DFB diode laser | $NH_3$ | 1.55 | 6,450 | Miklós and Fehér (1996) |
| Optical parametric oscillator | $CH_3OH$, $CH_3CH_2OH$, $CH_4$, $C_5H_{12}$, $C_6H_6$, $C_7H_8$ | 2.5–4.5 | 4,000–2,220 | Bohren and Sigrist (1997) |
| Optical parametric oscillator | $C_2H_6$ | 3.34 | 2,990 | Kühnemann et al. (1998) |
| DFB diode laser | $CH_4$ | 1.653–1.657 | 6,050–6,035 | Schäfer et al. (1998a) |
| Diode laser | $C_6H_6$, $C_7H_8$, $C_8H_{10}$, $H_2O$ | 1.31, 1.67 | 5,990, 7,630 | Beenen and Niessner (1999a) |
| Diode laser | $C_6H_6$, $C_7H_8$, $C_8H_{10}$, $H_2O$ | 1.31, 1.67 | 5,990, 7,630 | Beenen and Niessner (1999b) |
| Quantum-cascade DFB laser | $NH_3$, $H_2O$ | 8.49–8.52 | 1,173–1,178 | Paldus et al. (1999) |
| DFB diode laser | $NH_3$ | 1.53 | 6,529 | Miklós et al. (1999a) |
| External cavity diode laser | $CH_4$ | 1.32–1.34 | 7,590–7,450 | Miklós et al. (1999b) |
| Raman-shifted dye laser | HNCO | 0.62–1.5 | 16,200–6,750 | Coffey et al. (1999) |
| Ti:sapphire ring laser | $AsH_3$, $H_2Se$ | 0.85–0.87, 0.77–0.79 | 11,650–11,500, 12,925–12,600 | Hao et al. (2000) |
| F-center laser | $N_2$–DCCH, OC–DCCH | 3.00 | 3,337–3,334 | Hünig et al. (2000) |
| Difference-frequency generation | HCHO | 3.53 | 2,833 | Seiter and Sigrist (2000) |
| DFB diode laser | HF | 1.30 | 7,665 | Wolff and Harde (2000) |
| Ti:sapphire ring laser | HCN | 0.77–0.88 | 13,020–11,390 | Lecoutre et al. (2000a) |
| Ti:sapphire ring laser | $HSiF_3$ | 0.91–0.92, 0.77–0.78 | 10,960–10,900, 12,925–12,875 | Lecoutre et al. (2000b) |
| Ti:sapphire ring laser | $H_3SiD$ | 0.81–0.85 | 12,278–11,797 | Bürger et al. (2001) |
| Difference-frequency generation | $CH_4$ | 3.2–3.7 | 3,125–2,703 | Fischer et al. (2001) |
| Quantum-cascade DFB laser | $CO_2$, $CH_3OH$, $NH_3$ | 10.33–10.36, 10.16–10.19 | 968–965, 984–981 | Hofstetter et al. (2001) |

[a] Various lasers used as infrared sources.

[b] DFB, distributed feedback.

the CO laser. Moreover, because the OPO is tunable from 2.3 to 4 μm, PA spectra of many other gases should be obtainable by this technique. In the other publication, Hünig et al. (2000) described high-resolution PA spectra of the $N_2$–DCCH and OC–DCCH complexes. This article emphasizes the interpretation of the PA spectra rather than the experimental technique that was employed. Indeed, the reader might interpret this situation as a confirmation of the maturity of the PA infrared spectroscopy of gases.

As shown in Table 6.4, several different types of near-infrared lasers have been used since 1994 to obtain PA spectra of gases. Among these, distributed-feedback (DFB) diode lasers with emission wavelengths greater than 1 μm have been utilized by the largest number of researchers; publications describing this work will be summarized next. PA spectra obtained using other near-infrared lasers will then be considered.

Ammonia has been the subject of several PA investigations that utilized DFB diode near-infrared lasers. Fehér et al. (1994) constructed a system using a laser diode that emitted at about 1.53 μm (6529 $cm^{-1}$). Frequency modulation was achieved over a very narrow range by adjusting the drive current of the laser. With this apparatus, the authors detected $NH_3$ at pressures down to $5.3 \times 10^{-5}$ torr; the sensitivity of the PA system was estimated as 8 ppb. In subsequent related work, two of these authors (Miklós and Fehér, 1996) confirmed this detection limit and showed that it was approximately three orders of magnitude better than that obtainable by absorption measurements with a 1-m cell.

The analysis of $NH_3$ is complicated by the tendency of the gas to adsorb on the walls of the detector and the tubing of the gas transfer system. This situation prompted Miklós et al. (1999a) to develop a differential flow-through system based on the same 1.53-μm 5-mW DFB laser. In this improved arrangement, the gas flowed through two tubes, one of which was irradiated collinearly. The second tube was then used to subtract the background signal from the flowing gas. The PA signal from $NH_3$ was observed to vary linearly with concentration down to 1 ppm with this apparatus. The sensitivity of this PA system could be further improved through the use of a more powerful laser.

Other DFB diode lasers have also been used to obtain PA near-infrared spectra of gases. For example, Schäfer et al. (1998a) used a 1.65-μm DFB laser to detect $CH_4$ at concentrations as low as 60 ppm. These researchers found that absorption spectroscopy actually yielded significantly better sensitivity than their PA measurements, which were limited by the low output power of the laser. On the other hand, much higher sensitivity was attained by Wolff and Harde (2000), who employed a 1.30-μm DFB laser to detect gaseous HF. The detection limit in this experiment was about 80 ppb.

Beenen and Niessner (1999a,b) utilized near-infrared (NIR) laser diodes and a resonant PA cell to detect benzene, toluene, xylene, and water vapor. The long-term objective of this work was the development of portable apparatus for the detection of gases. In this laboratory-based study, the laser diodes were coupled to a flow-through cell with optical fibers. Diode wavelengths of 0.908, 0.911, and ~1.67 μm were used for the hydrocarbons, whereas water was monitored at 1.31 μm. These wavelengths can be tuned over narrow ranges by varying the temperature of the diode, a capability that was utilized to distinguish between different compounds. For example, $C_6H_6$ was monitored at 1.67 μm (5988 $cm^{-1}$), while toluene was analyzed at a slightly different wavelength of 1.68 μm (5952 $cm^{-1}$). The sensitivities achieved in this work were 70 μg/L ($C_6H_6$), 100 μg/L (toluene), and 160 μg/L (xylene) and are sufficient for environmental monitoring.

The literature described in the preceding paragraphs shows that PA infrared spectroscopy is indeed capable of extremely sensitive gas analysis in the near-infrared region. However, it should be kept in mind that the wavelength ranges covered by the diode lasers used in these studies are extremely narrow; to state the obvious, PA detection of trace gases is feasible only when their characteristic absorption bands coincide with the available laser lines.

Table 6.4 lists several publications in which a titanium:sapphire ring laser was used to obtain near-infrared PA spectra of gases. Three articles resulted from collaborations between Lecoutre, Huet, and their colleagues at l'Université de Sciences et Technologies de Lille and at several other institutions. These investigations involved the measurement—and more particularly the interpretation—of high-resolution spectra of HCN (Lecoutre et al., 2000a), $HSiF_3$ (Lecoutre et al., 2000b), and $H_3SiD$ (Bürger et al., 2001). The high sensitivity of the PA system allowed the authors to study very weak bands, using only a small volume of gas, and to obtain research-quality results.

Hao et al. (2000) used a similar system to obtain Doppler-limited spectra of $AsH_3$ and $H_2Se$. A minimum detectable absorption coefficient of $6.35 \times 10^{-9}$ $cm^{-1}$ was calculated for water vapor in this work, which examined noise sources and cell resonance in considerable detail.

Other near-infrared lasers have also been used in the measurement of PA spectra of gases. Hornberger et al. (1995) employed a 1.5-μm color center laser and a multipass cell to record high-resolution spectra of acetylene. The minimum detectable absorption coefficient with this apparatus is on the order of $10^{-10}$ $cm^{-1}$. Miklós et al. (1999b) utilized an external cavity diode laser (mentioned above in another context) to obtain PA spectra of $CH_4$ in the 7450–7590 $cm^{-1}$ range. Close agreement between PA and absorption spectra was demonstrated for the Q branch of the methane combination band centered near 7510 $cm^{-1}$. Finally, Coffey et al. (1999) used a Raman-shifted dye laser in the measurement of overtone spectra of HNCO gas. The PA detection scheme used by these authors is discussed very briefly in this publication.

### 6.9.4. FTIR Spectra of Gases

The preceding discussion of the recent literature on the PA infrared spectroscopy of gases reviewed work in which various mid- and near-infrared lasers were used for excitation. As shown by the large number of references in Tables 6.2–6.4, a considerable amount of research has been carried out in this area within a comparatively short period. By contrast, only a few articles that describe the measurement of PA spectra of gases with an FTIR spectrometer are known to exist. This literature, which also spans a longer period of time, is briefly summarized in this section.

PA infrared spectra of solid carbonyl compounds are discussed later in this chapter. As will be shown, the research of I. S. Butler and his group comprises most of the published work in this area. During their investigations, these researchers also observed gas-phase spectra. For example, experiments on the group VIB metal chalcocarbonyl complexes $M(CO)_6$ ($M = Cr$, Mo, and W) and $Cr(CO)_5CS$ by Xu et al. (1986) yielded spectra of the corresponding vapors instead of the expected spectra of the solid complexes. Indeed, the PA spectra agreed with published spectra of the gaseous complexes and differed significantly from absorption spectra of the same complexes in KBr disks. The observation of gas-phase spectra in this work was put down to the heating of the samples by the infrared beam.

Several years later, Butler et al. (1992) reported gas-phase PA infrared spectra of $CH_3Mn(CO)_5$ and $CH_3Re(CO)_5$. The spectrum of the former compound is depicted in Figure 6.33. This result shows typical examples of the near-infrared bands that are clearly visible in these spectra; similar results were obtained in the previous study and in the investigations of the solid complexes described below. The authors assigned the bands near 4000 $cm^{-1}$ to overtones and combinations of the fundamental transitions that were identified at lower frequencies.

Recently, Olafsson et al. (1999) described a multipass cell optimized for the detection of gases in PA FTIR spectroscopy. Two factors ensured that the signal arising from the cell wall was negligible, even though reflection losses were relatively high. First, the thermal time constant of the cell wall was much less than the width of the interferogram center burst; and second, the heat capacity of the cell wall was much greater than that of the gas. An impressive detection limit of about 0.1 ppm was achieved in this experiment, corresponding to an absorption coefficient of $10^{-5}$ $cm^{-1}$. These values com-

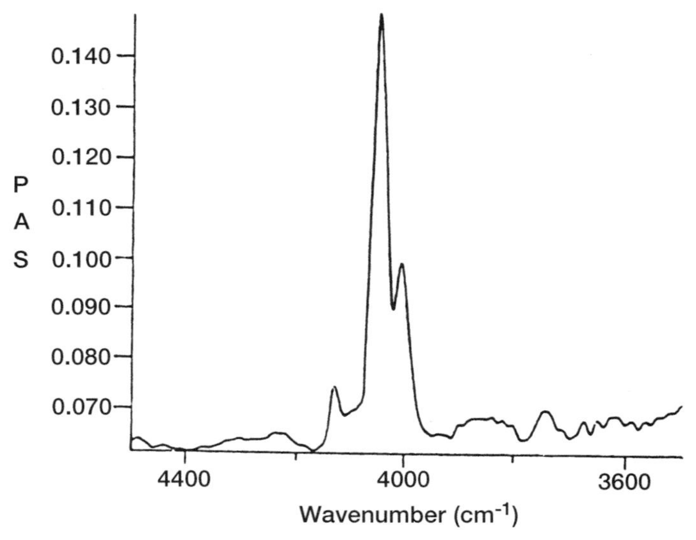


**Figure 6.33.** PA infrared spectrum of gaseous $CH_3Mn(CO)_5$. (Reproduced from Butler, I. S. et al., *Appl. Spectrosc.* **46**: 1605–1607, by permission of the Society for Applied Spectroscopy; copyright © 1992.)

pare favorably with those mentioned earlier with regard to laser-excited PA spectra of gases.

As stated at the beginning of this section, some of the most important achievements in PA spectroscopy have occurred with respect to gas analysis. The high sensitivity afforded by PA detection of gas-phase spectra is particularly impressive and has led to the development of many successful analytical applications and the publication of an extensive amount of literature. This field of research is expected to continue to grow in the future.

## 6.10. FOOD PRODUCTS

Food products generally consist of proteins, fats, carbohydrates, and water. To the analyst, it is rather obvious that many foods are not particularly amenable to traditional infrared sample preparation techniques. Fortunately, near- and mid-infrared PA spectroscopy provide a welcome opportunity for both qualitative and quantitative analyses of foods that otherwise might not exist.

Several research groups have successfully studied the PA infrared spectra of foods, food products, and packaging materials. Some of the earlier literature in this subject area was reviewed by McQueen et al. (1995a), who compared near- and mid-infrared PA techniques with both ATR infrared spectroscopy and wet chemical methods. In general, previously published analyses of food products by PA spectroscopy have addressed the following problems: (a) the near-infrared determination of moisture content, (b) the quantitation of fats and edible oils, (c) the elucidation of molecular structures in foods through the assignment of mid-infrared bands to particular functional groups, and (d) the characterization of coatings and microorganisms on foods. The existing literature on these topics is summarized below. Some additional examples of PA infrared spectra of typical food products are also illustrated.

### 6.10.1. Early PA Studies of Foods

The capability of PA spectroscopy for the quantitation of moisture in single-cell protein was demonstrated by Castleden et al. (1980). These authors obtained dispersive near-infrared PA spectra (1.3–2.3 μm, approximately 7700–4350 $cm^{-1}$) of protein samples with a range of particle sizes and found that the combination band due to water at about 1.9 μm (5265 $cm^{-1}$) could be used to quantify moisture at levels from 0 to 12%. Bands due to —NH and —CH groups were also identified in the spectra. However, quantitative analysis was not feasible unless the samples were first separated into fractions on the basis of particle size since PA intensities are usually greater for smaller particles. Castleden et al. (1980) performed such fractionations and were consequently able to carry out successful moisture determinations for specific particle size distributions. These experiments were continued by Jin et al. (1982), who described PA near-infrared spectra of both protein samples and milk substitutes. PA intensities of bands due to water were again found to vary linearly with concentration in the ~0–10% range in this study.

During the 1980s, P. S. Belton and his colleagues at the Institute of Food Research and the University of East Anglia in Norwich, United Kingdom, published a series of studies on near- and mid-infrared PA spectra of various food products. These authors were particularly interested in the use of PA spectroscopy for the quantitative analysis of specific constituents in foods. A brief review of PA spectroscopy, which mentions possible applications of the technique to the study of foods, was written during this period by Belton (1984).

An early investigation by this research group involved the PA near-infrared (1–2.5 μm, or 10,000–4000 $cm^{-1}$) determination of the moisture content of potato starch (Belton and Tanner, 1983). Dispersive PA spectra were obtained in this study. Instrumentation included an Xe lamp, a monochromator, chopper, and lock-in amplifier. It was observed that the PA intensity due to water absorbed on starch did not increase linearly with the amount of water present at higher concentrations. Hence this finding disagreed with the above-mentioned results of Castleden et al. (1980) and Jin et al. (1982), who had reported a linear dependence of PA near-infrared intensity on water concentration; however, a wider concentration range was investigated by Belton and Tanner (1983). In fact, the nonlinear variation of PA intensity with analyte concentration led these authors to investigate the theory in which the reciprocal of PA intensity is predicted to be proportional to the reciprocal of analyte concentration. Their experimental results agreed well with this model, which is discussed in more detail in Chapter 7.

Following their initial near-infrared study, this research group published a series of studies on mid-infrared PA spectra of foods. Belton et al. (1987a) compared PA spectra with results obtained using the ATR and diffuse reflectance techniques. The objective of this investigation was the quantitative analysis of protein–starch mixtures, prepared using either wheat gluten or casein together with potato starch. The authors found that PA and ATR spectra gave the best results, whereas the diffuse reflectance spectra were less satisfactory. Problems with sample mixing and loading were encountered with both the ATR and diffuse reflectance methods. Indeed, PA spectroscopy was concluded to be the most suitable technique for the analysis of these mixtures.

In a related investigation, Belton et al. (1987b) described a detailed PA study of sucrose, sucrose–KBr mixtures, and carbon black. The main goal of this work was to elucidate the influence of particle size on quantitative analysis in PA infrared spectroscopy. Spectra of sucrose powders with different particle sizes were shown but not discussed in detail in this study. This work is also mentioned in the discussion of experimental methods in PA spectroscopy (Chapter 3). Sucrose and its chars were also described above in the section on PA spectra of carbons.

The final study in this series (Belton et al., 1988a) compares PA and ATR infrared spectroscopies for the analysis of chocolate and cocoa liquors. These methods were selected because the samples were unsuitable for ordinary infrared transmission measurements. Both types of spectra shown in this work are of good quality, although the bands in the PA spectrum are rather broad because it was obtained at the comparatively low resolution of 16 $cm^{-1}$. Despite this limitation, about a dozen bands are clearly identifiable in the PA spectrum. In fact, the authors concluded that PA spectroscopy was the best method for the quantitative analysis of fat in chocolate. This study is also mentioned with regard to quantitative analysis in Chapter 7.

### 6.10.2. Typical Example: Milk Chocolate

The capability of PA infrared spectroscopy for the characterization of chocolate is further demonstrated by a previously unpublished result for ordinary

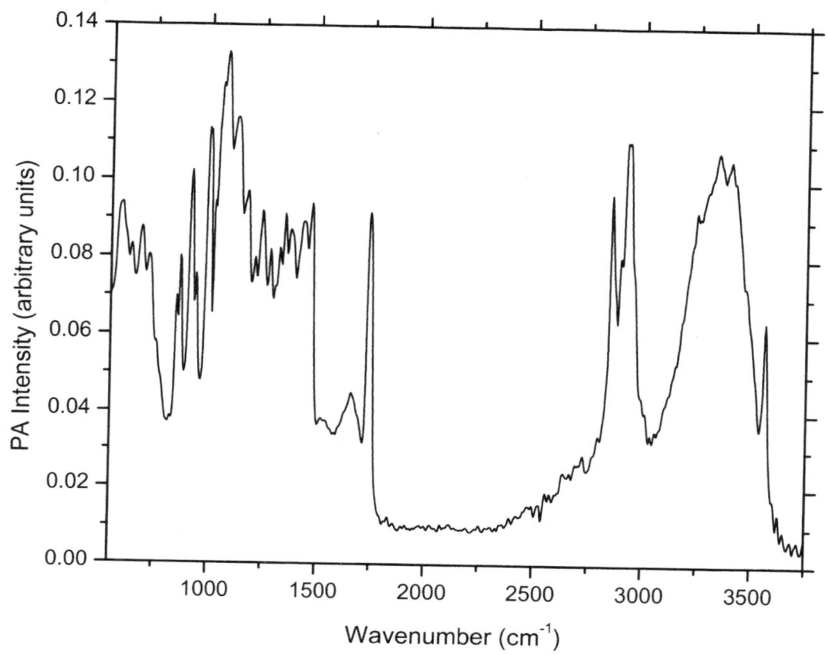


**Figure 6.34.** PA infrared spectrum of milk chocolate, obtained using a Bruker IFS 113v FTIR and Princeton Applied Research Corporation 6003 PA cell.

milk chocolate, depicted in Figure 6.34. This PA spectrum, which was recorded in the author's laboratory in an acquisition time of about 25 min, displays about 25 well-defined bands. Indeed, food scientists would probably agree that this infrared spectrum conveys a considerable amount of useful information regarding the composition of this chocolate sample. Perhaps even more importantly, the reader should also recognize this demonstration of the capability for the measurement of a research-quality infrared spectrum of a soft solid without any sample preparation. This point, of course, is relevant to samples other than foods.

### 6.10.3. Agricultural Grains

PA infrared spectroscopy can be used to characterize agricultural grains as well as prepared foods. For example, Greene et al. (1992) obtained PA and diffuse reflectance infrared spectra of corn in the course of a study of fungal contamination, which can potentially be a serious health hazard. These spectroscopic techniques were used because they permitted convenient analysis of the relevant solid samples. In this study, PA infrared spectra of corn infected with *Fusarium moniliforme* or *Aspergillus flavus* were observed to be greatly different from the spectrum of uninfected corn: Major differences were observed in the amide I (1650-$cm^{-1}$) and amide II (1550-$cm^{-1}$) bands, implying a significant increase in protein or acetylated sugar content as a consequence of the infection. Differences between diffuse reflectance and PA spectra were put down to the sample grinding required in the former technique. However, a minor drawback to this use of PA spectroscopy was noted: Only one kernel of corn could be analyzed in each experiment! The examination of a large number of samples by this technique would therefore undoubtedly require significant time and effort. In fact, these investigators proposed an alternative to PA spectroscopy several years later; emission spectra of corn kernels on a moving conveyor belt were obtained using a technique referred to as transient infrared spectroscopy (Gordon et al., 1999). This effectively increased the amount of corn that could be analyzed in a limited period of time.

A related study by this group (Gordon et al., 1997) compared results from PA infrared spectroscopy with those from a technique known as BGYF (bright greenish-yellow fluorescence), an accepted analytical method for the detection of pathogenic fungi on corn kernels. The authors noted 10 specific changes that occur in the PA spectra of corn as a consequence of infection; these arise from phenomena such as an increase in the number of COOH groups, greater $CO_2$ evolution, and a decrease in carbohydrates, all of which are caused by the effect of the fungi on the corn. The observation of these changes in the PA spectra of a series of blind samples agreed completely with positive results in separate BGYF tests. Hence these experiments validated the use of PA infrared spectroscopy for the detection of fungal infection in corn. Because of the high information content in the PA infrared spectra of corn, the authors proposed that the PA results be combined with knowledge-based pattern recognition techniques in an expert system.

The number of features in the PA spectra that can be used to discriminate between healthy corn and samples infected with mycotoxigenic fungi was eventually increased to 12 (Gordon et al., 1998). In this later work, an artificial neural network was trained to distinguish contaminated from uncontaminated corn by pattern recognition of the infrared spectra. Although the use of individual bands was generally not adequate for discriminating between the two types of corn, the authors found that the simultaneous examination of eight or more features in the PA spectra led to a high success rate for the neural network. The question as to whether PA spectroscopy is capable of distinguishing among infections caused by different fungal species was not resolved in this work.

### 6.10.4. The PA Infrared Spectrum of Flour

The use of dispersive and FT near-infrared spectroscopies for the determination of the moisture and protein contents in wheat and other grains is now well established. In the context of the present discussion, it is relevant to ask whether PA infrared spectroscopy might usefully be employed for the analysis of these agricultural grains and products derived from them. To partially address this question, a PA infrared spectrum was recorded for commercially available whole wheat flour. The result is shown in Figure 6.35; it should also be mentioned that this spectrum closely resembles one measured for ordinary white flour under similar conditions. In the fingerprint region, the well-known amide I and II bands (see above) are clearly visible, as are a number of features arising from various other functional groups. The broad band near 3400 $cm^{-1}$ can readily be attributed to hydrogen-bonded species. Thus the information obtainable from the PA infrared spectrum of flour is analogous to that derived from the corresponding near-infrared spectrum. The ease with which a PA spectrum such as this may be recorded, as well

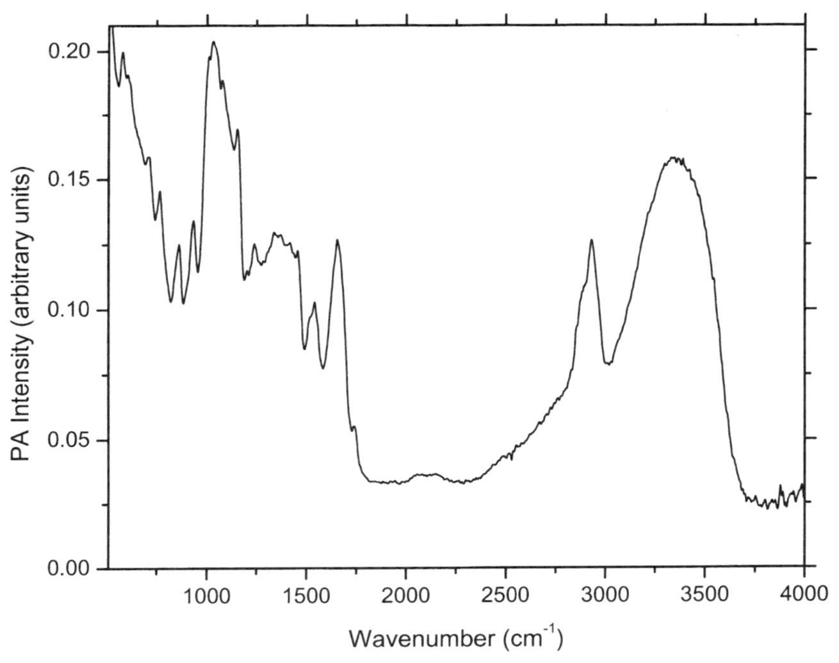


**Figure 6.35.** PA infrared spectrum of whole wheat flour, recorded under conditions similar to those for Fig. 6.34.

as the intrinsically greater information content that is conveyed by the mid-infrared region, imply that PA infrared spectroscopy could certainly be used instead of the established near-infrared techniques for the analysis of wheat, other grains, and products such as flour.

### 6.10.5. Recent PA Studies of Food Products

McQueen et al. (1995b) utilized optothermal near-infrared spectroscopy for the analysis of an extensive series of cheese samples. The optothermal technique is described in Chapter 3. In this work, measurements were carried out at three wavelengths [1740 nm (5750 $cm^{-1}$), 1935 nm (5170 $cm^{-1}$), and 2180 nm (4590 $cm^{-1}$)] that were selected by broad-bandpass optical filters. Protein, fat, and moisture contents were also measured using standard analytical techniques. Correlation coefficients ranging from 0.93 to 0.96 were calculated for the spectroscopic and wet chemical data. These results confirm the suitability of the optothermal method for cheese analysis. Because the spectroscopic data were recorded at only three wavelengths, no spectra were plotted in this publication.

The rapid quantitative analysis of the main components of pea seeds (starch, proteins, and lipids) was investigated by Letzelter et al. (1995). Both PA infrared spectroscopy and established wet chemical methods were employed in this work. To assess the relationship of PA band intensities with concentration, the authors analyzed mixtures of starch in KBr, as well as protein in starch. In the former case, the PA intensity of the broad carbohydrate band at about 1000 $cm^{-1}$ exhibited partial saturation at concentrations of about 50%; by contrast, the strong amide I protein band varied linearly with concentration in the second experiment. Thus band intensities are not the only factor that determines whether saturation affects the PA spectra of these samples.

Proceeding to the analysis of single pea seeds, the authors found that the use of PA infrared spectra and partial least squares (PLS), a common multivariate regression algorithm, gave reliable results with regard to the compositions of the samples. They concluded that the PA data were adequate for the reliable prediction of the contents of the three components mentioned above, which were required for breeding purposes. This work is also mentioned in Chapter 7 with regard to quantitative analysis.

Near-infrared optothermal spectroscopy was mentioned above. This technique can also be implemented at mid-infrared wavelengths, provided that a suitable gas laser or thermal source is used as a source of infrared radiation. For example, Favier et al. (1996) used the 966-$cm^{-1}$ line of a $CO_2$ laser in optothermal experiments designed to detect trans fatty acids (TFAs) in margarines. This frequency is diagnostic for the carbon–carbon bonds in these acids. TFA concentrations between about 6–65% were determined in this work. The accuracy of the optothermal results was confirmed by their agreement with data obtained from two different types of chromatography and from transmission infrared spectroscopy.

Favier et al. (1998) next employed a tunable $CO_2$ laser in optothermal window experiments carried out to detect contaminants in extra-virgin olive oil. $CO_2$ laser lines at 931, 953, 966, 1041, and 1079 $cm^{-1}$ were used in this work. Olive oil does not absorb significantly in this region. On the other hand, common adulterants such as the safflower and sunflower oils utilized in this study both absorb light at one or more of these frequencies. Therefore, optothermal window (OW) determinations of these contaminants in olive oil are feasible. Favier et al. (1998) showed that the observed OW signals were proportional to the amount of safflower oil or sunflower oil that was present in mixtures with olive oil. The limit of detection for either of these contaminants was approximately 5%.

A brief overview of infrared photothermal detection schemes—including several hyphenated techniques—that are suitable for quantitation of TFAs in margarines and edible oils was presented by Bićanić et al. (1999). No spectra were included in this study. The methods discussed include dual-beam thermal lensing (DBTL) spectrometry, the OW technique, gas chromatography combined with either PA or photopyroelectric (PPE) detection, and high-performance liquid chromatography (HPLC)/DBTL. The latter method has not yet been extended to infrared wavelengths.

The DBTL and OW measurements were based on the above-mentioned absorption of 966-$cm^{-1}$ $CO_2$ laser radiation by a trans carbon–carbon double bond. The sensitivity of DBTL was found to be about two orders of magnitude better than that in conventional infrared spectroscopy, with an impressive detection limit of 0.002%. The OW method [mentioned above with regard to the work of McQueen et al. (1995b) and discussed in detail in Chapter 3] was used at higher concentrations, displaying an operating range of 4–60% TFA.

The gas chromtography (GC) methods were developed to quantify TFAs in the vapor state. Initial experiments with a flow-through, high-temperature PA cell using $CO_2$ laser radiation achieved limited success because of problems with temperature instability and varying TFA vapor pressure. Photopyroelectric detection was more satisfactory, although its 2-μg sensitivity was still inferior to that of a conventional GC detector.

### 6.10.6. Research of Irudayaraj's Group

A substantial series of studies describing applications of PA infrared spectroscopy to the characterization of foods has recently been published by J.

Irudayaraj and his colleagues at Pennsylvania State University. The salient points from these articles are discussed in the final part of this section.

Yang and Irudayaraj (2000a) compared PA and ATR spectroscopies in the analysis of semisolid fat and edible oils, choosing butter, soybean oil, and lard as representative materials. The PA and ATR methods were selected because they require little sample preparation, obviously an important consideration in view of the rather problematic physical characteristics of fats and oils. Indeed, both sampling techniques yielded satisfactory spectra, as illustrated by the figures in this article. Frequencies and assignments of a total of 18 bands observed in the PA infrared spectra of soybean oil and lard were also given by these authors.

An important limitation of PA infrared spectroscopy was noted during this investigation. Heat generated by the absorption of infrared radiation tended to evaporate the water contained in butter, leading to the appearance of rotation-vibration bands due to water vapor in the 1450–1650 and 3100–3700 $cm^{-1}$ regions. These sharp features tended to obscure the broader, more relevant bands arising from the condensed-phase samples. This phenomenon, of course, is well known to many researchers who use PA infrared spectroscopy since it occurs frequently during the acquisition of PA spectra of samples that contain significant amounts of water.

Irudayaraj et al. (2001) continued their exploratory comparison of the PA and ATR techniques through the measurement of spectra of lard, peanut butter, mayonnaise, and whipped topping. Commercially available samples of these common foods were analyzed as purchased and again after heating at either 60 or 90°C for periods as long as 32 days. Heating caused the elimination of water from the latter two samples, as well as the oxidation of the lipids in lard and peanut butter. The authors again concluded that ATR was preferable to PA infrared spectroscopy with regard to high-moisture samples (see above). Approximately 10–20 bands were identifiable in both types of spectra for these four food products.

Three works by this group (Irudayaraj and Yang, 2000, 2001; Yang and Irudayaraj, 2000b) reported depth profiling studies of cheese slices and their associated polymer-based packaging materials. The use of PA infrared spectroscopy in this research was prompted by the fact that cheese is certainly not very amenable to most infrared sample preparation methods and also by the possibility of successfully depth profiling both the wrapper and the cheese.

In-phase and quadrature step-scan data were used to calculate magnitude and phase PA spectra of both the packaging material and its contents. The use of widely separated phase modulation frequencies, ranging from 50 to 900 Hz, allowed the retrieval of spectra corresponding to greatly different sample depths. In this way, it was confirmed that two components (fat and protein) of the cheese had diffused into the surface layer of the wrapper during storage. On the other hand, bands arising from the package material mostly originated from deeper within the wrapper. Depth profiling of dried cheese confirmed the rather intuitive expectation that water had been eliminated from its surface, whereas moisture beneath the surface was less prone to evaporation.

Two of these articles described the use of generalized two-dimensional (G2D) spectral correlation analysis. This method, which is discussed in Chapter 4, can be used to confirm the ordering of a multilayered system. In favorable cases it may also assist in the identification of overlapping bands because spectral information is spread across an additional dimension. Indeed, Yang and Irudayaraj (2000b) found that G2D confirmed the diffusion of the amide I and II components from the cheese sample into the packaging. Moreover a 1461-cm$^{-1}$ C—H band, not detectable in one-dimensional spectra because of the existence of several neighboring bands, was clearly revealed in the G2D correlation spectrum. Depth profiling of a model three-layer protein/starch/polyethylene system was also demonstrated by these researchers, who used both the G2D technique and PA phase spectra (Irudayaraj and Yang, 2002).

This group has also used PA spectroscopy to study edible coatings and microorganisms on fruit surfaces. Yang et al. (2001) carried out depth profiling experiments on these systems, using either rapid-scan or step-scan PA infrared spectroscopy. Edible coatings were modeled using two-layer (protein/apple) and three-layer (starch/protein/apple, protein/starch/apple) samples, prepared by treating an apple skin with aqueous solutions of protein and starch. For the two-layer sample, a magnitude PA spectrum exhibited bands from both layers, even though the sampling depth was estimated to be only 6 μm. However, the corresponding phase angle spectrum facilitated assignments of the bands to the first or second layers. G2D spectral correlation analysis confirmed the known fact that the apple skin was below the layer of protein film. For the more complicated three-layer samples, phase angle spectra were used to confirm the ordering of the layers, as was G2D.

Microorganisms on fruit contain proteins, which may be identified using the well-known amide I and amide II bands. Yang et al. (2001) examined apple and honeydew melon surfaces covered with several different microbes in this work. Principal component and canonical variate analyses were then used to discriminate between spectra of samples containing microorganisms and those of untreated samples. Magnitude spectra were used for depth profiling; plausibly, the amide I and II bands were stronger at higher modulation frequencies because the microbes were on the surface of the fruit. The authors suggested that the combination of PA spectroscopy and multivariate analysis might eventually become an accepted method for the determination of microorganisms on the surfaces of food products. Indeed, Irudayaraj et al. (2002) subsequently showed that apple skin surfaces contaminated with different microorganisms could be distinguished using PA infrared spectroscopy and suitable numerical techniques. The authors also demonstrated that it was possible to distinguish between pathogenic and nonpathogenic *Escherichia coli* in these experiments.

The contamination of extra-virgin olive oil with other oils was discussed above with regard to the optothermal technique used by Favier et al. (1998). Yang and Irudayaraj (2001a) also studied this problem, employing near-infrared, FT-Raman, as well as mid-infrared (PA and ATR) spectroscopies. The adulterant in the more recent study was olive pomace oil, a fully refined product. FT-Raman spectra exhibited the highest correlation with the concentration of the contaminant. The PA and ATR spectra, which were practically identical, confirm the statement by Favier et al. (1998) that extra-virgin olive oil absorbs very weakly throughout the 931–1079 $cm^{-1}$ region in which the $CO_2$ laser lines occur.

The final publication by this group to be included in this discussion presents infrared spectra for different types of meats. Yang and Irudayaraj (2001b) compared ATR and PA infrared spectroscopies with regard to their capabilities for the characterization of both beef and pork. The use of these particular infrared techniques obviously presents a significant advantage with regard to earlier infrared methods for the analysis of meats, in which protein was first solvated, and fat was either emulsified or extracted and refined. Both the ATR and the PA spectra of the beef samples obtained in this work were of very good quality. PA depth profiling experiments confirmed that the surface of each sample had a lower moisture content than the interior (Fig. 6.36). Moreover, differences among different types of meats (beef, pork, chicken, turkey) were expected to be observable using PA infrared spectroscopy.

The considerable body of research described above clearly demonstrates that PA infrared spectroscopy is well suited for the analysis of food products. Indeed, the two principal advantages of PA spectroscopy—minimal sample preparation and the capability for depth profiling of layered samples—play an important role in much of this work. It might be anticipated that the use of PA infrared spectroscopy for the characterization of foods and food products will eventually become sufficiently common that it no longer merits special comment. For example, the authors of a recent publication on sodium caseinate/glycerol and sodium caseinate/polyethylene glycol edible coatings (Siew et al., 1999) utilized PA infrared spectroscopy and a number of other analytical techniques but did not find it necessary to display any infrared spectra.

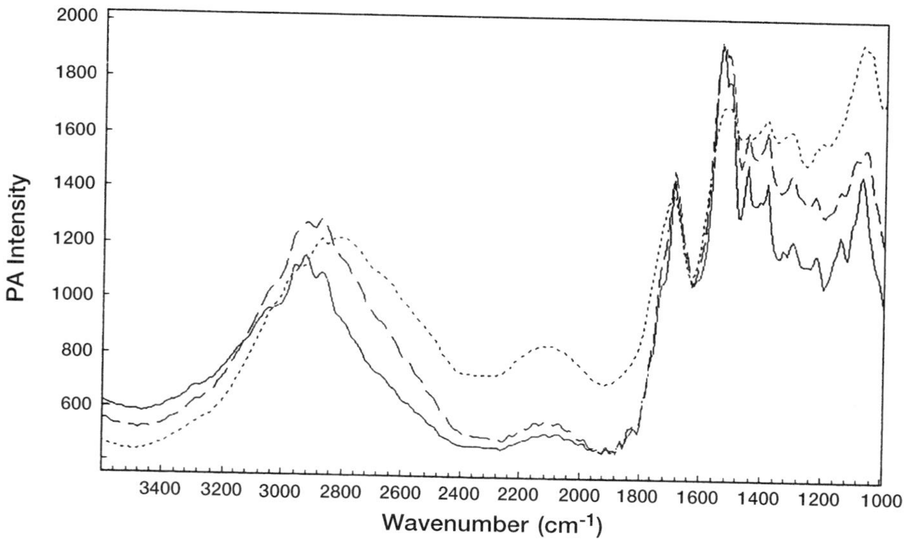


**Figure 6.36.** Depth profiling results for fresh beef. Phase modulation frequencies were as follows: solid curve, 900 Hz; dashed curve, 100 Hz; dotted curve, 10 Hz. (Reprinted from *Lebensm.-Wiss. u.-Technol.* **34**, Yang, H. and Irudayaraj, J., Characterization of beef and pork using Fourier transform infrared photoacoustic spectroscopy, 402–409, copyright © 2001, with permission from Elsevier Science.)

## 6.11. BIOLOGY AND BIOCHEMISTRY

Proteins and microorganisms such as bacteria and fungi are generally not very well suited to traditional infrared sample preparation techniques; hence, the virtual elimination of sample manipulation in PA spectroscopy offers an important advantage when the characterization of these rather difficult samples is mandated. Several typical examples of the use of PA infrared spectroscopy for the qualitative or quantitative analyses of biological and biochemical samples are reviewed in the following paragraphs.

An early application of dispersive PA spectroscopy to the characterization of proteins occurs in the work of Sadler et al. (1984). Near-infrared PA spectra of proteins, both in the solid state and in $D_2O$ solutions, were obtained in this investigation. PA spectroscopy was utilized because it facilitated the measurement of spectra for strongly scattering samples. Viable spectra were obtained for the proteins egg white lysozome and bovine pancreatic ribonuclease, various solid amino acids, proteins and polypeptides, as well as proteins in solution. The solid-state spectra generally contained much more detail than the solution spectra. This enabled Sadler et al. to propose detailed assignments of a number of bands to overtones and combinations of the fundamentals that occur in the mid-infrared.

The work of Gordon, Greene, and their colleagues at the U.S. Department of Agriculture on PA spectroscopy of grains and fungi was discussed in the previous section. Two additional publications by this group, which fall more naturally into the general subject area of biology, biotechnology, and biochemistry, should also be cited. In the first, PA infrared spectroscopy was used to monitor the growth of the filamentous fungus *Phanerochaete chrysosporium* on cellulose filter paper disks (Greene et al., 1988). The motivation for the study of this model system arises from the possible use of waste cellulose as a substrate for fermentation. As discussed below, PA spectroscopy provides a means for the measurement of biomass formed during this process, avoiding traditional analytical methods that require extensive sample handling and sometimes lead to the ultimate destruction of the sample.

The application of PA infrared spectroscopy to this system was demonstrated using a purified protein (bovine serum albumin), a purified phospholipid (asolectin), and a carbohydrate (cellulose): A mixture of the three is an approximate representation of a reconstituted microorganism. Indeed, the PA spectrum of this mixture qualitatively resembles that of *P. chrysosporium*. Next, known dry weights of the actual fungus were deposited on filter paper disks, and the intensities of the amide I bands in the PA spectra were determined. For relatively small weights, amide I intensity increased linearly with the amount of fungus; however, this intensity eventually reached a limiting value and did not increase further with weight. This saturation effect was attributed to mycelial layering, an already known phenomenon that is expected to limit optical penetration and thereby establish an upper limit to PA intensity. Finally, the growth of the fungus on filter paper disks was monitored by PA spectroscopy for several days and confirmed by a standard protein assay. These successful experiments clearly demonstrate the suitability of photoacoustic infrared spectroscopy for the analysis of *P. chrysosporium* on cellulose.

Gordon et al. (1990) continued this research, improving on the quantitation of protein biomass by PA infrared spectroscopy. Bands arising from amide groups were again utilized to monitor the increase in biomass corresponding to microorganism growth. Both pure proteins and microorganisms such as a bacterium, yeast, alga, and a fungus were studied in this investigation. Whereas the cellulose band at about 3338 $cm^{-1}$ had been used to normalize intensities in the previous investigation, the authors added a fixed amount of polyacrylonitrile (PAN) as an internal standard for quantitation in the later work. The nitrile absorption band at 2243 $cm^{-1}$, which is well separated from the features due to the samples or the cellulose disks, was used to scale these spectra. Four proteins and the four microorganisms mentioned above were studied in this way. Typical results are shown in

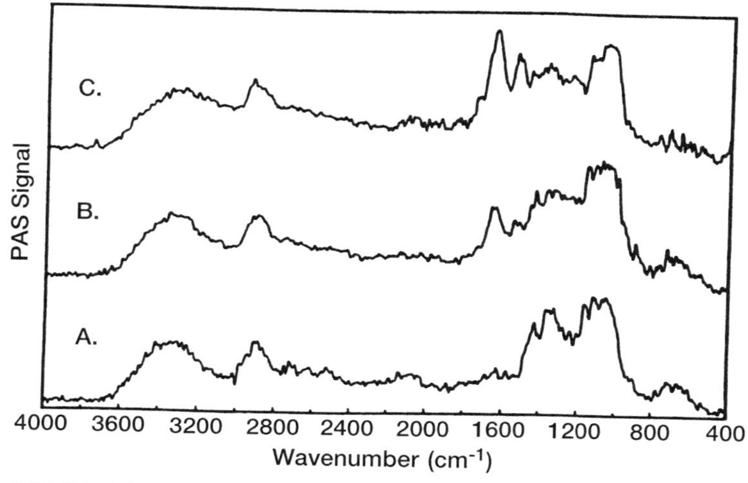


**Figure 6.37.** PA infrared spectra of *Phanerochaete chrysosporium* on cellulose filter paper. Curve A, filter paper only; curve B, 0.67-mg fungus on filter paper; curve c, 3.2-mg fungus on filter paper. (Reprinted from Measurement of protein biomass by Fourier transform infrared-photoacoustic spectroscopy, Gordon, S. H. et al., *Biotech. Appl. Biochem.* **12**: 1–10, copyright © 1990, with permission from Elsevier Science.)

Figure 6.37. The curves display saturation plateaus at high protein levels; the earlier onset of saturation for the microorganisms is a consequence of the fact that they contain about 70% nonprotein biomass, which contributes to the layering mentioned above and exacerbates saturation in the spectra.

An even more impressive result is shown in Figure 6.38, which displays the relationship between both normalized and unnormalized PA intensities and the weight of protein. It is obvious that the tendency for saturation is ameliorated when the spectra are scaled using the 2243-$cm^{-1}$ PAN band as an internal standard. These two studies are also discussed in Chapter 7 on quantitative analysis in PA infrared spectroscopy.

Other research groups have also used PA infrared spectroscopy to study proteins. For example, Luo et al. (1994) at Iowa State University obtained spectra of concanavalin A, hemoglobin, lysozome, and trypsin, which have different distributions of secondary structures. These structures were elucidated by calculating second derivatives and then curve fitting the amide I band centered at about 1660 $cm^{-1}$. In this way, it was found that seven components are required to account for the observed spectrum of lysozome. Samples were prepared as thin layers by evaporating 1–2 μL of protein solutions on Teflon membranes; in cases where saturation of the PA spectra was a problem, the thickness of the deposited layer was reduced. The authors noted the high sensitivity of PA spectroscopy, as well as the speed and ease with which the data were acquired. In fact PA spectroscopy

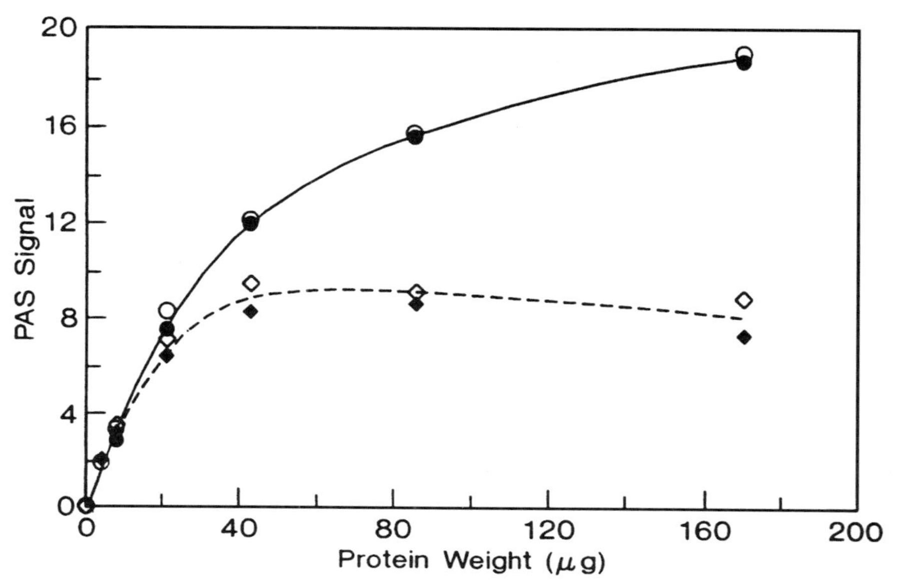


**Figure 6.38.** Effect of normalization on PA intensity of amide I (1660 $cm^{-1}$) band. Solid curve, data normalized using PAN intensity at 2243 $cm^{-1}$; dashed curve, data without normalization. Open and filled symbols refer to two different experiments. (Reprinted from Gordon, S. H. et al., Measurement of protein biomass by Fourier transform infrared-photoacoustic spectroscopy, *Biotech. Appl. Biochem.* **12**: 1–10, copyright © 1990, with permission from Elsevier Science.)

was concluded to be superior to both traditional infrared spectroscopy and circular dichroism (CD) spectroscopy—the latter being commonly used to study protein structure.

This group also used PA infrared spectroscopy to study phosphoamino acids and phosphoproteins (Graves and Luo, 1994). The PA technique was found to be well suited for these experiments because there was no need for crystalline materials or for large quantities of sample: Typical amounts analyzed were between 30 and 40 μg. Solid samples were prepared by spreading 10 μL of protein solution or phosphate ester on a 7-mm polyethylene membrane disk and evaporating to dryness. Serine phosphate was distinguished from tyrosine phosphate, the former giving rise to a band at 984 $cm^{-1}$ and the latter yielding a 974-$cm^{-1}$ peak. These spectral features were attributed to dianions; by contrast, samples prepared at different pH values exhibited peaks due to monoanions. The authors also studied the interaction between Al ions and phosphate groups in phosvitin, a phosphoseryl-containing protein, and used the PA spectra to determine the corresponding p$K_a$ value.

Lotta et al. (1990) demonstrated another application of PA infrared spectroscopy that is relevant in the present context. These authors investigated the interaction between the well-known electron acceptor 7,7,8,8-tetracyanoquinodimethane (TCNQ) and phospholipids, specifically diacylphosphatidylcholines and diacylphosphatidylglycerols. These interactions produce colored charge-transfer complexes, which of course can be studied by visible and ultraviolet absorption spectroscopies. PA spectroscopy was used by these researchers because it does not require any sample preparation that might perturb the complexes. The infrared bands of neutral TCNQ disappeared upon complexation and new bands characteristic of the anion appeared in the PA spectra. Comparison of the spectra of the neutral and complexed species showed that a stretching band due to phosphatidylglycerol ester carbonyl groups was modified by complex formation. PA infrared spectroscopy was treated as an established analytical technique in this work; in other words, instrumental questions were not considered in detail.

The reader has probably noticed that the number of studies dealing with the PA infrared spectra of biological and biochemical systems is still relatively small, even though the quality of the published results is very good. These observations prompt the assertion that this particular application of PA infrared spectroscopy possesses the potential for significant future growth.

## 6.12. MEDICAL APPLICATIONS

PA infrared spectroscopy has been utilized in a variety of medical applications and by a significant number of research teams during the last two decades. Even a cursory reading of the relevant literature quickly reveals the rather impressive diversity of this work. To illustrate, it can be mentioned that PA spectroscopy has been employed in the following applications pertaining to human health: (1) qualitative and quantitative analysis of drugs, (2) drug transport in membranes, (3) the analysis of calcified tissue (teeth) and fingernails, (4) diagnostics associated with adverse medical conditions (gallstones, cancerous tissue), (5) the identification of bacteria, and (6) the determination of total body water by analysis of blood serum. Examples of the use of PA spectroscopy in each of these situations are given in the following paragraphs.

The reader may already be aware that the spectroscopic characterization of drugs (pharmaceuticals) has recently become a fairly routine matter. In particular, near-infrared and FT-Raman spectroscopies (sometimes facilitated by the use of fiber optics) are now being used for qualitative and quantitative analysis of drugs in both research laboratories and production environments with good success. At the same time, there is an obvious need for the use of mid-infrared spectroscopy in these circumstances; one could argue that this role is best filled by PA infrared spectroscopy since the virtual elimination of sample preparation afforded by the near-infrared and Raman techniques is retained. The advantages arising from the use of neat samples in PA spectroscopy are obvious in several of the publications discussed below.

An early example of the use of PA spectroscopy for quantitative analysis of a drug exists in the work of Castleden et al. (1982). These authors recorded dispersive near-infrared (1.3–2.5 μm, approximately 7700–4000 $cm^{-1}$) PA spectra of propranolol, a $\beta$-andrenergic receptor blocker. This investigation was motivated by the need to quantify the analyte without the usual time-consuming physical separation of excipients. Spectra were obtained for both intact and manually ground tablets in this study. The authors found that the aromatic C—H band at 2.21 μm (4525 $cm^{-1}$) arising from propranolol could be used for quantitation: This was accomplished by monitoring its intensity in spectra of samples containing known amounts of the drug, or by ratioing the band against the 1.65-μm (6060-$cm^{-1}$) band of lactose, the main excipient. Thus the capability of near-infrared PA spectroscopy for quantitative analysis of drugs was clearly demonstrated in this investigation. This study is mentioned again in Chapter 7.

After the pioneering study of Castleden et al. (1982), the attention of workers in this field shifted primarily to the mid-infrared. In a study that described a variety of applications of PA infrared spectroscopy, Krishnan (1981) presented spectra of a powdered mixture of acetylsalicylic acid and phenacetin, and of the undiluted acid. Subtraction of the second spectrum from the first yielded a spectrum that was practically identical to that of pure phenacetin. This result nicely demonstrates the additivity of PA spectra of drugs.

Several years later, the suitability of PA mid-infrared spectroscopy for characterization of drugs was confirmed by Belton et al. (1988b). In the context of a brief review of PA spectroscopy, these authors presented infrared spectra of two drug polymorphs; like the results of Krishnan (1981), these spectra are notable for both their complexity and their good quality. Indeed, Belton et al. (1988b) observed that PA spectroscopy could be superior to the diffuse reflectance technique in this particular application.

Huvenne and Lacroix (1988) also noted that PA infrared spectroscopy could be utilized to characterize drugs without the need for sample preparation. However, these authors observed that the relative intensities in the PA spectra of these samples were somewhat different from those in absorbance spectra and concluded that PA spectra could not be used for direct identification of drugs unless they were appropriately corrected. They obtained PA infrared spectra of flunitrazepam, dipyriamole, and lactose $\alpha$ monohydrate and empirically determined the proportionality constants that relate the PA signal intensity and the product $\beta\mu$ for a thermally thick sample.

The determination of drug content is even more challenging for semisolid (ointment and cream) formulations. Traditionally, this problem has been approached by first extracting the drug from the semisolid dispersant and then using standard analytical methods such as HPLC for quantitation. In an attempt to avoid this separation step, Neubert et al. (1997) utilized PA spectroscopy for quantitative analysis of dithranol and brivudin, prepared as mixtures with Vaseline. Both rapid- and step-scan PA spectra were obtained in this investigation; for equal measurement times (on the order of 20 min), the signal-to-noise ratios in the step-scan spectra were about an order of magnitude better than those in the corresponding rapid-scan spectra. A series of bands between 800 and 1800 $cm^{-1}$ in the step-scan magnitude spectra intensified as the weight percent of the drugs increased from 0.5 to 10%. With regard to quantitation, the integrated intensities of the bands between 1545 and 1770 $cm^{-1}$ for brivudin, or between 1560 and 1670 $cm^{-1}$ for dithranol, were found to vary linearly with the concentrations of these drugs as determined by capillary zone electrophoresis or HPLC. In fact, the instrumental errors in the PA spectra were significantly smaller than the variation observed in the results for the standard analytical methods. These encouraging findings suggest that PA infrared spectroscopy could eventually become the method of choice for analyzing drugs in semisolid formulations.

A subsequent investigation by this group (Schendzielorz et al., 1999) employed PA spectroscopy in a study of drug penetration from semisolid formulations into dodecanol-collodion membranes. The model system consisted of a 10% suspension of clotrimazole in Vaseline. The objective of this work was to elucidate the mechanism of drug transport through human skin. Membrane thicknesses of 31 and 17 μm were used; the bands due to the drug appeared in the PA spectra of the membrane/clotrimazole/Vaseline system as penetration progressed, on a time scale of approximately 4 h. The results in the high-frequency region (Fig. 6.39) show that the hydrogen bond system of the membrane is affected by the penetration of the drug. The sampling depth in the step-scan PA magnitude spectra was estimated as 15 μm. Diffusion coefficients were calculated from the PA data and compared with those obtained with both a multilayer membrane system and a modified liberation model. In fact, the accuracy of the results from PA infrared spectroscopy was found to be much better than that attainable by the other two methods. These findings, like those in the previous investigation by these authors, nicely illustrate the suitability of PA infrared spectroscopy for the study of drugs in semisolid dispersants.

The above discussion clearly shows that PA spectroscopy is well suited for the quantitative analysis of drugs in several different media. The second group of medical applications, alluded to in the list at the start of this section, is based on the characterization of various human tissues. As is

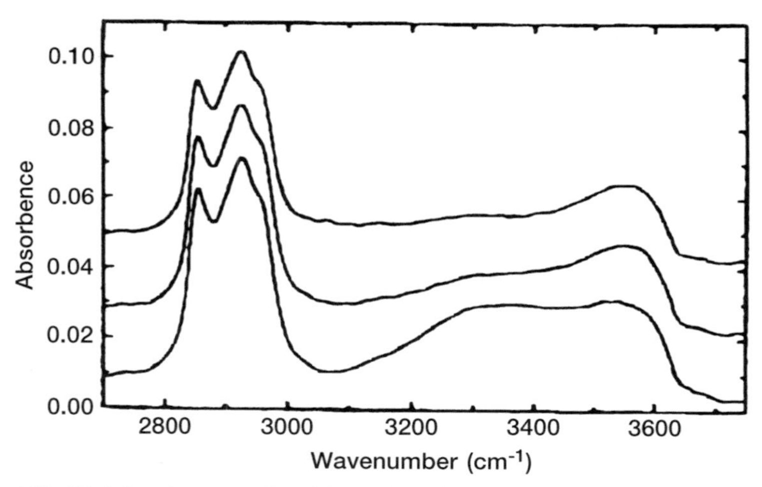


**Figure 6.39.** PA infrared spectra of model membrane/clotrimazole/vaseline system recorded at various times during penetration experiments. Bottom curve, initial spectrum; middle curve, elapsed time 66 min; top curve, elapsed time 242 min. (Reproduced from Schendzielorz, A. et al., *Pharm. Res.* **16**: 42–45, by permission of Kluwer Academic/Plenum Publishers; copyright © 1999.)

illustrated below, both the minimal sample preparation and the capability for depth profiling in PA spectroscopy have been an integral part of these studies.

An important early application of PA infrared spectroscopy to a sample derived from human tissue was based on a multiple-wavelength ($CO_2$ laser) experiment. Kanstad et al. (1981) obtained PA spectra of skin lipids in the 900–1100 $cm^{-1}$ region, using a series of approximately 40 $CO_2$ laser lines for illumination. This work was carried out at about the same time as a series of other studies by these authors, already described in Chapter 2. Samples were deposited as thin layers on 0.04-mm Al foils that were epoxied onto piezoelectric ceramic strain gage wafers. Cholesterol served as a model substance; the spectrum of a 14-μg sample contained prominent bands at 957 and 1055 $cm^{-1}$, although the available $CO_2$ laser lines did not permit the complete definition of either peak. Skin surface lipids obtained from a patient using a chloroform/methanol mixture yielded weaker bands, with the quantity of sample being estimated at about 5 μg. The sensitivity of PA detection is even more impressive when the results are compared with those from a transmission experiment on a KBr disk: Kanstad et al. observed that the amount of sample required for the transmission measurement was 100 times greater than that for the PA spectrum.

Sowa and Mantsch (1993, 1994) demonstrated the use of PA spectroscopy for the analysis of calcified tissue—specifically, an extracted but intact tooth. These authors carried out both rapid-scan and step-scan depth profiling experiments. To ameliorate the effects of saturation, they calculated linearized modulus (power) spectra; phase-modulated step-scan spectra also exhibited less saturation than did the corresponding rapid-scan PA spectra. In the step-scan spectra, depth profiling was accomplished by changing the modulation frequency or alternatively by varying the phase angle while keeping the phase modulation frequency fixed at 400 Hz. However, as pointed out by the authors, this latter experiment can be carried out more efficiently using phase synthesis. In this approach, spectra corresponding to various phase angles $\theta$ are calculated from the relationship

$$I(\theta) = I(0^\circ)\cos(\theta) + I(90^\circ)\sin(\theta) \tag{6.2}$$

where $I$ denotes intensity and the angles of $0^\circ$ and $90^\circ$ refer to the in-phase and quadrature spectra, respectively. This numerical method is discussed in Chapter 4. The success of the phase synthesis method in this experiment implies that the PA signal displays linear phase behavior.

The depth profiling studies of Sowa and Mantsch (1993, 1994) showed that the protein contribution increases relative to the mineral component as the sampling depth increases, a conclusion that is consistent with the known morphology of maturing tooth enamel. Moreover, crystallinity was found to be diminished for the subsurface apatitic structure. Even though the structure of the human tooth is relatively complex, the inorganic and protein distribution observed in this investigation is reminiscent of that in a two-component system.

This research group subsequently turned its attention to PA infrared spectra of finger nails (Sowa et al., 1995). Both mid- and near-infrared PA spectra of viable and clipped nails were obtained in this investigation. ATR, transmission, and diffuse reflectance spectra were also recorded and compared with the PA data. The objective of this work was to characterize the three layers that make up nails: From top to bottom, these are referred to as dorsal, intermediate, and ventral, respectively. Depth profiling was effected in step-scan phase modulation experiments. Modulus spectra were calculated at phase angles ranging from $90^\circ$ (near surface) to $180^\circ$ from in-phase and quadrature spectra, as described above. These spectra showed a decrease in the ratio of methylene to methyl groups as the sampling depth increased, consistent with the expectation that the lipid content in the intermediate layer should be lower than that in the dorsal layer. Saturation tended to affect the PA spectra, with the consequence that depth profiling was limited to the dorsal layer and the upper portion of the intermediate layer of the nails.

Near-infrared PA spectra, like the corresponding mid-infrared data, suggested a significant contribution from the intermediate layer in addition to that from the dorsal layer. On the other hand, the ATR spectra were concluded to arise from the near surface (approximately the first 10 μm) of the dorsal layer only. The near-infrared spectra indicated that the ordering of the protein structure in the intermediate layer is lower than that in the dorsal layer.

The infrared analysis of human hair by traditional techniques is problematic because of the ever-present need for sample preparation. For example, transmission measurements would ordinarily require dispersion of the hair sample in an infrared-transparent solid such as KBr. On the other hand, it is almost surprisingly straightforward to obtain PA infrared spectra of hair: Only a few individual hairs, cut to lengths on the order of 2–3 mm, are required for this experiment. To illustrate, Figure 6.40 shows a PA spectrum of approximately 10 hairs that was recently measured in the author's laboratory. About a dozen bands are clearly observable, the most prominent among these being the amide I and II peaks at about 1660 and 1540 $cm^{-1}$,

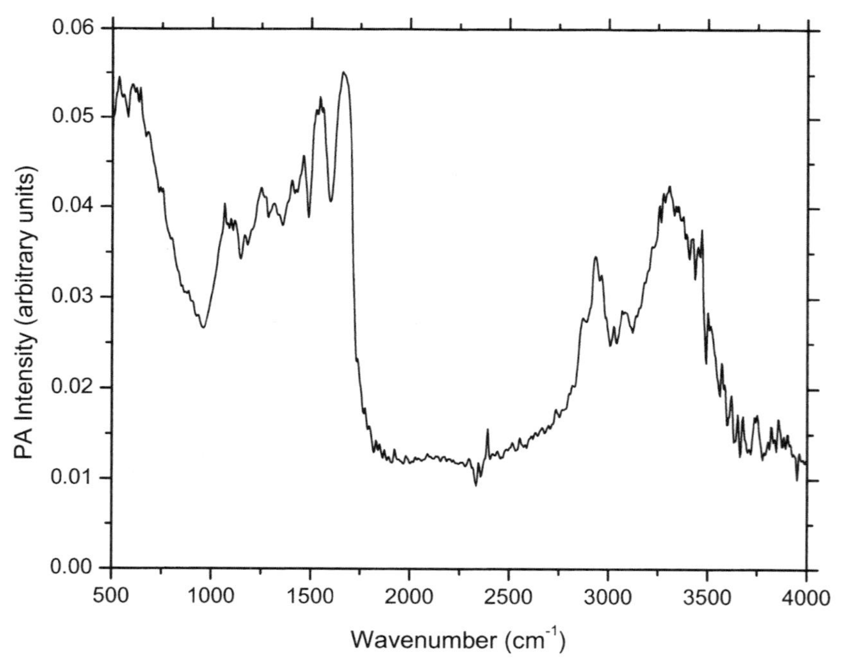


**Figure 6.40.** PA infrared spectrum of approximately 10 hairs (a mixture of brown and white), cut to length of about 3 mm. Sample was provided by the author.

respectively. PA infrared spectra of hair have been compared to spectra of textiles in the literature and are therefore also mentioned in the section on textiles. The PA spectrum of hair has also been obtained with a single-fiber accessory, and is discussed in Chapter 8.

The suitability of PA infrared spectroscopy for the characterization of other human tissues has recently been demonstrated by at least two research groups. For example, Wentrup-Byrne et al. (1997) described the use of PA spectroscopy for the categorization of human gallstones. PA spectroscopy was chosen for this work because it avoids the traditional crushing, extraction, and wet chemical techniques. It also examines a relatively large area (several square millimeters) of the gallstone surface, which is considered to be an advantage because it reduces the likelihood that any inhomogeneities might be overlooked.

The four major gallstone types can be described as cholesterol, mixed cholesterol and bile pigments, black, and brown stones. In this work, partial least-squares (PLS) was used to model cholesterol concentration using the results from PA spectroscopy. The model predictions yielded reasonable agreement between predicted and measured (by traditional, nonspectroscopic techniques) cholesterol concentration. Discrepancies were attributed to inhomogeneities in the stones. Grinding the stones would tend to distribute these inhomogeneities more evenly, but at the expense of structural integrity.

Recently, Schüle et al. (1999) utilized PA infrared spectroscopy and photothermal response to 10.6-μm $CO_2$ laser radiation to characterize skin, liver, and muscle tissue. This research addressed the problem that arises from the poor predictability of the layer thickness of tissue in cancer diagnostics. The measurement of thermal diffusion properties provides a possible means for the determination of this thickness.

Tissues can be distinguished by their photothermal response at a single wavelength; phase contrast experiments using the $CO_2$ laser were employed for this purpose. When samples were dried and embedded in polymethylmethacrylate, the phase response of the superficial temperature as a function of modulation frequency was about the same for the different tissues because the varying water contents that normally give rise to these contrasts in untreated tissues had been eliminated. On the other hand, untreated tissues were clearly distinguishable by this technique, suggesting that differentiation between healthy and malignant tissues might eventually be possible in future work.

Step-scan PA amplitude and phase spectra of the three tissues both showed a significant number of features between 1000 and 1800 $cm^{-1}$, with more detail appearing in the phase spectra. Indeed, the samples could be positively identified by their PA infrared spectra. These spectra were then simulated by the finite-difference method, using data from other absorption measurements and assumed thermal properties. Both the amplitude and phase spectra agreed well with the absorption spectrum, which was plotted as the wavelength (frequency) dependence of the absorption coefficient.

Traditional methods for the identification of bacteria can require up to 2 days to complete. This fact prompted Ardeleanu et al. (1992) to investigate the use of PA infrared spectroscopy for bacterial identification, in an attempt to develop a more rapid procedure. Four different bacteria were studied: *Staphylococcus aureus*, *Steptococcus pyogenes*, *Escherichia coli*, and *Proteus vulgaris*. In fact, the PA spectra of these bacteria were of very good quality and resembled published infrared spectra of bacteria obtained in transmission experiments. The information content of the PA infrared spectra was high, as is evidenced by a series of features. These included bands in the 900–1200 $cm^{-1}$ region, due to carbohydrates; between 1200 and 1500 $cm^{-1}$ bands were ascribed to proteins, fatty acids, and phosphates; the well-known amide I and II bands appeared at 1656 and 1547 $cm^{-1}$, respectively; ester C=O stretching was observed in the 1735–1745 $cm^{-1}$ region; and finally, several C—H stretching bands occurred between 2800 and 3000 $cm^{-1}$. Although the PA spectra displayed in this study were certainly viable, they did not differ greatly. Nevertheless, the authors concluded that the technique might eventually be used to identify different types of bacteria.

The optothermal window (OW) technique, discussed in Chapter 3, was used by Annyas et al. (1999) to measure total body water (TBW), an indicator of the fat content in human patients. In this experiment, $D_2O$ was used as a TBW tracer; blood serum samples from human subjects, containing between 150 and 400 ppm $D_2O$, were analyzed. Measurements were made at 4 μm since a large difference between the absorption coefficients of $D_2O$ and $H_2O$ exists at this wavelength. In the OW method, absorption by the sample generates heat in a supporting disk. The incident radiation is chopped, leading to a periodic expansion and contraction of the disk that can be detected with a piezoelectric transducer. A detection limit of 30 ppm for $D_2O$, which is comparable to that in infrared transmission experiments, was obtained with this technique. The authors also mentioned a series of potential improvements that were expected to improve the sensitivity of the OW method by as much as an order of magnitude.

The results described in this section show that PA infrared spectroscopy is very well suited for a wide range of medical applications. These have already been shown to include analysis of drugs and tissue samples; medical researchers may indeed suggest additional problems that are amenable to the PA techniques described in this and other chapters of this book. The implementation of these suggestions will further enhance the role of PA spectroscopy in medical applications, where several additional photothermal methods are already in active use.

## 6.13. CARBONYL COMPOUNDS

PA infrared spectroscopy has been successfully applied to the characterization of carbonyl compounds by at least two groups of researchers. In the work reviewed here, the lack of sample preparation was cited as the principal justification for the selection of the technique. Moreover, the need for nondestructive analysis of small samples also favored the choice of PA spectroscopy. Both mid- and near-infrared PA spectroscopy of carbonyl compounds are discussed in this section.

Natale and Lewis (1982) reported dispersive near-infrared PA spectra of $Mo(CO)_6$, $Ir_4(CO)_{12}$, $[RhCl(CO)_2]_2$ and $(C_6H_6)Cr(CO)_3$. The spectral region covered in this work extended from 1.0 to 2.6 μm (10,000–3850 $cm^{-1}$). The metal carbonyls each displayed two or three bands near the low-frequency end of the near-infrared region: These peaks were attributed to overtones and combinations of the fundamental carbonyl stretching vibrations that occur near 2000 $cm^{-1}$. The authors found that it was necessary to utilize both Raman and infrared transmission spectra in order to fully characterize the fundamental vibration frequencies, as neither spectrum exhibited all of the bands in this region. Thus the PA near-infrared spectrum of each of these metal carbonyls effectively combines information from two experimental techniques in a single measurement.

The assignments of the near-infrared bands of the carbonyl compounds studied by Natale and Lewis (1982) might well be considered as tentative in view of the band overlap and noise that occur in the PA spectra of the first two compounds. In contrast, seven well-defined near-infrared PA bands were observed for $(C_6H_6)Cr(CO)_3$; most of these features were readily assigned to overtones and combinations of the fundamental vibrations of the benzene moiety.

Both mid- and near-infrared PA spectra of a wide series of organometallic complexes were discussed in a series of studies published between 1986 and 1993 by I. S. Butler and his research team at McGill University. In the initial PA investigation by this group (Xu et al., 1986) both spectral regions were examined for group VIB chalcocarbonyls, $M(CO)_6$ ($M = Cr$, Mo, W), and $Cr(CO)_5CS$. To their surprise, the authors observed gas-phase PA spectra of these metal chalcocarbonyl complexes—an observation that can be put down to sample volatility and the very high sensitivity of the PA effect in gases. Indeed, the PA spectra obtained in this work resembled published infrared spectra of the gaseous complexes and moreover differed significantly from absorption spectra of the solid complexes in KBr disks. Thus the PA experiment yielded bands due to the vapor trapped in the PA cell directly above the solid complexes; as mentioned in several contexts in this book, this is not an uncommon phenomenon in PA infrared spectroscopy and is readily attributed to heating of the sample by the infrared beam. Xu et al. (1986) recognized this situation and turned it to their advantage, presenting a detailed assignment of the PA infrared bands of the gas-phase chalcocarbonyls. Several years later, a similar PA investigation of $CH_3Mn(CO)_5$ and $CH_3Re(CO)_5$ in this laboratory (Butler et al., 1992) also reported the observation of bands from vapor-phase complexes.

Butler et al. (1987a) described PA infrared spectra of 10 different solid organometallic complexes of chromium, manganese, rhenium, and iron. Near- and mid-infrared spectra were obtained in this work, and detailed band assignments were given for both frequency ranges. In the relatively narrow 3800–4600 $cm^{-1}$ near-infrared region alone, about 15 bands were observed for $[CpMo(CO)_3]_2$, $[CpFe(CO)_2]_2$, and $Fe_2(CO)_9$ ($Cp = \eta^5$-$C_5H_5$). For $CpM(CO)_2(CX)$ (M = Mn, Re; X = O, S), approximately 20 near-infrared bands were identified. As illustrated in Figure 6.41, these near-infrared spectra generally display very good signal-to-noise ratios. A number of mid-infrared PA bands were also observed in this investigation, and assigned to the fundamental vibrations of these complexes.

PA infrared spectroscopy was combined with micro-Raman and traditional infrared sampling methods in an investigation of $Ru_3(CO)_{12}$ by Butler et al. (1987b). Both $^{12}C$ and $^{13}C$ isotopomers were included in this study.

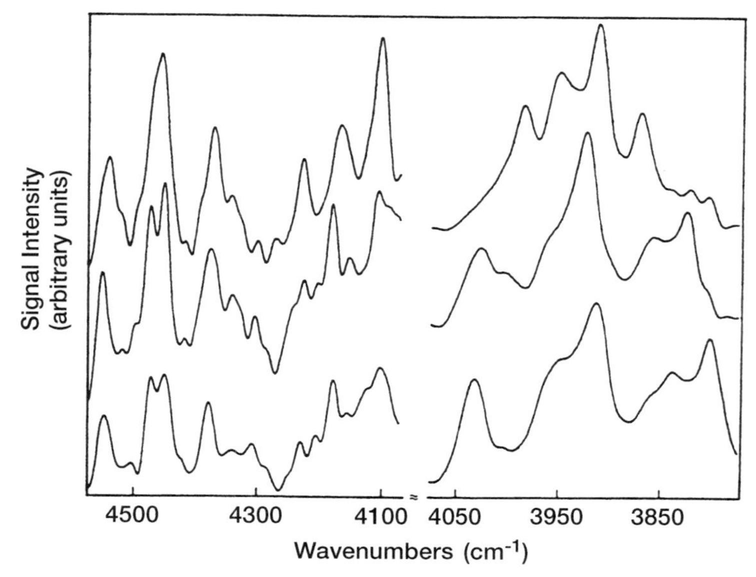


**Figure 6.41.** PA infrared spectra of $CpMn(CO)_2(CS)$, top curve; $CpMn(CO)_3$, middle curve; and $CpRe(CO)_3$, bottom curve. (Reproduced from Butler, I. S. et al., *Appl. Spectrosc.* **41**: 149–153, by permission of the Society for Applied Spectroscopy; copyright © 1987.)

PA and solution infrared spectroscopies were used to examine the 3800–4200 $cm^{-1}$ region, whereas micro-Raman and other infrared spectra were employed to characterize the lower frequencies. The PA spectra in this work exhibited poorer signal-to-noise ratios than those in the previous works by these authors. Despite this fact, a number of weak peaks that appeared only in the PA spectra were assumed to be genuine and attributed to factor group splitting since they were observed only in the solid state. The authenticities of some of these extra peaks appear to require confirmation by PA spectroscopy or another infrared technique.

In a subsequent investigation that also combined PA infrared spectroscopy with other techniques, Butler et al. (1991) obtained near-, mid-, and far-infrared spectra of solid organoiron(II) carbonyl complexes $[CpFe(CO)_2R]BF_4$ ($R = C_2H_4$, $CH_2C(CH_3)_2$, 3-methylthiophene, and $CpFe(CO)_2I$. PA spectroscopy was found to be the most sensitive technique in the 4000–4600 $cm^{-1}$ near-infrared region, with as many as 10 bands being identified in the data. In the mid-infrared, ATR and transmission spectra exhibited similar relative band intensities, except that the low-frequency bands in the ATR spectrum were stronger due to the greater depths of penetration that are characteristic of the technique at longer wavelengths. Relative intensities were somewhat different in the PA infrared spectra, where the stronger bands are broadened and appear to be partly saturated. While this saturation has a deleterious effect on the profiles of some mid-infrared bands, it also results in the beneficial intensification of the weaker near-infrared bands and thereby facilitate their analysis.

The final two studies in this series continued the approach described above, specifically the comparison of PA infrared spectra with data obtained by other sampling methods, and the use of PA near-infrared spectra to characterize the region near 4000 $cm^{-1}$. First, Li and Butler (1992) measured mid-infrared PA spectra of manganese(I)carbonyl halides, $Mn(CO)_5X$, and $[Mn(CO)_4X]_2$ ($X = Cl$, Br, I), complementing these results with infrared spectra obtained using both KBr pellets and solutions of these compounds. Even though the fundamental C=O stretching bands near 2000 $cm^{-1}$ in the PA spectra of these carbonyl halides may be partly saturated, the concomitant intensification of the near-infrared bands is advantageous because it reveals features that are completely absent from spectra recorded for the KBr pellets. The occurrence of these near-infrared bands in the PA spectra—although fortuitous—allowed the authors to calculate the anharmonicities of the carbonyl stretching vibrations. Similarly, the near-infrared PA spectra of mixed carbonyl-*t*-butylisocyanide complexes, $M(CO)_{6-n}(CN^tBu)_n$ ($M = Cr$, Mo, W; $n = 1–3$), were the first ever reported (Li and Butler, 1993) for this type of mixed-ligand complex. While the near-infrared features were easily detected in the PA spectra of the latter complexes, the corresponding mid-infrared PA spectra showed little indication of saturation.

## 6.14. TEXTILES

The analysis of textiles by infrared spectroscopy presents yet another situation in which traditional sample preparation techniques tend to be problematic. PA infrared spectroscopy nicely obviates this difficulty; moreover, because textile fibers are commonly coated to improve both their performance and appearance, the depth profiling capability of PA spectroscopy is also relevant to this application. Both of these facts have been recognized by at least half a dozen groups during the last two decades. The principal results of this research are summarized in the following discussion.

PA infrared spectroscopy was first used to study textiles about 20 years ago, at the time when the technique was still evolving. As a consequence, some of the initial reports on PA spectra of textiles were also concerned with issues regarding instrumentation that have now been resolved. For example, Teramae and Tanaka (1981) first described a home-built PA cell and then compared PA infrared spectra of carbon black with the more familiar "empty instrument" spectrum obtained with a triglycine sulfate (TGS) detector; the Helmholtz resonance of the PA cell was also characterized. To demonstrate the capabilities of their apparatus, these authors reported PA spectra of cotton and nylon cloths; cellulose bands at about 1100, 2900, and 3300 $cm^{-1}$ were observed in the former case, whereas nylon displayed characteristic polyamide bands at 1550, 1650, 2900, and 3350 $cm^{-1}$. These results were discussed in more detail in a companion work (Teramae et al., 1982) on PA infrared spectra of polymers. Rubbers and resins were also discussed in the second publication.

Exploratory PA studies of textiles were also carried out at shorter wavelengths. Davidson and King (1983) reported near-infrared (1.2–2.4 μm, or about 8300–4200 $cm^{-1}$) PA spectra of wool and polyester-fiber fabrics, and of wool–polyester blends. The PA spectra of wool and polyester were different, with each containing about a dozen bands. As might be expected, several features due to OH, NH, and CH groups were observed in these spectra. The PA near-infrared spectrum of cotton fabric was also measured in this investigation and found to resemble the spectrum of wool. The often-mentioned insensitivity of PA spectra to sample morphology was noticed in this work: Various types of wool fabric (challis, serge, flannel, and knitted) yielded very similar PA spectra. Moreover, the application of a fluorescent whitening agent and subsequent prolonged irradiation did not cause changes to the near-infrared spectra, even though this treatment caused the samples to turn yellow.

This near-infrared investigation was followed by a related PA mid-infrared study of wool, polyester, and nylon (Davidson and Fraser, 1984). As shown in Figure 6.42, the PA infrared spectra of wool and nylon yarns are distinctly different, with the nylon spectrum containing the larger number of bands. The authors confirmed the additivity of the PA spectra of textiles by examining results for blends of different yarns. For example, the spectrum of polyester was successfully retrieved by subtracting the spectrum of a wool fabric from that of a wool/polyester mixture. Similarly, a suitably weighted coaddition of wool and nylon PA spectra closely resembled the spectrum of a known mixture. Poorer agreement was obtained when this experiment was extended to ternary mixtures, primarily because of band overlap. Nevertheless, this research clearly demonstrated the capabilities of PA FTIR spectroscopy for both qualitative and quantitative analysis of textile mixtures.

The oxidation of wool was studied by McKenna et al. (1985), who employed PA infrared spectroscopy primarily because of the ease of sample preparation. The PA infrared spectrum of untreated wool displayed a number of amide bands (labeled A, B, I, II, and III; see Fig. 6.43), in addition to several well-known features near 2900 $cm^{-1}$ due to C—H stretching. Oxidation with dichloroisocyanuric acid resulted in the appearance of an additional band at 1022 $cm^{-1}$, which the authors attributed to a sulfoxide or sulfinic acid (RSOOH). By contrast, oxidation of wool in a corona discharge did not cause any observable changes in its infrared spectrum, which suggests a low concentration of oxidation products.

Keratins are fibrous proteins that occur in hair, feathers, hooves, and horns. PA infrared spectra of the surface (cuticle) and interior (cortex) of intact keratin fibers, specifically three types of wool and human hair, were investigated by Jurdana et al. (1994). Diffuse reflectance and ATR infrared spectroscopies were also utilized by these authors because of the physical nature of the samples and the particular need for characterization of their surfaces. It should also be recalled that the PA infrared spectrum of human hair is discussed in the section on medical applications.

All of the bands observed by McKenna et al. (1985) in the PA spectrum of wool were also detected in the spectra of wool and human hair by Jurdana et al. (1994); the amide I and II bands were studied in detail in an attempt to distinguish between the surface and interior of the fibers. Thermal diffusion lengths were calculated using a thermal conductivity of $9.05 \times 10^{-5}$ cal $°C^{-1}$ $s^{-1}$ $cm^{-1}$ and already known values for the density and heat capacity. The value of $\mu_s$ ranged from 0.8 to 22.9 μm, depending on the

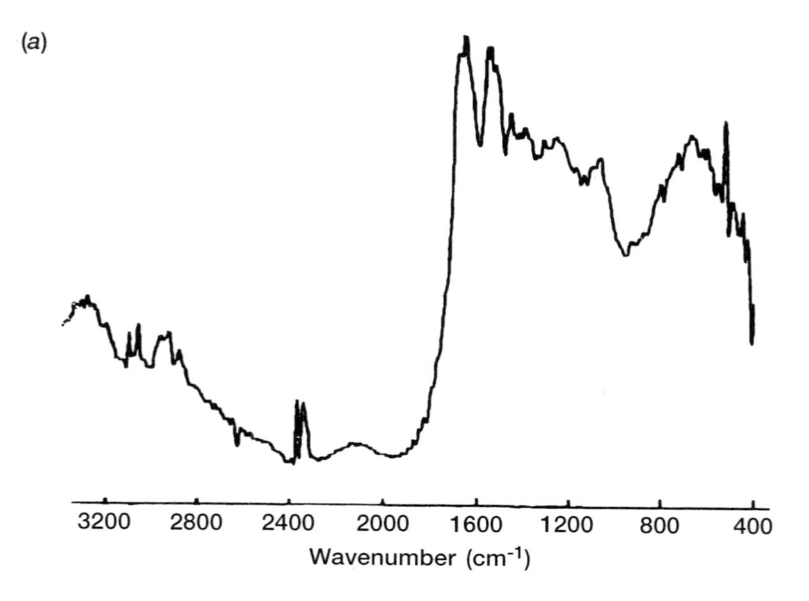


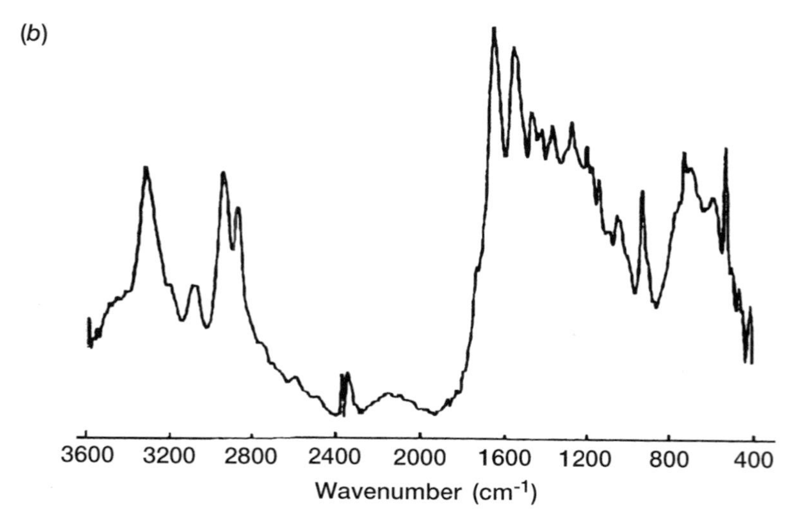


**Figure 6.42.** PA infrared spectra of (*a*) wool and (*b*) nylon yarns. (Davidson, R. S. and Fraser, G. V. Copyright © 1984. *J. Soc. Dyers Colour.* **100**: 167–170, reproduced by permission of the Society of Dyers and Colourists.)

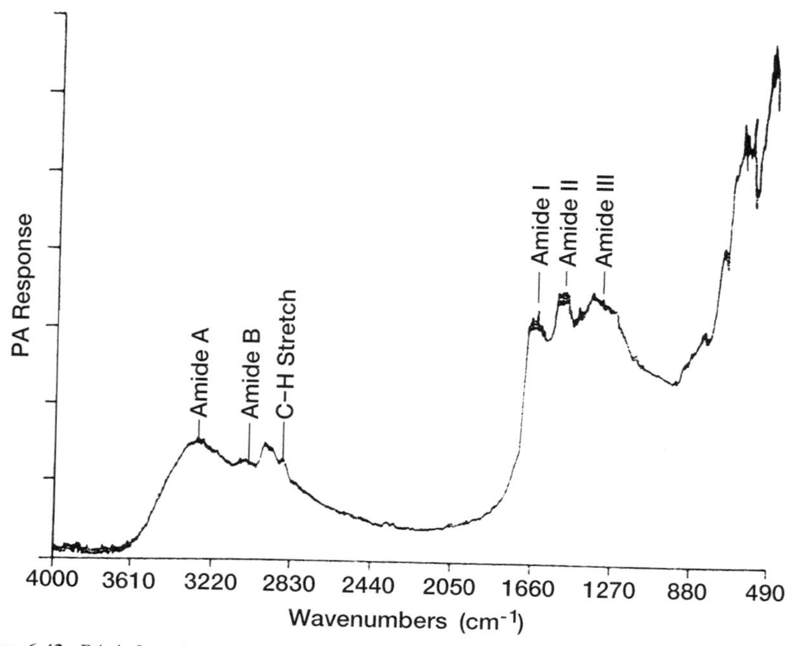


**Figure 6.43.** PA infrared spectrum of untreated wool. (Reprinted from McKenna, W. P. et al., *Spectrosc. Lett.* **18**: 115–122, by permission of Marcel Dekker, Inc.: copyright © 1985.)

rapid-scan mirror velocities that were utilized and the wavenumber region under consideration. This made it possible to distinguish the cuticle sheath of the wool from the cortex since the former has a thickness between 1 and 10 μm.

This investigation also showed that the frequency separation between the amide I and II bands varied with the sampling depth. This effect is due to differences between the chemical compositions of the cuticle and the cortex and is analogous to previously published results from ATR and diffuse reflectance studies. Some of the PA spectra measured at the lower velocities were partly saturated, making it impossible to obtain reliable intensities for the amide I and II bands. The diffuse reflectance spectra displayed relatively broad bands that were thought to be attributable to the composition of the fiber interiors. By contrast, the bands in the ATR spectra were narrower and due to the cuticle.

These researchers extended their studies in a subsequent publication (Jurdana et al., 1995). Both wool and hair were again examined, with an emphasis being placed on a comparison of the rapid-scan and step-scan PA techniques. The origins of the amide I and II bands were determined in step-scan phase modulation experiments, by changing the phase angle of the lock-in amplifier in small increments. Plots of the amide I band position and the amide I/II intensity ratio vs. phase lag, which could ideally be used to locate the interface between cuticle and cortex, showed discontinuities at phase angles between 126° and 153°. Subtraction of a phase error of 45° corrected these angles to 81° and 108° and yielded $\mu_s$ values between 5.2 and 9 μm. The transition from cuticle to cortex was thus determined in a rather elegant manner in this work. In general, saturation effects were found to be more problematic in the rapid-scan PA spectra of wool and hair than in the step-scan spectra.

Church and Evans (1995) used both PA and ATR infrared spectroscopies to study the surface treatment of wool with fluorocarbon polymer finishes. These authors observed that the PA spectrum of a thin free-standing film of the polymer closely resembles the absorption spectrum obtained in a transmission experiment; the ATR spectrum, after correction for the standard wavelength dependence of the sampling depth, also displayed similar relative intensities. The quantitative analysis of fluorocarbon polymers on wool was one of the objectives of this investigation. Although the PA spectra of the treated wool were influenced by saturation, it was still found that the integrated intensities of the C—F stretching bands were proportional to the known concentration of the polymer on the wool. While the detection limit in ATR was lower than that in PA spectroscopy, the ATR calibration deviated from linearity at higher concentrations. In this sense, the PA results could be concluded to be as good or better than those obtained by ATR.

The comparison of PA and ATR infrared spectroscopies with respect to the characterization of surface-treated wool was continued by Carter et al. (1996). Wool samples that had been chlorinated, or chlorinated and neutralized with bisulfite for shrinkproofing, were studied in this second investigation. Wool treated with a fluoropolymer was also examined in this study. Rapid-scan PA spectra obtained at different mirror velocities confirmed that chlorination led to surface oxidation, specifically the conversion of cystine to cysteic acid. PA spectra recorded at higher velocities qualitatively resembled the ATR spectrum, although ATR was found to yield better signal-to-noise ratios. Depth profiling experiments on wool samples treated with the fluoropolymer showed that it collected on the surface and did not penetrate into the fiber.

An extensive series of studies on the PA infrared spectroscopy of fabrics, fibers, and related polymeric materials was published between 1985 and 1991 by Yang, Fateley, and a number of their collaborators. The first stage of this research was carried out at Kansas State University, while the latter part was completed at Marshall University. Some of the major findings of this work are discussed in the following paragraphs.

The intial publications in this series emphasized depth profiling and surface sensitivity in rapid-scan PA infrared spectroscopy. A clear example of both effects with regard to sized (stiffened) cotton fibers was given by Yang et al. (1985); the 1735-$cm^{-1}$ band due to the polyurethane sizing agent was significantly stronger in the PA spectrum than in the transmission spectrum, the latter having been obtained for a ground sample. Similarly, the intensity of this band was diminished in the PA spectrum of a powdered sample. Grinding obviously mixes the surface and the bulk of the fibers, thereby decreasing the relative intensities of the PA bands due to the sizing agent on the surface. Depth profiling was also clearly demonstrated by measuring PA spectra of an intact sample at different mirror velocities: The characteristic 1735-$cm^{-1}$ band became proportionately stronger as the sampling depth was reduced, which is consistent with the presence of polyurethane on the fiber surface and the diminished contribution of the bulk of the fiber to the spectrum.

Both surface sensitivity and depth profiling were again observed for cotton fabrics treated by means of conventional padding and foam finishing techniques, where the distribution of the finishing agents is correlated with wrinkle recovery properties (Yang and Fateley, 1987; Yang et al., 1989). The wavenumber dependence of the sampling depth in rapid-scan PA spectroscopy was also quite noticeable in these experiments. In a model system comprising a glass fiber coated with polyvinyl acetate, the C—H stretching band was absent in PA spectra measured at higher velocities (corresponding to reduced sampling depths), while the 1735-$cm^{-1}$ band persisted; this was due to the greater thermal diffusion length at the lower infrared frequency and the penetration of the coating into the fiber (Yang et al., 1987).

The concentration of polymeric sizing agents near the surface of cotton yarns was further examined by Yang and Bresee (1987). PA infrared spectroscopy showed that desizing with a boiling NaOH solution removed the sizing agent from the surface but tended to leave it in the bulk. Although the PA spectra of both cotton yarn sized with polyurethane and unsized yarn appeared to be nearly the same, subtraction dramatically recovered the spectrum of the sizing agent from that of the sized yarn (Fig. 6.44). It is clear that the sizing agent was present mainly on the surface of the yarn. On the other hand, X-ray photoelectron spectroscopy (XPS) revealed that a copolymer finish on polyethylene terephthalate fibers was inhomogeneous (Yang et al., 1990). Scanning electron microscope (SEM) images were consistent with this conclusion. The apparent discrepancy between these later results and those from PA infrared spectroscopy probably arises from the differences in spatial resolution among the various techniques.

The oxidation of cotton cellulose by ultraviolet radiation was studied by Yang and Freeman (1991). After just 4 h of irradiation at 254 nm, a band

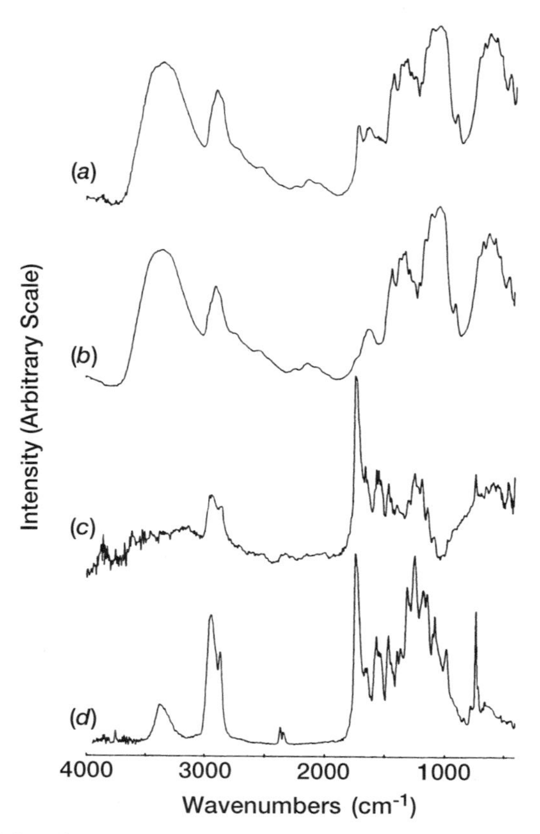


**Figure 6.44.** PA infrared spectra of (*a*) cotton yarn sized with polyurethane; (*b*) unsized cotton yarn; (*c*) difference between spectra in (*a*) and (*b*); (*d*) polyurethane. (Reproduced from Yang, C. Q. and Bresee, R. R., *J. Coated Fabrics* **17**: 110–128, by permission of Sage Publications Ltd; copyright © 1987.)

due to carboxylic acids appeared at 1735 $cm^{-1}$ in the PA infrared spectrum. This band intensified during prolonged (118-h) exposure to UV light. Treatment with NaOH converted the acid to a carboxylate salt, causing the 1735-$cm^{-1}$ peak to be replaced by another feature at 1618 $cm^{-1}$. Finally, acidification with HCl caused the reappearance of the carboxylic acid.

The final work in this series (Yang, 1991) presented a thorough comparison of PA, transmission, and diffuse reflectance spectra for fabrics, fibers, and films. The PA spectra of fabrics displayed the surface sensitivity that has already been described in detail. Diffuse reflectance spectra were obtained for a treated cotton fabric before and after grinding. Band definition was better than in the PA spectra and similar to that in the transmission spectra;

this result could, of course, be taken as an indication of partial saturation in the PA spectra. In general, diffuse reflectance spectroscopy was concluded to be less surface-sensitive than PA infrared spectroscopy. A comparison of PA and diffuse reflectance spectra obtained for a polypropylene fiber treated with a finishing solution supported this conclusion.

The above discussion illustrates the suitability of PA infrared spectroscopy for the analysis of textiles. Depth profiling of textile coatings, clearly demonstrated a number of years ago using rapid-scan PA spectroscopy, could undoubtedly be improved in suitable step-scan experiments. Textile and coatings researchers may propose other promising research topics for the PA infrared spectroscopist.

## 6.15. CATALYSTS

The requirement for infrared spectra of solid-state catalysts often presents yet another significant challenge to the analyst. This is partly due to the fact that many heterogeneous catalysts are supported on substrates such as alumina, silica, and metal oxides that strongly absorb or scatter light of various wavelengths; this complicates acquisition of their vibrational (both infrared and Raman) spectra. Moreover, it is highly desirable that these catalyst samples not be ground or otherwise perturbed prior to their analysis. Fortunately, PA infrared spectroscopy affords the opportunity for the characterization of these systems, while minimizing any potentially destructive sample preparation. The fact that PA spectra are generally characteristic of the surfaces of these samples provides additional motivation for the use of this technique, as does the potential for depth profiling.

Some of the earliest demonstrations of the feasibility of PA infrared spectroscopy included investigations on catalysts. A number of relevant articles have already been discussed in different parts of this book. Examples of this type occur in the research of Low and Parodi (1978), who obtained PA spectra of a chromia-alumina catalyst above 2000 $cm^{-1}$ and were able to detect the substitution of hydroxyls by methoxy groups. The silanization of silica was also reported in this study. Several years later, this group used PBDS to study the chemisorption of CO on Ni—C and other metal-carbon catalysts (Low et al., 1982b).

PA infrared spectra of catalysts were also described in two studies presented at the 1981 conference on Fourier Transform infrared spectroscopy. Vidrine (1981) examined a partially poisoned catalyst and was able to demonstrate depth profiling for this sample by varying the mirror velocity in a series of rapid-scan spectra. Mehicic et al. (1981) briefly discussed their PA investigations on metal oxide catalysts and put forward the useful suggestion that support disks of KBr or KCl could be sprinkled with the metal oxide of interest to facilitate the measurement of its PA infrared spectrum.

Dispersive near-infrared PA spectra of catalysts were reported in two early publications by Lochmüller and Wilder (1980a,b). These studies were discussed in Chapter 2. The first investigation demonstrated the suitability of PA spectroscopy for the qualitative analysis of chemically modified silica gel, whereas the second study showed that the surface coverage of silica by various organic compounds could additionally be quantified by this technique. The authors noted that the intrinsic intensity of an infrared band associated with a particular functional group in the adsorbate is frequently modified by its interaction with the substrate; this implies that the calibration plot used for quantitation of adsorbed species must be developed using intensities observed under similar conditions.

Kinney and Staley (1983) designed and constructed a PA cell for use with an FTIR spectrometer. These authors went on to investigate the infrared spectra of a variety of catalyst systems, including palmitic acid deposited on silica, CO on alumina-supported platinum, and Ag on alumina after reaction with gaseous $(CN)_2$. A typical result for the reaction of CO with Pt on alumina is shown in Figure 6.45. The authors reported that PA

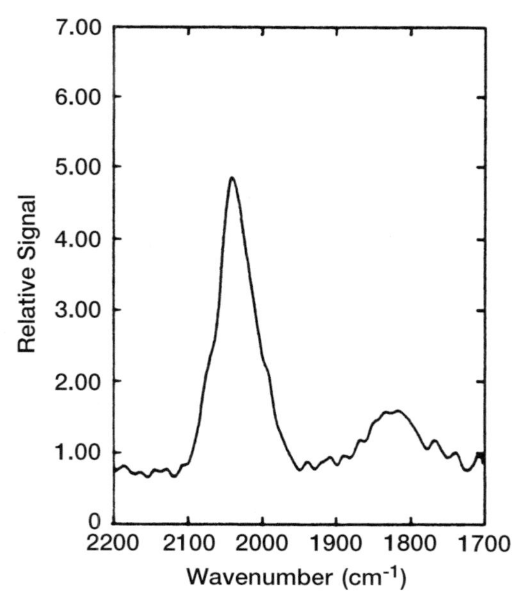


**Figure 6.45.** PA infrared spectrum of CO adsorbed on 5% Pt on alumina. (Reprinted with permission from Kinney, J. B. and Staley, R. H., *Anal. Chem.* **55**: 343–348. Copyright © 1983 American Chemical Society.)

spectra of good quality were generally obtained within a few minutes, although scattering and absorption by the support occasionally proved to be troublesome.

The pioneering work of E. M. Eyring and his group on various aspects of PA infrared spectroscopy has been discussed in several places in this book. During the first decade of widespread activity in this field, these researchers published several articles on the study of catalysts that are pertinent to the present discussion. For example, the quantitative determination of surface adsorption sites in silica-alumina and $\gamma$-alumina was described by Riseman et al. (1982). Bronsted and Lewis acid sites were characterized by their effect on the infrared spectrum of adsorbed pyridine. Because of its emphasis on quantitative analysis, this study is discussed again in the next chapter, which is dedicated to that subject.

The adsorption of CO on dehydroxylated $SiO_2$ was investigated by Gardella et al. (1983b). Both gaseous and adsorbed CO were quantified using observed PA intensities; 40% of the sites on the catalyst were found to be active. Similarly, the adsorption of CO on a $Ni/SiO_2$ catalyst was studied by Gardella et al. (1983a). A noteworthy result in the latter work deserves special mention: The adsorbed CO was found to be gradually converted to gaseous $CO_2$. In fact, the authors of this investigation demonstrated that the infrared source in the FTIR spectrometer was responsible for the promotion of this reaction.

More recently, Wang et al. (1991) used PA, diffuse reflectance, and transmission infrared spectroscopies to study the adsorption of pyridine on the zeolites HZSM-5 and HY. The researchers observed an unexpected enhancement of the PA intensities when He was used as the transducing gas and interpreted their results in terms of the model in which the PA signal is generated by means of the volumetric expansion of the interstitial gas (McGovern et al., 1985). This mechanism, which is separate and distinct from the usual thermal conduction to the carrier gas above the sample, can be important in powders and other highly porous samples such as zeolites. Wang et al. reported intensity enhancements by a factor of at least 2 for adsorbed pyridine when He was used in their experiments. The effect was smaller when nitrogen was employed in the PA cell because this gas possesses a lower thermal conductivity and speed of sound than does helium.

The research of B. S. H. Royce and his colleagues at Princeton University, briefly alluded to in the previous paragraph, comprised another important contribution to the study of catalysts during the 1980s. This work began with a brief report (Royce et al., 1980) in which the exposure of zeolite to moist air for 30 min prior to the measurement of the PA spectra resulted in the appearance of bands due to both $CO_2$ and water vapor. By contrast, the infrared spectrum of a fresh zeolite sample did not exhibit these bands. This

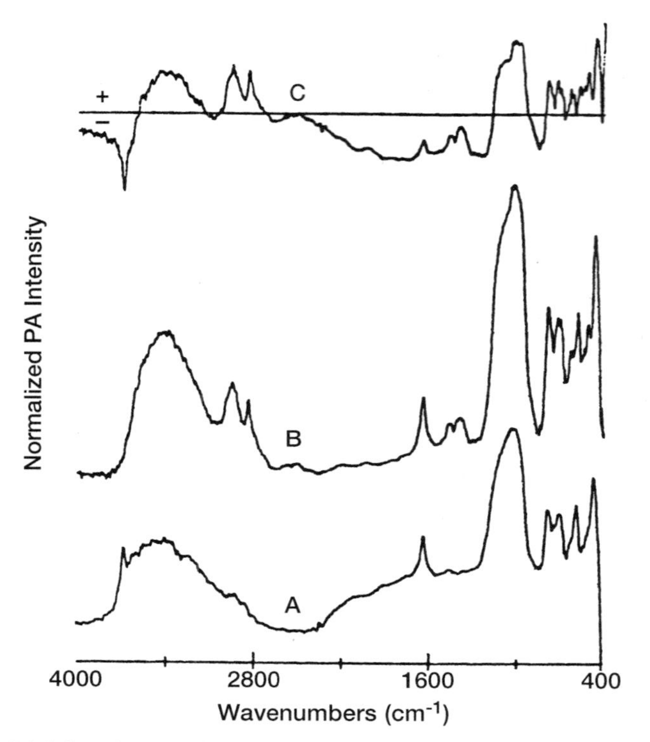


**Figure 6.46.** PA infrared spectra depicting methanol adsorption on Na-Y zeolite. Curve A, adsorbed sample; curve B, dried zeolite; curve C, difference, $(a - b) \times 2$. (Reprinted from *Appl. Surf. Sci.* **18**, Infrared photoacoustic spectroscopy of adsorption on powders, 401–413, copyright © 1984 with permission from Elsevier Science.)

experiment confirmed the suitability of the technique for these samples and, moreover, illustrated the surface activity of the zeolite.

In subsequent research, this group utilized PA infrared spectroscopy to characterize the methoxylation of silica, methanol adsorption on Na-Y zeolite, and the adsorption of CO on platinum black and $Pt/Al_2O_3$ (McGovern et al., 1984). Figure 6.46 illustrates the high quality of the results obtained for the adsorption of methanol on zeolite. One noteworthy result in this study was the general absence of saturation in the PA spectra. The powdered samples exhibited the following optical and thermal characteristics: (a) The optical absorption coefficient is given by the product of the absorption coefficient of the solid phase and its volume fraction; (b) the thermal conductivity is that of the interstitial gas; and (c) the heat capacity is that of the solid. This behavior distinguishes fine powders from bulk solids.

PA infrared spectra of silica, methoxylated silica, aluminum oxides, and hydroxides were described in detail by Benziger et al. (1985). The emphasis in this work was primarily on the information revealed by the spectra, which were again of very good quality. For example, adsorbed water was found to affect the lattice vibrations of silica, whereas particle–particle interactions influenced the vibrations of the surface species. The results obtained for the aluminum-containing compounds (some of which resemble the clays and minerals discussed previously in this chapter) demonstrated the suitability of the method for the study of structural (phase) transformations, insofar as these compounds are interrelated. The PA infrared method and the results obtained in earlier studies were then thoroughly reviewed by Royce and Benziger (1986), who presented additional results for supported metal oxide catalysts. In situ measurements at temperatures up to 400°C were also described in this publication.

PA infrared spectroscopy has also been used to study heteropoly catalysts. Highfield and Moffat (1985a) studied the interaction of methanol with 12-tungstophosphoric acid, $H_3PW_{12}O_{40}$. Brønsted acidity was found to be essential for catalyzing the elimination of water from methanol, a process thought to be related to the formation of hydrocarbons. In related work, the same authors described the quantitation of pyridine and ammonia on $H_3PW_{12}O_{40}$ (Highfield and Moffat, 1985b). This research is summarized in Chapter 7. The use of PA intensities for quantitative analysis in these systems was feasible only at low concentrations since saturation tended to limit the band intensities at higher concentrations.

Recent publications that describe the use of PA infrared spectroscopy for the characterization of catalyst systems (Ryczkowski, 1994; Mohamed, 1995) primarily discussed the interactions between the adsorbed species and the catalyst. From the perspective of these authors, the use of PA detection was somewhat incidental to the main objective of catalyst characterization. Indeed, the reader may conclude that this situation resembles that described above with regard to the PA spectra of gases: The routine use of a technique can be interpreted as a sign of its maturity.

## REFERENCES

Adams, M. J., Beadle, B. C., and Kirkbright, G. F. (1978). Optoacoustic spectrometry in the near-infrared region. *Anal. Chem.* **50**: 1371–1374.

Angle, C. W., Donini, J. C., and Hamza, H. A. (1988). The effect of ultrasonication on the surface properties, ionic composition and electrophoretic mobility of an aqueous coal suspension. *Colloids Surfaces* **30**: 373–385.

Annyas, J., Bićanić, D., and Schouten, F. (1999). Novel instrumental approach to the measurement of total body water: Optothermal detection of heavy water in the blood serum. *Appl. Spectrosc.* **53**: 339–343.

Ardeleanu, M., Morisset, R., and Bertrand, L. (1992). Fourier transform infrared photoacoustic spectra of bacteria. In: *Photoacoustic and Photothermal Phenomena III.* D. Bićanić (ed.). Springer, Berlin, pp. 81–84.

Bajic, S. J., Jones, R. W., McClelland, J. F., Hames, B. R., and Meglen, R. R. (1998). Rapid analysis of wood using transient infrared spectroscopy and photoacoustic spectroscopy with PLS regression. *AIP Conf. Proc.* **430**: 466–469.

Beenen, A., and Niessner, R. (1999a). Development of a photoacoustic trace gas sensor based on fiber-optically coupled NIR laser diodes. *Appl. Spectrosc.* **53**: 1040–1044.

Beenen, A., and Niessner, R. (1999b). Trace gas analysis by photoacoustic spectroscopy with NIR laser diodes. *AIP Conf. Proc.* **463**: 211–213.

Belton, P. S. (1984). Photoacoustic spectroscopy (PAS). In: *Biophysical Methods in Food Research.* H. W.-S. Chan (ed.). Blackwell Scientific, Oxford, pp. 123–135.

Belton, P. S., and Tanner, S. F. (1983). Determination of the moisture content of starch using near infrared photoacoustic spectroscopy. *Analyst* **108**: 591–596.

Belton, P. S., Saffa, A. M., and Wilson, R. H. (1987a). Use of Fourier transform infrared spectroscopy for quantitative analysis: A comparative study of different detection methods. *Analyst* **112**: 1117–1120.

Belton, P. S., Wilson, R. H., and Saffa, A. M. (1987b). Effects of particle size on quantitative photoacoustic spectroscopy using a gas-microphone cell. *Anal. Chem.* **59**: 2378–2382.

Belton, P. S., Saffa, A. M., and Wilson, R. H. (1988a). The potential of Fourier transform infrared spectroscopy for the analysis of confectionary products. *Food Chem.* **28**: 53–61.

Belton, P. S., Saffa, A. M., and Wilson, R. H. (1988b). In: *Analytical Applications of Spectroscopy*. C. S. Creaser and A. M. C. Davies (eds.). The Royal Society of Chemistry, Cambridge, pp. 245–250.

Bensebaa, F., Majid, A., and Deslandes, Y. (2001). Step-scan photoacoustic Fourier transform and X-rays photoelectron spectroscopy of oil sands fine tailings: new structural insights. *Spectrochim. Acta A* **57**: 2695–2702.

Benziger, J. B., McGovern, S. J., and Royce, B. S. H. (1985). IR photoacoustic spectroscopy of silica and aluminum oxide. In: *Catalyst Characterization Science*, pp. 449–463.

Bertrand, L., Monchalin, J.-P., and Lepoutre, F. (1982). Magnitude and phase photoacoustic spectra of chrysotile asbestos, a powdered sample. *Appl. Opt.* **21**: 248–252.

Bertrand, L. (1988). Advantages of phase analysis in Fourier transform infrared photoacoustic spectroscopy. *Appl. Spectrosc.* **42**: 134–138.

Bićanić, D., Fink, T., Franko, M., Močnik, G., van de Bovenkamp, P., van Veldhuizen, B., and Gerkema, E. (1999). Infrared photothermal spectroscopy in the science of human nutrition. *AIP Conf. Proc.* **463**: 637–639.

Bijnen, F. G. C., Zuckermann, H., Harren, F. J. M., and Reuss, J. (1998). Multicomponent trace-gas analysis by three intracavity photoacoustic cells in a CO laser: Observation of anaerobic and postanaerobic emission of acetaldehyde and ethanol in cherry tomatoes. *Appl. Opt.* **37**: 3345–3353.

Bohren, A., and Sigrist, M. W. (1997). Optical parametric oscillator based difference frequency laser source for photoacoustic trace gas spectroscopy in the 3 μm mid-IR range. *Infrared Phys. Technol.* **38**: 423–435.

Bouzerar, R., Amory, C., Zeinert, A., Benlahsen, M., Racine, B., Durand-Drouhin, O., and Clin, M. (2001). Optical properties of amorphous hydrogenated carbon thin films. *J. Non-Cryst. Solids* **281**: 171–180.

Bürger, H., Lecoutre, M., Huet, T. R., Breidung, J., Thiel, W., Hänninen, V., and Halonen, L. (2001). The ($n$00), $n = 3, 4$, and 6, local mode states of $H_3SiD$: Fourier transform infrared and laser photoacoustic spectra and *ab initio* calculations of spectroscopic parameters. *J. Chem. Phys.* **114**: 8844–8854.

Butler, I. S., Xu, Z. H., Werbowyj, R. S., and St.-Germain, F. (1987a). FT-IR photoacoustic spectra of some solid organometallic complexes of chromium, manganese, rhenium, and iron. *Appl. Spectrosc.* **41**: 149–153.

Butler, I. S., Xu, Z. H., Darensbourg, D. J., and Pala, M. (1987b). FT-IR, photoacoustic and micro-Raman spectra of the dodecacarbonyltriruthenium(0) complexes $Ru_3(^{13}CO)_{12}$ and $Ru_3(CO)_{12}$. *J. Raman Spectrosc.* **18**: 357–363.

Butler, I. S., Li, H., and Gao, J. P. (1991). Comparison of photoacoustic, attenuated total reflection, and transmission infrared spectra of crystalline organoiron(II) carbonyl complexes. *Appl. Spectrosc.* **45**: 223–226.

Butler, I. S., Gilson, D. F. R., and Lafleur, D. (1992). Infrared photoacoustic spectra of gaseous pentacarbonyl(methyl)manganese(I) and pentacarbonyl(methyl)-rhenium(I). *Appl. Spectrosc.* **46**: 1605–1607.

Calasso, I. G., and Sigrist, M. W. (1999). Selection criteria for microphones used in pulsed nonresonant gas-phase photoacoustics. *Rev. Sci. Instrum.* **70**: 4569–4578.

Carter, R. O., and Wright, S. L. (1991). Evaluation of the appropriate sample position in a PAS/FT-IR experiment. *Appl. Spectrosc.* **45**: 1101–1103.

Carter, R. O., Paputa Peck, M. C., and Bauer, D. R. (1989a). The characterization of polymer surfaces by photoacoustic Fourier transform infrared spectroscopy. *Polym. Degrad. Stabil.* **23**: 121–134.

Carter, R. O., Paputa Peck, M. C., Samus, M. A., and Killgoar, P. C. (1989b). Infrared photoacoustic spectroscopy of carbon black filled rubber: Concentration limits for samples and background. *Appl. Spectrosc.* **43**: 1350–1354.

Carter, E. A., Fredericks, P. M., and Church, J. S. (1996). Fourier transform infrared photoacoustic spectroscopy of surface-treated wool. *Textile Res. J.* **66**: 787–794.

Castleden, S. L., Kirkbright, G. F., and Menon, K. R. (1980). Determination of moisture in single-cell protein utilising photoacoustic spectroscopy in the near-infrared region. *Analyst* **105**: 1076–1081.

Castleden, S. L., Kirkbright, G. F., and Long, S. E. (1982). Quantitative assay of propranolol by photoacoustic spectroscopy. *Can. J. Spectrosc.* **27**: 244–248.

Chalmers, J. M., Stay, B. J., Kirkbright, G. F., Spillane, D. E. M., and Beadle, B. C. (1981). Some observations on the capabilities of photoacoustic Fourier transform infrared spectroscopy (PAFTIR). *Analyst* **106**: 1179–1186.

Chien, P.-L., Markuszewski, R., and McClelland, J. F. (1985a). Comparison of Fourier transform infrared-photoacoustic spectroscopy (FTIR-PAS) and conventional methods for analysis of coal oxidation. *Preprints, ACS Div. Fuel Chem.* **30**: 13–20.

Chien, P.-L., Markuszewski, R., Araghi, H. G., and McClelland, J. F. (1985b). Study of coal oxidation kinetics by Fourier transform infrared-photoacoustic spectroscopy (FTIR-PAS). In: *Proceedings, 1985 International Conference on Coal Science*. Pergamon Press, Sydney, pp. 818–821.

Choquet, M., Rousset, G., and Bertrand, L. (1985). Phase analysis of infrared Fourier transform photoacoustic spectra. *SPIE* **553**: 224–225.

Choquet, M., Rousset, G., and Bertrand, L. (1986). Fourier-transform photoacoustic spectroscopy: A more complete method for quantitative analysis. *Can. J. Phys.* **64**: 1081–1085.

Church, J. S., and Evans, D. J. (1995). The quantitative analysis of fluorocarbon polymer finishes on wool by FT-IR spectroscopy. *J. Appl. Polymer Sci.* **57**: 1585–1594.

Coffey, M. J., Berghout, H. L., Woods, E., and Crim, F. F. (1999). Vibrational spectroscopy and intramolecular energy transfer in isocyanic acid (HNCO). *J. Chem. Phys.* **110**: 10850–10862.

Cook, L. E., Luo, S. Q., and McClelland, J. F. (1991). Fourier transform infrared photoacoustic spectroscopy of polymers adsorbed from solution by *gamma*-iron oxide. *Appl. Spectrosc.* **45**: 124–126.

Davidson, R. S., and King, D. (1983). A new method of distinguishing wool from polyester-fibre and cotton fabrics. *J. Textile Inst.* **74**: 382–384.

Davidson, R. S., and Fraser, G. V. (1984). The analysis of textile mixtures using Fourier transform infra-red photoacoustic spectroscopy. *J. Soc. Dyers Colour.* **100**: 167–170.

Dittmar, R. M., Palmer, R. A., and Carter, R. O. (1994). Fourier transform photoacoustic spectroscopy of polymers. *Appl. Spectrosc. Rev.* **29**: 171–231.

Donini, J. C., and Michaelian, K. H. (1985). Near-, mid-, and far-infrared photoacoustic spectroscopy. *SPIE* **553**: 344–345.

Donini, J. C., and Michaelian, K. H. (1986). Near infrared photoacoustic FTIR spectroscopy of clay minerals and coal. *Infrared Phys.* **26**: 135–140.

Donini, J. C., and Michaelian, K. H. (1988). Low frequency photoacoustic spectroscopy of solids. *Appl. Spectrosc.* **42**: 289–292.

Eckhardt, H., and Chance, R. R. (1983). Defect states in polyacetylene: A photoacoustic and diffuse reflectance study. *J. Chem. Phys.* **79**: 5698–5704.

Factor, A., Tilley, M. G., and Codella, P. J. (1991). Determination of residual unsaturation in photo-cured acrylate formulations using photoacoustic FT-IR spectroscopy. *Appl. Spectrosc.* **45**: 135–138.

Favier, J. P., Bićanić, D., van de Bovenkamp, P., Chirtoc, M., and Helander, P. (1996). Detection of total trans fatty acids content in margarine: an intercomparison study of GLC, GLC + TLC, FT-IR, and optothermal window (open photoacoustic cell). *Anal. Chem.* **68**: 729–733.

Favier, J. P., Bićanić, D., Cozijnsen, J., van Veldhuizen, B., and Helander, P. (1998). $CO_2$ laser infrared optothermal spectroscopy for quantitative adulteration studies in binary mixtures of extra-virgin olive oil. *J. Am. Oil Chem. Soc.* **75**: 359–362.

Fehér, M., Jiang, Y., Maier, J. P., and Miklós, A. (1994). Optoacoustic trace-gas monitoring with near-infrared diode lasers. *Appl. Opt.* **33**: 1655–1658.

Fischer, C., Sigrist, M. W., Yu, Q., and Seiter, M. (2001). Photoacoustic monitoring of trace gases by use of a diode-based difference frequency laser source. *Opt. Lett.* **26**: 1609–1611.

Friesen, W. I., and Michaelian, K. H. (1986). Fourier deconvolution of photoacoustic FTIR spectra. *Infrared Phys.* **26**: 235–242.

Friesen, W. I., and Michaelian, K. H. (1991). Deconvolution and curve-fitting in the analysis of complex spectra. The CH stretching region in infrared spectra of coal. *Appl. Spectrosc.* **45**: 50–56.

Gagarin, S. G., Gladun, T. G., Friesen, W. I., and Michaelian, K. H. (1993). Simulating infrared spectra of macerals and estimation of petrographic composition of coals by spectra of fractions with various density. *Coke Chem.* 9–15.

Gagarin, S. G., Friesen, W. I., Michaelian, K. H., and Gladun, T. G. (1994). Reactivity of coal based on photoacoustic IR spectroscopic investigations. *Solid Fuel Chem.* **28**: 35–42.

Gagarin, S. G., Friesen, W. I., and Michaelian, K. H. (1995a). Estimation of the heat of combustion of coal by infrared spectroscopy. *Coke Chem.* 6–16.

Gagarin, S. G., Friesen, W. I., and Michaelian, K. H. (1995b). Prediction of the Roga index of coal blends based on data of IR-spectroscopy. *Coke Chem.* 23–28.

Gardella, J. A., Jr., Eyring, E. M., Klein, J. C., and Carvalho, M. B. (1982). Surface spectroscopic analysis of $HCN/NH_3$ corrosion products on iron by Fourier transform infrared photoacoustic spectrospcopy and X-ray photoelectron spectroscopy. *Appl. Spectrosc.* **36**: 570–573.

Gardella, J. A., Jiang, D.-Z., McKenna, W. P., and Eyring, E. M. (1983a). Applications of Fourier transform infrared photoacoustic spectroscopy (FT-IR/PAS) to surface and corrosion phenomena. *Appl. Surf. Sci.* **15**: 36–49.

Gardella, J. A., Jiang, D.-Z., and Eyring, E. M. (1983b). Quantitative determination of catalytic surface adsorption sites by Fourier transform infrared photoacoustic spectroscopy. *Appl. Spectrosc.* **37**: 131–133.

Gardella, J. A., Grobe, G. L., Hopson, W. L., and Eyring, E. M. (1984). Comparison of attenuated total reflectance and photoacoustic sampling for surface analysis of polymer mixtures by Fourier transform infrared spectroscopy. *Anal. Chem.* **56**: 1169–1177.

Garcia, J. A., Mandelis, A., Marinova, M., Michaelian, K. H., and Afrashtehfar, S. (1998). Quantitative photothermal radiometric and FT-IR photoacoustic measurements of specialty papers. *Appl. Spectrosc.* **52**: 1222–1229.

Garcia, J. A., Mandelis, A., Marinova, M., Michaelian, K. H., and Afrashtehfar, S. (1999). Quantitative photothermal radiometric and FTIR photoacoustic measurments of specialty papers. *AIP Conf. Proc.* **463**: 395–397.

Gerson, D. J., McClelland, J. F., Veysey, S., and Markuszewski, R. (1984). Characterization of coal using Fourier transform infrared photoacoustic spectroscopy. *Appl. Spectrosc.* **38**: 902–904.

Gondal, M. A. (1997). Laser photoacoustic spectrometer for remote monitoring of atmospheric pollutants. *Appl. Opt.* **36**: 3195–3201.

Goodarzi, F., and McFarlane, R. A. (1991). Chemistry of fresh and weathered resinites—An infrared photoacoustic spectroscopic study. *Int. J. Coal Geol.* **19**: 283–301.

Gordon, S. H., Greene, R. V., Freer, S. N., and James, C. (1990). Measurement of protein biomass by Fourier transform infrared-photoacoustic spectroscopy. *Biotech. Appl. Biochem.* **12**: 1–10.

Gordon, S. H., Schudy, R. B., Wheeler, B. C., Wicklow, D. T., and Greene, R. V. (1997). Identification of Fourier transform infrared photoacoustic spectral features for detection of *Aspergillus flavus* infection in corn. *Int. J. Food Microbiol.* **35**: 179–186.

Gordon, S. H., Wheeler, B. C., Schudy, R. B., Wicklow, D. T., and Greene, R. V. (1998). Neural network pattern recognition of photoacoustic FTIR spectra and knowledge-based techniques for detection of mycotoxigenic fungi in food grains. *J. Food Prot.* **61**: 221–230.

Gordon, S. H., Jones, R. W., McClelland, J. F., Wicklow, D. T., and Greene, R. V. (1999). Transient infrared spectroscopy for detection of toxigenic fungi in corn: Potential for on-line evaluation. *J. Agric. Food Chem.* **47**: 5267–5272.

Gosselin, F., Di Renzo, M., Ellis, T. H., and Lubell, W. D. (1996). Photoacoustic FTIR spectroscopy, a nondestructive method for sensitive analysis of solid-phase organic chemistry. *J. Org. Chem.* **61**: 7980–7981.

Graves, D. J., and Luo, S. (1994). Use of photoacoustic Fourier-transform infrared spectroscopy to study phosphates in proteins. *Biochem. Biophys. Res. Comm.* **205**: 618–624.

Greene, R. V., Freer, S. N., and Gordon, S. H. (1988). Determination of solid-state fungal growth by Fourier transform infrared-photoacoustic spectroscopy. *FEMS Microbiol. Lett.* **52**: 73–78.

Greene, R. V., Gordon, S. H., Jackson, M. A., Bennett, G. A., McClelland, J. F., and Jones, R. W. (1992). Detection of fungal contamination in corn: Potential of FTIR-PAS and -DRS. *J. Agric. Food Chem.* **40**: 1144–1149.

Gurnagul, N., St.-Germain, F. G. T., and Gray, D. G. (1986). Photoacoustic Fourier transform infrared measurements on paper. *J. Pulp Paper Sci.* **12**: J156–J159.

Halttunen, M., Tenhunen, J., Saarinen, T., and Stenius, P. (1999). Applicability of FTIR/PAS depth profiling for the study of coated papers. *Vibr. Spectrosc.* **19**: 261–269.

Hamza, H. A., Michaelian, K. H., and Andersen, N. E. (1983). A fundamental approach to beneficiation of fine oxidized coals. In: *Proceedings, 1983 International Conference on Coal Science*. Center for Conference Management, Pittsburgh, pp. 248–251.

Hao, L.-Y., Han, J.-X., Shi, Q., Zhang, J.-H., Zheng, J.-J., and Zhu, Q.-S. (2000). A highly sensitive photoacoustic spectrometer for near infrared overtone. *Rev. Sci. Instrum.* **71**: 1975–1980.

Harbour, J. R., Hopper, M. A., Marchessault, R. H., Dobbin, C. J., and Anczurowski, E. (1985). Photoacoustic spectroscopy of cellulose, paper and wood. *J. Pulp Paper Sci.* **11**: J42–J47.

Herres, W., and Zachmann, G. (1984). FT-IR photoakustische Spektroskopie in der Feststoffanalytik. *LaborPraxis* 632–638.

Highfield, J. G., and Moffat, J. B. (1985a). Elucidation of the mechanism of dehydration of methanol over 12-tungstophosphoric acid using infrared photoacoustic spectroscopy. *J. Catal.* **95**: 108–119.

Highfield, J. G., and Moffat, J. B. (1985b). The influence of experimental conditions in quantiative analysis of powdered samples by Fourier transform infrared photoacoustic spectroscopy. *Appl. Spectrosc.* **39**: 550–552.

Hocking, M. B., Syme, D. T., Axelson, D. E., and Michaelian, K. H. (1990a). Water-soluble imide-amide copolymers. I. Preparation and characterization of poly[acrylamide-co-sodium *N*-(4-sulfophenyl) maleimide]. *J. Polym. Sci.: Part A: Polym. Chem.* **28**: 2949–2968.

Hocking, M. B., Syme, D. T., Axelson, D. E., and Michaelian, K. H. (1990b). Water-soluble imide-amide copolymers. II. Preparation and characterization of poly(acrylamide-co-*p*-maleimidobenzoic acid). *J. Polym. Sci.: Part A: Polym. Chem.* **28**: 2969–2982.

Hocking, M. B., Syme, D. T., Axelson, D. E., and Michaelian, K. H. (1990c). Water-soluble imide-amide copolymers. III. Preparation and characterization of poly(acrylamide-co-*N*,*N*-diallylaniline) and poly(acrylamide-co-sodium-N,N-diallylsulfanilate). *J. Polym. Sci.: Part A: Polym. Chem.* **28**: 2983–2996.

Hofstetter, D., Beck, M., Faist, J., Nägele, M., and Sigrist, M. W. (2001). Photoacoustic spectroscopy with quantum cascade distributed-feedback lasers. *Opt. Lett.* **26**: 887–889.

Hornberger, Ch., König, M., Rai, S. B., and Demtröder, W. (1995). Sensitive photoacoustic overtone spectroscopy of acetylene with a multipass photoacoustic cell and a colour centre laser at 1.5 μm. *Chem. Phys.* **190**: 171–177.

Hünig, I., Oudejans, L., and Miller, R. E. (2000). Infrared optothermal spectroscopy of $N_2$– and OC–DCCH: The C—H stretching region. *J. Mol. Spectrosc.* **204**: 148–152.

Huvenne, J. P., and Lacroix, B. (1988). Spectres d'absorption détectés par effet photoacoustique en infrarouge par transformée de Fourier—applications à quelques médicaments. *Spectrochim. Acta A* **44**: 109–113.

Irudayaraj, J., and Yang, H. (2000). Analysis of cheese using step-scan Fourier transform infrared photoacoustic spectroscopy. *Appl. Spectrosc.* **54**: 595–600.

Irudayaraj, J., and Yang, H. (2001). Depth profiling of a heterogeneous food-packaging model using step-scan Fourier transform infrared photoacoustic spectroscopy. *J. Food Eng.* **55**: 25–33.

Irudayaraj, J., Sivakesava, S., Kamath, S., and Yang, H. (2001). Monitoring chemical changes in some foods using Fourier transform photoacoustic spectroscopy. *J. Food Sci.* **66**: 1416–1421.

Irudayaraj, J., Yang, H., and Sakhamuri, S. (2002). Differentiation and detection of microorganisms using Fourier transform infrared photoacoustic spectroscopy. *J. Mol. Struct.* **606**: 181–188.

Jalink, H., Bićanić, D., Franko, M., and Bozóki, Z. (1995). Vapor-phase spectra and the pressure-temperature dependence of long-chain carboxylic acids studied by a CO laser and the photoacoustic heat-pipe detector. *Appl. Spectrosc.* **49**: 994–999.

Jasse, B. (1989). Fourier-transform infrared photoacoustic spectroscopy of synthetic polymers. *J. Macromol. Sci.-Chem. A* **26**: 43–67.

Jin, Q., Kirkbright, G. F., and Spillane, D. E. M. (1982). The determination of moisture in some solid materials by near infrared photoacoustic spectroscopy. *Appl. Spectrosc.* **36**: 120–124.

Jurdana, L. E., Ghiggino, K. P., Leaver, I. H., Barraclough, C. G., and Cole-Clarke, P. (1994). Depth profile analysis of keratin fibers by FT-IR photoacoustic spectroscopy. *Appl. Spectrosc.* **48**: 44–49.

Jurdana, L. E., Ghiggino, K. P., Leaver, I. H., and Cole-Clarke, P. (1995). Application of FT-IR step-scan photoacoustic phase modulation methods to keratin fibers. *Appl. Spectrosc.* **49**: 361–366.

Kanstad, S. O., Nordal, P.-E., Hellgren, L., and Vincent, J. (1981). Infrared photoacoustic spectroscopy of skin lipids. *Naturwiss.* **68**: 47–48.

Kästle, R., and Sigrist, M. W. (1996a). CO laser photoacoustic spectroscopy of acetic, deuterated acetic and propionic acid molecules. *Spectrochim. Acta A* **52**: 1221–1228.

Kästle, R., and Sigrist, M. W. (1996b). Temperature-dependent photoacoustic spectroscopy with a Helmholtz resonator. *Appl. Phys. B* **63**: 389–397.

Kinney, J. B., and Staley, R. H. (1983). Photoacoustic cell for Fourier transform infrared spectrometry of surface species. *Anal. Chem.* **55**: 343–348.

Kirkbright, G. F. (1978). Analytical optoacoustic spectrometry. *Optica Pura y Aplicada* **11**: 125–136.

Koenig, J. L. (1985). Recent advances in FT-IR (Fourier transform infrared spectroscopy) of polymers. *Pure Appl. Chem.* **57**: 971–976.

Krishnan, K. (1981). Some applications of Fourier transfrom infrared photoacoustic spectroscopy. *Appl. Spectrosc.* **35**: 549–557.

Krishnan, K., Hill, S., Hobbs, J. P., and Sung, C. S. P. (1982). Orientation measurements from polymer surfaces using Fourier transform infrared photoacoustic spectroscopy. *Appl. Spectrosc.* **36**: 257–259.

Kühnemann, F., Schneider, K., Hecker, A., Martis, A. A. E., Urban, W., Schiller, S., and Mlynek, J. (1998). Photoacoustic trace-gas detection using a cw single-frequency parametric oscillator. *Appl. Phys. B* **66**: 741–745.

Kuo, M.-L., McClelland, J. F., Luo, S., Chien, P.-L., Walker, R. D., and Hse, C.-Y. (1988). Applications of infrared photoacoustic spectroscopy for wood samples. *Wood Fiber Sci.* **20**: 132–145.

Larsen, J. W. (1988). The macromolecular structure of bituminous coals: Macromolecular anisotropy, aromatic-aromatic interactions, and other complexities. *Preprints, ACS Div. Fuel Chem.* **33**: 400–406.

Lecoutre, M., Huet, T. R., Mkadmi, E. B., and Bürger, H. (2000a). High-resolution laser photoacoustic spectroscopy of $HSiF_3$: The $5\nu_1$ and $6\nu_1$ overtone bands. *J. Mol. Spectrosc.* **202**: 207–212.

Lecoutre, M., Rohart, F., Huet, T. R., and Maki, A. G. (2000b). Photoacoustic detection of new bands of HCN between 11 390 and 13 020 $cm^{-1}$. *J. Mol. Spectrosc.* **203**: 158–164.

Letzelter, N. S., Wilson, R. H., Jones, A. D., and Sinnaeve, G. (1995). Quantitative determination of the composition of individual pea seeds by Fourier transform infrared photoacoustic spectroscopy. *J. Sci. Food Agric.* **67**: 239–245.

Lewis, L. N. (1982). The analysis of the near IR of some organic and organo-metallic compounds by photoacoustic spectroscopy. *J. Organometall. Chem.* **234**: 355–365.

Li, H., and Butler, I. S. (1992). Infrared photoacoustic spectra of the solid manganese(I) carbonyl halides, $Mn(CO)_5X$ and $[Mn(CO)_4X]_2$ (X = Cl, Br, I). *Appl. Spectrosc.* **46**: 1785–1789.

Li, H., and Butler, I. S. (1993). Solid-state infrared photoacoustic spectra of group 6B metal mixed carbonyl-*t*-butylisocyanide complexes, $M(CO)_{6-n}(CN^t Bu)_n$ ($M$ = Cr, Mo, W; $n = 1$–3). *Appl. Spectrosc.* **47**: 218–221.

Lochmüller, C. H., and Wilder, D. R. (1980a). Qualitative examination of chemically-modified silica surfaces by near-infrared photoacoustic spectroscopy. *Anal. Chim. Acta* **116**: 19–24.

Lochmüller, C. H., and Wilder, D. R. (1980b). Quantitative photoacoustic spectroscopy of chemically-modified silica surfaces. *Anal. Chim. Acta* **118**: 101–108.

Lotta, T. I., Tulkki, A. P., Virtanen, J. A., and Kinnunen, P. K. J. (1990). Interaction of 7,7,8,8-tetracyanoquinodimethane with diacylphosphatidylcholines and -phosphatidylglycerols. A photoacoustic Fourier transform infrared study. *Chem. Phys. Lipids* **52**: 11–27.

Low, M. J. D. (1993a). Unusual bands in the infrared spectra of some oxidized coals and coal chars. *Spectrosc. Lett.* **26**: 453–459.

Low, M. J. D. (1983b). Normalizing infrared photoacoustic spectra of solids. *Spectrosc. Lett.* **16**: 913–922.

Low, M. J. D. (1984). Infrared spectra of infrared detectors. *Spectrosc. Lett.* **17**: 455–461.

Low, M. J. D. (1985). Hexane soot as standard for normalizing infrared photothermal spectra of solids. *Spectrosc. Lett.* **18**: 619–625.

Low, M. J. D. (1986). Some practical aspects of FT-IR/PBDS. Part I: vibrational noise. *Appl. Spectrosc.* **40**: 1011–1019.

Low, M. J. D., and Glass, A. S. (1989). The assignment of the 1600 $cm^{-1}$ mystery band of carbons. *Spectrosc. Lett.* **22**: 417–429.

Low, M. J. D., and Lacroix, M. (1982). An infrared photothermal beam deflection Fourier transform spectrometer. *Infrared Phys.* **22**: 139–147.

Low, M. J. D., and Morterra, C. (1983). IR studies of carbons—I. IR photothermal beam deflection spectroscopy. *Carbon* **21**: 275–281.

Low, M. J. D., and Morterra, C. (1985). IR studies of carbons—V. Effects of NaCl on cellulose pyrolysis and char oxidation. *Carbon* **23**: 311–316.

Low, M. J. D., and Morterra, C. (1986). The infrared examination of carbons with beam deflection spectroscopy. *IEEE Trans. UFFC* **33**: 585–589.

Low, M. J. D., and Morterra, C. (1989). Infrared spectra of carbons. X. The spectral profile of medium-temperature chars. In: *Structure and Reactivity of Surfaces*. C. Morterra, A. Zecchina, and G. Costa (eds.). Elsevier, Amsterdam, pp. 601–609.

Low, M. J. D., and Parodi, G. A. (1978). Infrared photoacoustic spectra of surface species in the 4000–2000 $cm^{-1}$ region using a broad band source. *Spectrosc. Lett.* **11**: 581–588.

Low, M. J. D., and Parodi, G. A. (1980a). An infrared photoacoustic spectrometer. *Infrared Phys.* **20**: 333–340.

Low, M. J. D., and Parodi, G. A. (1980b). Carbon as reference for normalizing infrared photoacoustic spectra. *Spectrosc. Lett.* **13**: 663–669.

Low, M. J. D., and Parodi, G. A. (1980c). Infrared photoacoustic spectra of solids. *Spectrosc. Lett.* **13**: 151–158.

Low, M. J. D., and Parodi, G. A. (1980d). Infrared photoacoustic spectroscopy of solids and surface species. *Appl. Spectrosc.* **34**: 76–80.

Low, M. J. D., and Tascon, J. M. D. (1985). An approach to the study of minerals using infrared photothermal beam deflection spectroscopy. *Phys. Chem. Minerals* **12**: 19–22.

Low, M. J. D., and Wang, N. (1990). A relation between the spectral profile and the infrared continuum of spectra of medium-temperature carbons. *Spectrosc. Lett.* **23**: 983–990.

Low, M. J. D., Lacroix, M., and Morterra, C. (1982a). Infrared photothermal beam deflection Fourier transform spectroscopy of solids. *Appl. Spectrosc.* **36**: 582–584.

Low, M. J. D., Morterra, C., and Severdia, A. G. (1982b). Infrared photothermal deflection spectroscopy of carbon-supported metal catalysts. *Spectrosc. Lett.* **15**: 415–421.

Low, M. J. D., Morterra, C., and Severdia, A. G. (1984). An approach to the infrared study of materials by photothermal beam deflection spectroscopy. *Mater. Chem. Phys.* **10**: 519–528.

Low, M. J. D., Politou, A. S., Varlashkin, P. G., and Wang, N. (1990). Unusual bands in the infrared spectra of some chars. *Spectrosc. Lett.* **23**: 527–531.

Luo, S., Huang, C.-Y. F., McClelland, J. F., and Graves, D. J. (1994). A study of protein secondary structure by Fourier transform infrared/photoacoustic spectroscopy and its application for recombinant proteins. *Anal. Biochem.* **216**: 67–76.

Lynch, B. M., and MacPhee, J. A. (1989). In: *Chemistry of Coal Weathering*. C. R. Nelson (ed.). Elsevier, Amsterdam, pp. 83–106.

Lynch, B. M., MacEachern, A. M., MacPhee, J. A., Nandi, B. N., Hamza, H., and Michaelian, K. H. (1983). Photoacoustic infrared Fourier transform (PAIFT) and diffuse reflectance infrared Fourier transform (DRIFT) spectroscopies in studies of oxidation, conversion, and derivatization reactions of bituminous and sub-bituminous coals. In: *Proceedings, 1983 International Conference on Coal Science*. Center for Conference Management, Pittsburgh, pp. 653–654.

Lynch, B. M., Lancaster, L., and Fahey, J. T. (1986). Detection by photoacoustic infrared Fourier transform spectroscopy of surface peroxide species in chemically and thermally modified coals. *Preprints, ACS Division of Fuel Chemistry*, **31**: 43–48.

Lynch, B. M., Lancaster, L.-I., and MacPhee, J. A. (1987a). Characterization of surface functionality of coals by photoacoustic FTIR (PAIFT) spectroscopy, reflectance infrared microspectrometry, and X-ray photoelectron spectroscopy (XPS). *Preprints, ACS Div. Fuel Chem.* **32**: 138–145.

Lynch, B. M., Lancaster, L.-I., and MacPhee, J. A. (1987b). Carbonyl groups from chemically and thermally promoted decomposition of peroxides on coal surfaces. Detection of specific types using photoacoustic infrared Fourier transform spectroscopy. *Fuel* **66**: 979–983.

Lynch, B. M., Lancaster, L., and MacPhee, J. A. (1988). Detection of carbonyl functionality of oxidized, vacuum-dried coals by photoacoustic infrared Fourier transform (PAIFT) spectroscopy: Correlations with added oxygen and with plastic properties. *Energy Fuels* **2**: 13–17.

Manzanares, C., Blunt, V. M., and Peng, J. (1993). Vibrational spectroscopy of nonequivalent C—H bonds in liquid *cis*- and *trans*-3-hexene. *Spectrochim. Acta A* **49**: 1139–1152.

McAskill, N. A. (1987). Near-infrared photoacoustic spectroscopy of coals and shales. *Appl. Spectrosc.* **41**: 313–317.

McClelland, J. F. (1983). Photoacoustic spectroscopy. *Anal. Chem.* **55**: 89A–105A.

McFarlane, R. A., Gentzis, T., Goodarzi, F., Hanna, J. V., and Vassallo, A. M. (1993). Evolution of the chemical structure of Hat Creek resinite during oxidation; a combined FT-IR photoacoustic, NMR and optical microscopic study. *Int. J. Coal Geol.* **22**: 119–147.

McGovern, S. J., Royce, B. S. H., and Benziger, J. B. (1984). Infrared photoacoustic spectroscopy of adsorption on powders. *Appl. Surf. Sci.* **18**: 401–413.

McGovern, S. J., Royce, B. S. H., and Benziger, J. B. (1985). The importance of interstitial gas expansion in infrared photoacoustic spectroscopy of powders. *J. Appl. Phys.* **57**: 1710–1718.

McKenna, W. P., Gale, D. J., Rivett, D. E., and Eyring, E. M. (1985). FT-IR photoacoustic spectroscopy investigation of oxidized wool. *Spectrosc. Lett.* **18**: 115–122.

McQueen, D. H., Wilson, R., and Kinnunen, A. (1995a). Near and mid-infrared photoacoustic analysis of principal components of foodstuffs. *Trends Anal. Chem.* **14**: 482–492.

McQueen, D. H., Wilson, R., Kinnunen, A., and Jensen, E. P. (1995b). Comparison of two infrared spectroscopic methods for cheese analysis. *Talanta* **42**: 2007–2015.

Mead, D. G., Lowry, S. R., Vidrine, D. W., and Mattson, D. R. (1979). Infrared spectroscopy using a photoacoustic cell. Fourth International Conference on Infrared and Millimeter Waves and Their Applications, p. 231. Miami Beach, FL, Dec. 10–15, 1979.

Mehicic, M., Kollar, R. G., and Grasselli, J. G. (1981). Analytical applications of photoacoustic spectroscopy using Fourier transform infrared (FTIR). *SPIE* **289**: 99–101.

Merker, U., Engels, P., Madeja, F., Havenith, M., and Urban, W. (1999). High-resolution CO-laser sideband spectrometer for molecular-beam optothermal spectroscopy in the 5–6.6 μm wavelength region. *Rev. Sci. Instrum.* **70**: 1933–1938.

Michaelian, K. H. (1987). Signal averaging of photoacoustic FTIR data. I. Computation of spectra from double-sided low resolution interferograms. *Infrared Phys.* **27**: 287–296.

Michaelian, K. H. (1989a). Depth profiling and signal saturation in photoacoustic FT-IR spectra measured with a step-scan interferometer. *Appl. Spectrosc.* **43**: 185–190.

Michaelian, K. H. (1989b). Interferogram symmetrization and multiplicative phase correction of rapid-scan and step-scan photoacoustic FT-IR data. *Infrared Phys.* **29**: 87–100.

Michaelian, K. H. (1990). Step-scan photoacoustic infrared spectra of kaolinite. *Infrared Phys.* **30**: 181–186.

Michaelian, K. H. (1990). Data treatment in photoacoustic FT-IR spectroscopy. In: *Vibrational Spectra and Structure*, J. R. Durig (ed.). Elsevier, Amsterdam, Vol. 18, pp. 81–126.

Michaelian, K. H. (1991). Depth profiling of oxidized coal by step-scan photoacoustic FT-IR spectroscopy. *Appl. Spectrosc.* **45**: 302–304.

Michaelian, K. H., and Friesen, W. I. (1990). Photoacoustic FT-IR spectra of separated western Canadian coal macerals. Analysis of the CH stretching region by curve-fitting and deconvolution. *Fuel* **69**: 1271–1275.

Michaelian, K. H., Bukka, K., and Permann, D. N. S. (1987). Photoacoustic infrared spectra (250–10,000 $cm^{-1}$) of partially deuterated kaolinite #9. *Can. J. Chem.* **65**: 1420–1423.

Michaelian, K. H., Yariv, S., and Nasser, A. (1991). Study of the interactions between caesium bromide and kaolinite by photoacoustic and diffuse reflectance infrared spectroscopy. *Can. J. Chem.* **69**: 749–754.

Michaelian, K. H., Ogunsola, O. I., and Bartholomew, R. J. (1995a). Infrared spectroscopy of thermally treated low-rank coals. II. Mid- and near-infrared photoacoustic spectra of coals heated in oxidizing and inert atmospheres. *Can. J. Appl. Spectrosc.* **40**: 94–99.

Michaelian, K. H., Friesen, W. I., Zhang, S. L., Gentzis, T., Crelling, J. C., and Gagarin, S. G. (1995b). FT-IR spectroscopy of western Canadian coals and macerals. In: *Coal Science*. J. A. Pajares and J. M. D. Tascón (eds.). Elsevier, Amsterdam, pp. 255–258.

Michaelian, K. H., Akers, K. L., Zhang, S. L., Yariv, S., and Lapides, I. (1997). Far-infrared spectra of kaolinite/alkali metal halide complexes. *Mikrochim. Acta* **14**(Suppl.): 211–212.

Michaelian, K. H., Lapides, I., Lahav, N., Yariv, S., and Brodsky, I. (1998). Infrared study of the intercalation of kaolinite by caesium bromide and caesium iodide. *J. Coll. Interf. Sci.* **204**: 389–393.

Michaelian, K. H., Zhang, S. L., Hall, R. H., and Bulmer, J. T. (2001). Photoacoustic infrared spectroscopy of distillation fractions from Syncrude Sweet Blend. Curve-fitting and integration of C—H stretching bands. *Can. J. Anal. Sci. Spectrosc.* **46**: 10–22.

Michaelian, K. H., Hall, R. H., and Bulmer, J. T. (2002). Photoacoustic infrared spectroscopy and thermophysical properties of Syncrude cokes. *J. Therm. Anal. Calorim.* **69**: 135–147.

Michaelian, K. H., Hall, R. H., and Bulmer, J. T. (2003). FT-Raman and photoacoustic infrared spectroscopy of Syncrude heavy gas oil distillation fractions. *Spectrochim. Acta A*. in press.

Miklós, A., and Fehér, M. (1996). Optoacoustic detection with near-infrared diode lasers: Trace gases and short-lived molecules. *Infrared Phys. Technol.* **37**: 21–27.

Miklós, A., Hess, P., Mohácsi, Á., Sneider, J., Kamm, S., and Schäfer, S. (1999a). Improved photoacoustic detector for monitoring polar molecules such as ammonia with a 1.53 μm DFB diode laser. *AIP Conf. Proc.* **463**: 126–128.

Miklós, A., Hess, P., Romolini, A., Spada, C., Lancia, A., Kamm, S., and Schäfer, S. (1999b). Measurement of the $(\nu_2 + 2\nu_3)$ band of methane by photoacoustic and long path absorption spectroscopy. *AIP Conf. Proc.* **463**: 217–219.

Mikula, R. J., Axelson, D. E., and Michaelian, K. H. (1985). Oxidation and weathering of stockpiled western Canadian coals. In: *Proceedings, 1985 International Conference on Coal Science*. Pergamon, Sydney, pp. 495–498.

Mohamed, M. M. (1995). Fourier-transform infrared/photoacoustic study of pyridine adsorbed on silica supported copper-molybdenum catalysts. *Spectrochim. Acta A* **51**: 1–9.

Monchalin, J.-P., Gagné, J.-M., Parpal, J.-L., and Bertrand, L. (1979). Photoacoustic spectroscopy of chrysotile asbestos using a cw HF laser. *Appl. Phys. Lett.* **35**: 360–363.

Morterra, C., and Low, M. J. D. (1982). The nature of the 1600 $cm^{-1}$ band of carbons. *Spectrosc. Lett.* **15**: 689–697.

Morterra, C., and Low, M. J. D. (1983). IR studies of carbons—II. The vacuum pyrolysis of cellulose. *Carbon* **21**: 283–288.

Morterra, C., and Low, M. J. D. (1985a). IR studies of carbons—IV. The vacuum pyrolysis of oxidized cellulose and the characterization of the chars. *Carbon* **23**: 301–310.

Morterra, C., and Low, M. J. D. (1985b). IR studies of carbons—VI. The effects of $KHCO_3$ on cellulose pyrolysis and char oxidation. *Carbon* **23**: 335–341.

Morterra, C., and Low, M. J. D. (1985c). IR studies of carbons—VII. The pyrolysis of a phenol-formaldehyde resin. *Carbon* **23**: 525–530.

Morterra, C., and Low, M. J. D. (1985d). IR studies of carbons. 8. The oxidation of phenol-formaldehyde chars. *Langmuir* **1**: 320–326.

Morterra, C., and Low, M. J. D. (1985e). An infrared spectroscopic study of some carbonaceous materials. *Mater. Chem. Phys.* **12**: 207–233.

Morterra, C., Low, M. J. D., and Severdia, A. G. (1984). IR studies of carbons—III. The oxidation of cellulose chars. *Carbon* **22**: 5–12.

Morterra, C., O'Shea, M. L., and Low, M. J. D. (1988). Infrared studies of carbons. IX. The vacuum pyrolysis of non-oxygen-containing materials: PVC. *Mater. Chem. Phys.* **20**: 123–144.

Nägele, M., and Sigrist, M. W. (2000). Mobile laser spectrometer with novel resonant multipass photoacoustic cell for trace-gas sensing. *Appl. Phys. B* **70**: 895–901.

Natale, M., and Lewis, L. N. (1982). Applications of PAS for the investigation of overtones and combinations in the near IR. *Appl. Spectrosc.* **36**: 410–413.

Nelson, J. H., MacDougall, J. J., Baglin, F. G., Freeman, D. W., Nadler, M., and Hendrix, J. L. (1982). Characterization of Carlin-type gold ore by photoacoustic, Raman, and EPR spectroscopy. *Appl. Spectrosc.* **36**: 574–576.

Neubert, R., Collin, B., and Wartewig, S. (1997). Direct determination of drug content in semisolid formulations using step-scan FT-IR photoacoustic spectroscopy. *Pharm. Res.* **14**: 946–948.

Nishikawa, Y., Kimura, K., Matsuda, A., and Kenpo, T. (1992). Surface characterization of gamma-iron oxides for magnetic memory media using Fourier transform infrared photoacoustic spectroscopy. *Appl. Spectrosc.* **46**: 1695–1698.

Olafsson, A., Hansen, G. I., Loftsdottir, A. S., and Jakobsson, S. (1999). FTIR photoacoustic trace gas detection. *AIP Conf. Proc.* **463**: 208–210.

O'Shea, M. L., Low, M. J. D., and Morterra, C. (1989). Spectroscopic studies of carbons. XI. The vacuum pyrolysis of non-oxygen-containing materials: PVBr. *Mater. Chem. Phys.* **23**: 499–516.

O'Shea, M. L., Morterra, C., and Low, M. J. D. (1990a). Spectroscopic studies of carbons. XIV. The vacuum pyrolysis of non-oxygen containing materials: PVF. *Mater. Chem. Phys.* **25**: 501–521.

O'Shea, M. L., Morterra, C., and Low, M. J. D. (1990b). Spectroscopic studies of carbons. XVII. Pyrolysis of polyvinylidene fluoride. *Mater. Chem. Phys.* **26**: 193–209.

O'Shea, M. L., Morterra, C., and Low, M. J. D. (1991a). Spectroscopic studies of carbons. XXII. The oxidation of polyvinyl halide chars. *Mater. Chem. Phys.* **28**: 9–31.

O'Shea, M. L., Morterra, C., and Low, M. J. D. (1991b). Spectroscopic studies of carbon. XX. The pyrolysis of polyvinylidene chloride and of Saran. *Mater. Chem. Phys.* **27**: 155–179.

Paldus, B. A., Spence, T. G., Zare, R. N., Oomens, J., Harren, F. J. M., Parker, D. H., Gmachl, C., Cappasso, F., Sivco, D. L., Baillargeon, J. N., Hutchinson, A. L., and Cho, A. Y. (1999). Photoacoustic spectroscopy using quantum-cascade lasers. *Opt. Lett.* **24**: 178–180.

Papendorf, U., and Riepe, W. (1989). Identification of adsorbed species on activated carbon by photoacoustic spectroscopy. In: Proceedings, 1989 International Conference on Coal Science, Vol. II., pp. 1111–1113. Tokyo, Oct. 23–27, 1989.

Pelletier, M., Michot, L. J., Barrès, O., Humbert, B., Petit, S., and Robert, J.-L. (1999). Influence of KBr conditioning on the infrared hydroxyl-stretching region of saponites. *Clay Miner.* **34**: 439–445.

Persijn, S. T., Veltman, R. H., Oomens, J., Harren, F. J. M., and Parker, D. H. (1999). A CO laser based photoacoustic system applied to the detection of trace gases emitted by conference pears stored at high $CO_2$ and low $O_2$ levels. *AIP Conf. Proc.* **463**: 609–611.

Persijn, S. T., Veltman, R. H., Oomens, J., Harren, F. J. M., and Parker, D. H. (2000). CO laser absorption coefficients for gases of biological relevance: $H_2O$, $CO_2$, ethanol, acetaldehyde, and ethylene. *Appl. Spectrosc.* **54**: 62–71.

Petkovska, L. T., and Miljanić, Š. S. (1997). $CO_2$-laser photoacoustic spectroscopy of deuterated ammonia. *Infrared Phys. Technol.* **38**: 331–336.

Politou, A. S., Morterra, C., and Low, M. J. D. (1990a). Infrared studies of carbons. XII. The formation of chars from a polycarbonate. *Carbon* **28**: 529–538.

Politou, A. S., Morterra, C., and Low, M. J. D. (1990b). Infrared studies of carbons. XIII. The oxidation of polycarbonate chars. *Carbon* **28**: 855–865.

Politou, A. S., Morterra, C., and Low, M. J. D. (1991). Infrared studies of carbons. XVI. The carbonization of an aliphatic allyl polycarbonate. *Polym. Degrad. Stab.* **32**: 331–356.

Radak, B. B., Pastirk, I., Ristić, G. S., and Petkovska, L. T. (1998). Pressure effects on $CO_2$-laser coincidences with ethene and ammonia investigated by photoacoustic detection. *Infrared Phys. Technol.* **39**: 7–13.

Raveh, A., Martinu, L., Domingue, A., Wertheimer, M. R., and Bertrand, L. (1992). Fourier transform infrared photoacoustic spectroscopy of amorphous carbon films. In: *Photoacoustic and Photothermal Phenomena III*. D. Bićanić (ed.). Springer, Berlin, pp. 151–154.

Repond, P., and Sigrist, M. W. (1994). Photoacoustic spectroscopy on gases with high pressure continuously tunable $CO_2$ laser. *J. Phys. IV, Colloq. C7* **4**: C7-523–C7-526.

Riseman, S. M., and Eyring, E. M. (1981). Normalizing infrared FT photoacoustic spectra of solids. *Spectrosc. Lett.* **14**: 163–185.

Riseman, S. M., Yaniger, S. I., Eyring, E. M., MacInnes, D., MacDiarmid, A. G., and Heeger, A. J. (1981). Infrared photoacoustic spectroscopy of conducting polymers. I. Undoped and *n*-doped polyacetylene. *Appl. Spectrosc.* **35**: 557–559.

Riseman, S. M., Massoth, F. E., Dhar, G. M., and Eyring, E. M. (1982). Fourier transform infrared photoacoustic spectroscopy of pyridine adsorbed on silica-alumina and $\gamma$-alumina. *J. Phys. Chem.* **86**: 1760–1763.

Royce, B. S. H., and Benziger, J. B. (1986). Fourier transform photoacoustic spectroscopy of solids. *IEEE Trans. UFFC* **33**: 561–572.

Rockley, M. G., and Devlin, J. P. (1980). Photoacoustic infrared spectra (IR-PAS) of aged and fresh-cleaved coal surfaces. *Appl. Spectrosc.* **34**: 407–408.

Rockley, N. L., and Rockley, M. G. (1987). The FT-IR analysis by PAS and KBr pellet of cation-exchanged clay mineral and phosphonate complexes. *Appl. Spectrosc.* **41**: 471–475.

Royce, B. S. H., Teng, Y. C., and Enns, J. (1980). Fourier transform infrared photoacoustic spectroscopy of solids. *Ultrasonics Symposium Proceedings*, pp. 652–657.

Rockley, M. G., Richardson, H. H., and Davis, D. M. (1982). The Fourier-transformed infrared photoacoustic spectroscopy of asbestos fiber. *J. Photoacoustics* **1**: 145–149.

Rockley, M. G., Ratcliffe, A. E., Davis, D. M., and Woodard, M. K. (1984). Examination of a C-black by various FT-IR spectroscopic methods. *Appl. Spectrosc.* **38**: 553–556.

Ryczkowski, J. (1994). FTIR and FTIR PAS applications in qualitative analyses of EDTA adsorption on alumina surface. *SPIE* **2089**: 182–183.

Sadler, A. J., Horsch, J. G., Lawson, E. Q., Harmatz, D., Brandau, D. T., and Middaugh, C. R. (1984). Near-infrared photoacoustic spectroscopy of proteins. *Anal. Biochem.* **138**: 44–51.

Salnick, A. O., and Faubel, W. (1995). Photoacoustic FT-IR spectroscopy of natural copper patina. *Appl. Spectrosc.* **49**: 1516–1524.

Salnick, A., and Faubel, W. (1996). Photoacoustic FT-IR spectroscopy of natural copper patina. *Prog. Nat. Sci.* **6**(suppl.): S14–S17.

Santosa, E., te Lintel Hekkert, S., Harren, F. J. M., and Parker, D. H. (1999). The $\Delta v = 2$ CO laser: Photoacoustic trace gas detection. *AIP Conf. Proc.* **463**: 612–614.

Sarma, T. V. K., Sastry, C. V. R., and Santhamma, C. (1987). The photoacoustic spectra of substituted benzenes in the near infrared region—I. *Spectrochim. Acta A* **43**: 1059–1065.

Saucy, D. A., Simko, S. J., and Linton, R. W. (1985a). Comparison of photoacoustic and attenuated total reflectance sampling depths in the infrared region. *Anal. Chem.* **57**: 871–875.

Saucy, D. A., Cabaniss, G. E., and Linton, R. W. (1985b). Surface reactivities of polynuclear aromatic adsorbates on alumina and silica particles using infrared photoacoustic spectroscopy. *Anal. Chem.* **57**: 876–879.

Schäfer, S., Mashni, M., Sneider, J., Miklós, A., Hess, P., Pitz, H., Pleban, K.-U., and Ebert, V. (1998a). Sensitive detection of methane with a 1.65 μm diode laser by photoacoustic and absorption spectroscopy. *Appl. Phys. B* **66**: 511–516.

Schäfer, S., Miklós, A., Pusel, P., and Hess, P. (1998b). Absolute measurement of gas concentrations and saturation behavior in pulsed photoacoustics. *Chem. Phys. Lett.* **285**: 235–239.

Schendzielorz, A., Hanh, B. D., Neubert, R. H. H., and Wartewig, S. (1999). Penetration studies of clotrimazole from semisolid formulation using step-scan FT-IR photoacoustic spectroscopy. *Pharm. Res.* **16**: 42–45.

Schüle, G., Schmitz, B., and Steiner, R. (1999). Spectral photothermal tissue diagnostics with step-scan FT-IR technique. *AIP Conf. Proc.* **463**: 615–617.

Seaverson, L. M., McClelland, J. F., Burnet, G., Andregg, J. W., and Iles, M. K. (1985). Investigation of water and hydroxyl groups associated with coal fly ash by thermal desorption and Fourier transform infrared photoacoustic spectroscopies. *Appl. Spectrosc.* **39**: 38–45.

Seiter, M., and Sigrist, M. W. (2000). Trace-gas sensor based on mid-IR difference-frequency generation in PPLN with saturated output power. *Infrared Phys. Technol.* **41**: 259–269.

Shoval, S., Yariv, S., Michaelian, K. H., Boudeuille, M., and Panczer, G. (1999a). Hydroxyl stretching bands "A" and "Z" in Raman and IR spectra of kaolinites. *Clay Miner.* **34**: 551–563.

Shoval, S., Yariv, S., Michaelian, K. H., Lapides, I., Boudeuille, M., and Panczer, G. (1999b). A fifth OH-stretching band in IR spectra of kaolinites. *J. Coll. Interf. Sci.* **212**: 523–529.

Shoval, S., Yariv, S., Michaelian, K. H., Boudeulle, M., and Panczer, G. (2001). Hydroxyl-stretching bands in curve-fitted micro-Raman, photoacoustic and transmission infrared spectra of dickite from St. Claire, Pennsylvania. *Clays Clay Miner.* **49**: 347–354.

Shoval, S., Yariv, S., Michaelian, K. H., Boudeulle, M., and Panczer, G. (2002a). Hydroxyl stretching bands in polarized micro-Raman spectra of orientated single-crystal Keokuk kaolinite. *Clays Clay Miner.* **50**: 56–62.

Shoval, S., Michaelian, K. H., Boudeulle, M., Panczer, G., Lapides, I., and Yariv, S. (2002b). Study of thermally treated dickite by curve-fitted transmission infrared, photoacoustic and micro-Raman spectra. *J. Thermal Anal. Calorim.* **69**: 205–225.

Siew, D. C. W., Heilmann, C., Easteal, A. J., and Cooney, R. P. (1999). Solution and film properties of sodium caseinate/glycerol and sodium caseinate/polyethylene glycol edible coating systems. *J. Agric. Food Chem.* **47**: 3432–3440.

Solomon, P. R., and Carangelo, R. M. (1982). FTIR analysis of coal. 1. Techniques and determination of hydroxyl concentrations. *Fuel* **61**: 663–669.

Sowa, M. G., and Mantsch, H. H. (1993). Phase modulated-phase resolved photoacoustic FT-IR study of calcified tissues. *SPIE* **2089**: 128–129.

Sowa, M. G., and Mantsch, H. H. (1994). FT-IR step-scan photoacoustic phase analysis and depth profiling of calcified tissue. *Appl. Spectrosc.* **48**: 316–319.

Sowa, M. G., Wang, J., Schultz, C. P., Ahmed, M. K., and Mantsch, H. H. (1995). Infrared spectroscopic investigation of in vivo and ex vivo human nails. *Vib. Spectrosc.* **10**: 49–56.

St.-Germain, F. G. T., and Gray, D. G. (1987). Photoacoustic Fourier transform infrared spectroscopic study of mechanical pulp brightening. *J. Wood Chem. Technol.* **7**: 33–50.

Teramae, N., and Tanaka, S. (1981). Structure elucidation of textile fabrics by Fourier transform infrared photoacoustic spectroscopy. *Spectrosc. Lett.* **14**: 687–694.

Teramae, N., and Tanaka, S. (1984). Fourier transform infrared photoacoustic spectroscopy of film-like samples. *Bunseki Kagaku* **33**: E397–E400.

Teramae, N., and Tanaka, S. (1985). Subsurface layer detection by Fourier transform infrared photoacoustic spectroscopy. *Appl. Spectrosc.* **39**: 797–799.

Teramae, N., and Tanaka, S. (1987). Fourier transform infrared photoacoustic spectroscopy of films. In: *Fourier Transform Infrared Characterization of Polymers*. H. Ishida (ed.). Plenum, New York, pp. 315–340.

Teramae, N., and Tanaka, S. (1988). Effect of heat from rear surface of a film sample on spectral features in Fourier transform infrared photoacoustic spectroscopy. *Anal. Chem.* **57**: 95–99.

Teramae, N., Hiroguchi, M., and Tanaka, S. (1982). Fourier transform infrared photoacoustic spectroscopy of polymers. *Bull. Chem. Soc. Jpn.* **55**: 2097–2100.

Varlashkin, P. G., and Low, M. J. D. (1986). FT-IR photothermal beam deflection spectroscopy of solids submerged in liquids. *Appl. Spectrosc.* **40**: 1170–1176.

Vidrine, D. W. (1980). Photoacoustic Fourier transform infrared spectroscopy of solid samples. *Appl. Spectrosc.* **34**: 314–319.

Vidrine, D. W. (1981). Photoacoustic Fourier transform infrared (FTIR) spectroscopy of solids. *SPIE* **289**: 355–360.

Vidrine, D. W. (1984). Photoacoustic and reflection FT-IR spectrometry: photometric approximations for practical quantitative analysis. *Preprints, ACS Div. Fuel Chem.* **25**: 147–148.

Vidrine, D. W., and Lowry, S. R. (1983). Photoacoustic Fourier transform IR spectroscopy and its application to polymer analysis. *Adv. Chem. Ser. (Polym. Charact.)* **203**: 595–613.

Wahls, M. W. C., Kenttä, E., and Leyte, J. C. (2000). Depth profiles in coated paper: Experimental and simulated FT-IR photoacoustic difference magnitude spectra. *Appl. Spectrosc.* **54**: 214–220.

Wang, N., and Low, M. J. D. (1989). Spectroscopic studies of carbons. XV. The pyrolysis of a lignin. *DOE/PC/*79920–10, *DE*90 006239, pp. 1–14.

Wang, N., and Low, M. J. D. (1990a). Spectroscopic studies of carbons. XVIII. The charring of rice hulls. *Mater. Chem. Phys.* **26**: 117–130.

Wang, N., and Low, M. J. D. (1990b). Spectroscopic studies of carbons. XIX. The charring of sucrose. *Mater. Chem. Phys.* **26**: 465–481.

Wang, N., and Low, M. J. D. (1991). Spectroscopic studies of carbons. XXI. An infrared study of the charring of coconut shell. *Mater. Chem. Phys.* **27**: 359–374.

Wang, H. P., Eyring, E. M., and Huai, H. (1991). Photoacoustic enhancement of surface IR modes in zeolite channels. *Appl. Spectrosc.* **45**: 883–885.

Wentrup-Byrne, E., Rintoul, L., Gentner, J. M., Smith, J. L., and Fredericks, P. M. (1997). Photoacoustic spectroscopy vs. chemical categorization of human gallstones. *Mikrochim. Acta* **14**(Suppl.): 615–616.

Wetzel, D. L., and Carter, R. O. (1998). Synchrotron powered FT-IR microspectroscopic incremental probing of photochemically degraded polymer films. *AIP Conf. Proc.* **430**: 567–570.

Wolff, M., and Harde, H. (2000). Photoacoustic spectrometer based on a DFB-diode laser. *Infrared Phys. Technol.* **41**: 283–286.

Xu, Z. H., Butler, I. S., and St.-Germain, F. G. T. (1986). FT-IR photoacoustic spectra of gaseous group VIB metal chalocarbonyl complexes. *Appl. Spectrosc.* **40**: 1004–1009.

Yamada, O., Yasuda, H., Soneda, Y., Kobayashi, M., Makino, M., and Kaiho, M. (1996). The use of step-scan FT-IR/PAS for the study of structural changes in coal and char particles during gasification. *Preprints, ACS Div. Fuel Chem.* **41**: 93–97.

Yang, C. Q. (1991). Comparison of photoacoustic and diffuse reflectance infrared spectroscopy as near-surface analysis techniques. *Appl. Spectrosc.* **45**: 102–108.

Yang, C. Q., and Bresee, R. R. (1987). Studies of sized cotton yarns by FT-IR photoacoustic spectroscopy. *J. Coated Fabrics* **17**: 110–128.

Yang, C. Q., and Fateley, W. G. (1987). Fourier-transform infrared photoacoustic spectroscopy evaluated for near-surface characterization of polymeric materials. *Anal. Chim. Acta* **194**: 303–309.

Yang, C. Q., and Freeman, J. M. (1991). Photo-oxidation of cotton cellulose studied by FT-IR photoacoustic spectroscopy. *Appl. Spectrosc.* **45**: 1965–1698.

Yang, H., and Irudayaraj, J. (2000a). Characterization of semisolid fats and edible oils by Fourier transform infrared photoacoustic spectroscopy. *J. Am. Oil Chem. Soc.* **77**: 291–295.

Yang, H., and Irudayaraj, J. (2000b). Depth profiling Fourier transform analysis of cheese package using generalized two-dimensional photoacoustic correlation spectroscopy. *Trans. Am. Soc. Agric. Eng.* **43**: 953–961.

Yang, H., and Irudayaraj, J. (2001a). Comparison of near infrared, Fourier transform infrared, and Fourier tranform-Raman methods for determining olive pomace oil adulteration in extra virgin olive oil. *J. Am. Oil Chem. Soc.* **78**: 889–895.

Yang, H., and Irudayaraj, J. (2001b). Characterization of beef and pork using Fourier transform infrared photoacoustic spectroscopy. *Lebensm.-Wiss. u.-Technol.* **34**: 402–409.

Yang, C. Q., and Simms, J. R. (1993). Infrared spectroscopy studies of the petroleum pitch carbon fiber—I. The raw materials, the stabilization, and carbonization processes. *Carbon* **31**: 451–459.

Yang, C. Q., and Simms, J. R. (1995). Comparison of photoacoustic, diffuse reflectance and transmission infrared spectroscopy for the study of carbon fibres. *Fuel* **74**: 543–548.

Yang, C. Q., Ellis, T. J., Bresee, R. R., and Fateley, W. G. (1985). Depth profiling of FT-IR photoacoustic spectroscopy and its application for polymeric material studies. *Polymer. Mater. Sci. Eng.* **53**: 169–175.

Yang, C. Q., Bresee, R. R., and Fateley, W. G. (1987). Near-surface analysis and depth profiling by FT-IR photoacoustic spectroscopy. *Appl. Spectrosc.* **41**: 889–896.

Yang, C. Q., Perenich, T. A., and Fateley, W. G. (1989). Studies of foam finished cotton fabrics using FT-IR photoacoustic spectroscopy. *Textile Res. J.* **59**: 562–568.

Yang, C. Q., Bresee, R. R., and Fateley, W. G. (1990). Studies of chemically modified poly(ethylene terephthalate) fibers by FT-IR photoacoustic spectroscopy and X-ray photoelectron spectroscopy. *Appl. Spectrosc.* **44**: 1035–1039.

Yang, H., Irudayaraj, J., and Sakhamuri, S. (2001). Characterization of edible coatings and microorganisms on food surfaces using Fourier transform infrared photoacoustic spectroscopy. *Appl. Spectrosc.* **55**: 571–583.

Yaniger, S. I., Riseman, S. M., Frigo, T., and Eyring, E. M. (1982). Infrared photoacoustic spectroscopy of conducting polymers. II. P-doped polyacetylene. *J. Chem. Phys.* **76**: 4298–4299.

Yariv, S., Nasser, A., Michaelian, K. H., Lapides, I., Deutsch, Y., and Lahav, N. (1994). Thermal treatment of the kaolinite/CsCl/$H_2O$ intercalation complex. *Thermochim. Acta* **234**: 275–285.

Zachmann, G. (1984). FT-IR spectroscopy of solid surfaces. *J. Mol. Struct.* **115**: 465–468.

Zeninari, V., Tikhomirov, B. A., Ponomarev, Yu. N., and Courtois, D. (1998). Preliminary results on photoacoustic study of the relaxation of vibrationally excited ozone ($\nu_3$). *J. Quant. Spectrosc. Radiat. Transfer* **59**: 369–375.

Zeninari, V., Kapitanov, V. A., Coutois, D., and Ponomarev, Yu. N. (1999). Design and characteristics of a differential Helmholtz resonant photoacoustic cell for infrared gas detection. *Infrared Phys. Technol.* **40**: 1–23.

Zeninari, V., Tikhomirov, B. A., Ponomarev, Yu. N., and Courtois, D. (2000). Photoacoustic measurements of the vibrational relaxation of the selectively excited ozone ($\nu_3$) molecule in pure ozone and its binary mixtures with $O_2$, $N_2$, and noble gases. *J. Chem. Phys.* **112**: 1835–1843.

Zerlia, T. (1985). Fourier transform infrared photoacoustic spectroscopy of raw coal. *Fuel* **64**: 1310–1312.

Zerlia, T. (1986). Depth profile study of large-sized coal samples by Fourier transform infrared photoacoustic spectroscopy. *Appl. Spectrosc.* **40**: 214–217.

## RECOMMENDED READING

Bićanić, D., Jalink, H., Chirtoc, M., Sauren, H., Lubbers, M., Quist, J., Gerkema, E., van Asselt, K., Miklós, A., Sólyom, A., Angeli, Gy. Z., Helander, P., and Vargas, H. (1992). Interfacing photoacoustic and photothermal techniques for new hyphenated methodologies and instrumentation suitable for agricultural, environmental and medical applications. In: *Photoacoustic and Photothermal Phenomena III*. D. Bićanić (ed.). Springer, Berlin, pp. 20–27.

Burt, J. A., Michaelian, K. H., and Zhang, S. L. (1997). Photothermal absorption spectroscopy of mica. *Mikrochim. Acta* **14**(Suppl.): 173–174.

Farmer, V. C. (ed.). (1974). *The Infrared Spectra of Minerals.* The Mineralogical Society, London.

Farmer, V. C. (1998). Differing effects of particle size and shape in the infrared and Raman spectra of kaolinite. *Clay Miner.* **33**: 601–604.

Hess, P. (ed.). (1989). *Photoacoustic, Photothermal and Photochemical Processes in Gases*. Top. Curr. Phys., Vol. 46, Springer, Berlin.

Kaplanová, M., and Katuščáková, G. (1992). Photoacoustic study of the thermal effusivity of cellulose and paper. In: *Photoacoustic and Photothermal Phenomena III*. D. Bićanić (ed.). Springer, Berlin, pp. 180–182.

McClelland, J. F., Luo, S., Jones, R. W., and Seaverson, L. M. (1992). A tutorial on the state-of-the-art of FTIR photoacoustic spectroscopy. In: *Photoacoustic and Photothermal Phenomena III*. D. Bićanić (ed.). Springer, Berlin, pp. 113–124.

Michaelian, K. H. (2001). Photoacoustic infrared spectroscopy of bitumens and clays. In: *Frontiers in Science and Technology*. Stefan University Press, La Jolla, CA.

Monchalin, J.-P., Bertrand, L., and Rousset, G. (1984). Photoacoustic spectroscopy of thick powdered or porous samples at low frequency. *J. Appl. Phys.* **56**: 190–210.

Saffa, A. M., and Michaelian, K. H. (1993). Quantitative analysis of clay mixtures by photoacoustic FT-IR spectroscopy. *SPIE* **2089**: 566–567.

Saffa, A. M., and Michaelian, K. H. (1994). Quantitative analysis of kaolinite/silica and kaolinite/KBr mixtures by photoacoustic FT-IR spectroscopy. *Appl. Spectrosc.* **48**: 871–874.

Sigrist, M. W. (ed.). (1994). *Air Monitoring by Spectroscopic Techniques*, Chem. Anal. Vol. 127, Wiley (Interscience), New York.

Willing, B., Muralt, P., and Oehler, O. (1999). Infrared gas spectrometer based on a pyroelectric thin film array detector. *AIP Conf. Proc.* **463**: 277–279.

Yariv, S., and Cross, H. (2002). *Organo-Clay Complexes and Interactions.* Marcel Dekker, New York.

Zhang, S. L., Michaelian, K. H., Bulmer, J. T., Hall, R. H., and Hellman, J. L. (1996). Fourier transform Raman spectroscopy of fuels: Curve-fitting of C—H stretching bands. *Spectrochim. Acta A* **52**: 1529–1540.

Zhang, S. L., Michaelian, K. H., and Burt, J. A. (1997). Phase correction in piezoelectric photoacoustic Fourier transform infrared spectroscopy of mica. *Opt. Eng.* **36**: 321–325.

# CHAPTER 7
# QUANTITATIVE ANALYSIS

The two quantities of greatest interest in virtually any type of spectroscopy are, of course, band positions and intensities—the former generally conveying qualitative information, the latter quantitative. As regards PA infrared spectroscopy, the reader may have already concluded that a substantial amount of the published literature describes qualitative (or, alternatively, semiquantitative) applications of the technique. However, a different perspective is assumed in the present chapter: This narrative is given over to a description of the use of PA infrared spectroscopy for the quantitative analysis of condensed-phase materials. It should be noted that PA spectra of gases, which were briefly reviewed in the previous chapter, are not included in this discussion.

## 7.1. QUANTITATION IN PA NEAR-INFRARED SPECTROSCOPY

The potentiality for the utilization of near-infrared PA intensities for quantitative analysis was examined in several pioneering investigations by G. F. Kirkbright and his collaborators that were referred to in previous chapters. Indeed, it would appear that these authors considered the importance of this issue to be comparable to that of the reliable assignment of the bands in the PA spectra they obtained. This perspective is illustrated by the discussion in the following paragraphs.

Adams et al. (1978) showed that a characteristic near-infrared band at 2.2 μm (4545 $cm^{-1}$) could be used to identify aromatic hydrocarbons in the presence of aliphatics. PA spectra of mixtures of benzene in *n*-hexane were analyzed, and a linear variation in the intensity of this band with benzene concentration was established. Similar results were obtained when benzene was added to an Iranian crude oil at concentrations of 10 and 30%.

These findings are significant inasmuch as they demonstrate the feasibility of using near-infrared PA spectroscopy for the quantitation of total aromatics in hydrocarbon fuels. It can also be pointed out that this specific analysis will become increasingly important in the future as sources of conventional, light crude oils are depleted and the use of heavier (more aromatic) crudes becomes more prevalent. An alternative FT-Raman method for the determination of total aromatics in fuels already exists; however, Raman spectra are plagued by fluorescence when hydrocarbon samples with very high boiling points are analyzed. Fortunately, such a limitation does not occur in PA infrared spectroscopy.

The familiar, strong absorption by water in the mid- and near-infrared regions provides another opportunity for quantitative analysis by means of PA infrared spectroscopy. This is demonstrated by work on the determination of moisture content in single-cell protein samples and milk substitutes (Castleden et al., 1980; Jin et al., 1982). As discussed in the previous chapter, the quantitative determination of moisture in these samples was shown to be feasible at concentrations that ranged up to about 10%, provided that the samples were first subdivided according to particle size. When this essential preliminary step was carried out, linear correlations between the intensities of the 1.9-μm (5265-$cm^{-1}$) band and the amounts of moisture in the samples were obtained. It should also be recognized that this technique could be applied—at both near- and mid-infrared wavelengths—to other foods, grains, and various classes of industrial materials.

The final demonstration of quantitative analysis by near-infrared PA spectroscopy to be mentioned here involved the characterization of drug tablets (Castleden et al., 1982). This study also gave very encouraging results: Drug concentrations as high as 36% yielded linear calibrations relating the intensities of specific bands to the amount of drug present in each sample. Similar results were obtained using ultraviolet and visible PA spectra. To summarize the preceding comments on these three studies, the research of Kirkbright's group demonstrated that PA near-infrared spectroscopy is well suited for quantitative studies, as long as analyte concentrations were in the low-to-moderate range.

## 7.2. QUANTITATION IN PA MID-INFRARED SPECTROSCOPY

The evolution of a spectroscopic technique logically passes through several stages, including its initial demonstration, acceptance by the scientific community, and refinement; this tends to be followed by its subsequent use for qualitative and, eventually, quantitative analysis. Thus, after the original successful experiments showing the feasibility of PA infrared spectroscopy, a number of investigators became interested in the question as to whether the technique could realistically be used for quantitative measurements. It was already known that ultraviolet and visible PA spectra are sometimes subject to saturation: As this effect begins to influence a series of spectra, the likelihood of reliable quantitative analysis is diminished. On the other hand, saturation effects would generally be expected to be less important in the infrared, where absorption coefficients are typically orders of magnitude smaller than those at shorter wavelengths. In light of this situation, numerous research groups examined the possibility of using PA infrared spectroscopy for quantitative analysis in a variety of different applications. Some of the relevant publications on this subject will now be mentioned, beginning with the earlier investigations.

A series of publications on PA FTIR spectroscopy in the 1980s took up the issue of quantitative analysis. Of course, suitable samples must be examined so as to elucidate this issue. Rockley et al. (1980) first proposed that homogeneous mixtures prepared from molten naphthalene and benzophenone be used. Perhaps surprisingly, the authors found that the relative contributions to the PA spectra of the mixtures were not proportional to the known concentrations of the two components. It was suggested that this observation was due to the occurrence of eutectic mixtures of the hydrocarbons.

A more ideal system, consisting of intimate mixtures of $K^{14}NO_3$ and $K^{15}NO_3$, was next investigated. The spectra of these two salts can be distinguished by the shift of an absorption band from 825 to 800 $cm^{-1}$ that is caused by the increase in nitrogen mass; this makes the relative contributions of the two compounds to the observed spectra easily identifiable. Mixtures containing from 1 to 99% $K^{15}NO_3$ were analyzed, and a linear relationship between the relative contribution of this isotope to the spectrum and its abundance was observed (Rockley et al., 1981). This encouraging result is one of the first that suggested the feasibility of using mid-infrared PA intensities for quantitative analysis. However, these findings are accompanied by a caveat: The plot of relative intensity of the $K^{15}NO_3$ peak vs. the percentage of this species in the samples does not pass through zero. This situation may be attributable to inaccurate resolution of the two overlapping peaks or, alternatively, to the contamination of this salt with the more common $K^{14}NO_3$.

Shortly after the publication of the work by Rockley et al. (1981), E. M. Eyring and his colleagues at the University of Utah employed PA infrared spectroscopy for the quantitative analysis of catalytic surface adsorption sites. One of the objectives of this work was to determine the relative concentrations of Brønsted and Lewis acid sites on silica-alumina through the analysis of the PA spectrum of adsorbed pyridine. The results were compared with those for adsorption on $\gamma$-alumina, which has only Lewis acid sites. Riseman et al. (1982) showed that the strong low-frequency bands of the substrates could be used to normalize the spectra; this step removed the influence of particle size effects and various other experimental factors. The scaled PA spectra showed that 20% of the surface hydroxyls in silica-alumina interacted with pyridine, a result that was corroborated by other data. In subsequent work (Gardella et al., 1983) the amount of CO adsorbed onto an $Ni/SiO_2$ catalyst surface was calculated. A linear calibration, up to an added volume of 300 μL of CO, was obtained by plotting PA intensity vs. the volume of CO gas that was introduced. However, PA signal saturation occurred when larger amounts of CO were used. The PA intensities of the gas were again calculated after normalizing the spectra to constant silica band intensity in this work.

The occurrence of saturation in PA infrared spectra of adsorbed species was also investigated by Highfield and Moffat (1985). The sorption of $NH_3$ or pyridine on 12-tungstophosphoric acid, $H_3PW_{12}O_{40}$, was studied in this work. In one series of experiments, the $NH_3$ band at 1420 $cm^{-1}$ intensified linearly with the amount of gas that was added. On the other hand, the characteristic 1537-$cm^{-1}$ band of pyridine exhibited a linear dependence on the amount of analyte only at lower concentrations. The intensity of this band grew more gradually as increasing amounts of pyridine were adsorbed, suggesting that saturation was limiting the PA intensities. The authors used known thermal properties for compounds that resemble $H_3PW_{12}O_{40}$ to calculate a numerical value of 55 $cm^{-1}$ for $(\alpha/\pi f)^{-1/2}$, the reciprocal of the thermal diffusion length, for the pyridine band. The absorptivity of the 1537-$cm^{-1}$ band is known to be approximately 60 L $mol^{-1}$ $cm^{-1}$. The concentration of pyridine at the suspected onset of saturation was estimated as 1.9 mol $L^{-1}$; this in turn yielded an absorption coefficient of 114 $cm^{-1}$. Because this value is significantly greater than the 55 $cm^{-1}$ calculated for $(\alpha/\pi f)^{-1/2}$ in the PA infrared experiment, it was concluded that signal saturation was the likely cause of the nonlinear increase in PA intensity at higher concentrations.

The dependence of PA intensity on analyte concentration was also studied in an early single-wavelength PA experiment. Yokoyama et al. (1984) used the 1081-$cm^{-1}$ emission line of a $CO_2$ laser to monitor the PA signal arising from the $\nu_3$ sulfate band in $KAl(SO_4)_2 \cdot 12H_2O$, $CaSO_4 \cdot 2H_2O$, and $K_2SO_4$. These compounds were each divided into several size fractions and studied as binary mixtures with NaCl at concentrations that ranged from 0 to 100%. At very low concentrations (below 1%) PA intensity increased linearly with the amount of sulfate present in each mixture. However, the intensification of the PA signal was more gradual when the amount of analyte was greater. Yokoyama et al. noted that light scattering by the diluent can also contribute to PA intensity and attributed the nonlinear increase in PA intensity to a diminution of this scattering at higher analyte concentrations. Thermal interactions were also thought to play a role in the reduction of the PA intensity. Signal saturation was not mentioned in this study.

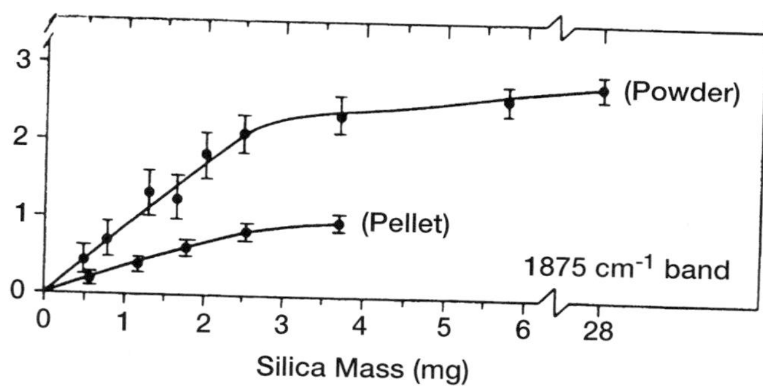


**Figure 7.1.** Variation of intensity of 1875-$cm^{-1}$ silica band in PA infrared spectra of powder (upper curve) and pellet (lower curve) forms of silica/kaolin mixtures. Nonlinear increase at higher concentrations was attributed to saturation. (Reproduced from Pandurangi, R. S. and Seehra, M. S., *Appl. Spectrosc.* **46**: 1719–1723, by permission of the Society for Applied Spectroscopy; copyright © 1992.)

In subsequent work, several research groups compared the suitability of PA FTIR spectroscopy with that of other infrared techniques for quantitative analysis. For example, Rosenthal et al. (1988) used PA and transmission infrared spectroscopies to analyze mixtures of acetylsalicylic acid, salicylic acid, and filler or binder used in the preparation of commercial analgesic tablets. Partial least-squares (PLS) analysis of the PA data gave correlation coefficients of 0.97 or higher for calibration spectra and at least 0.93 for validation samples. The latter results were obtained for concentrations between 1 and 5%. This study incorrectly states that only one previous study on the application of PA infrared spectroscopy to quantitative analysis existed at the time of publication.

PA infrared spectroscopy was compared with the diffuse reflectance and transmission techniques in the context of the determination of silica in silica–kaolin mixtures by Pandurangi and Seehra (1992). Both powders and pellets were examined in the PA and diffuse reflectance experiments. As shown in Figure 7.1, the PA intensities of several different silica bands increased in an approximately linear fashion at low concentrations; however, this intensification was much more gradual at higher concentrations. The latter results were attributed to saturation. By contrast, diffuse reflectance yielded a wider range of linearity than that in PA spectroscopy. It should also be noted that the onset of saturation for the weaker bands in the PA spectra occurred at higher concentrations.

Two examples of quantitation in PA infrared spectroscopy were mentioned in the previous chapter. In Section 6.11 on biology and biochemistry, the work of Gordon et al. (1990) on the growth of microorganisms on cellulose was discussed. As shown in Figure 6.38, the intensity of the protein amide I band in the PA spectra of *Saccharomyces cerevisiae* is less influenced by saturation if the spectra are normalized using a polyacrylonitrile band as an internal standard. It should be pointed out that the saturation referred to in this case arises from layering of the yeast on the cellulose substrate, a somewhat different phenomenon than the effect that occurs at high concentrations in the binary mixtures described above.

Saturation was much less troublesome in the PA infrared spectra of individual pea seeds (Letzelter et al., 1995). Mixtures of starch in KBr first showed evidence of PA saturation at concentrations of about 50%, whereas protein–starch mixtures exhibited a linear variation in the intensity of the amide I band over an even wider concentration range. The authors concluded that the use of PA spectra for quantitative determination of the starch, lipid, and protein contents of pea seeds was quite feasible.

It is important to include the work of J. F. McClelland and his colleagues at Iowa State University on quantitative analysis in this discussion. In a study of both qualitative and quantitative applications of PA infrared spectroscopy, McClelland et al. (1991) obtained spectra of copolymers of vinyl acetate in polyethylene over a wide concentration range. After normalizing the PA spectra to constant integrated intensity for the entire C—H stretching region, the authors found that the vinyl acetate concentrations predicted by factor analysis of three different spectral regions showed excellent agreement with the concentrations measured by titration, an accepted method of analysis.

These experiments were soon described again in a wide-ranging review article by the same authors (McClelland et al., 1992). Importantly, the latter work also reported results of an even more impressive experiment in which a multicomponent system comprising coal and three of its commonly associated minerals (kaolinite, pyrite, quartz) was examined using PA spectroscopy. The total mineral concentration in each of the 15 samples studied was 30% by weight, with the amount of each mineral varying from 0 to 30%. Excellent results were obtained in these experiments: Factor analysis gave standard errors of prediction between 1 and 2% for all three minerals when single samples were successively treated as unknowns and the others were used as a learning set.

Finally, Norton and McClelland (1997) investigated the feasibility of using PA infrared spectroscopy for process control in the industrial setting in which lime (CaO) is obtained by heating limestone ($CaCO_3$). The residual limestone from this commercial process is traditionally determined by an ignition test, which is relatively slow and also susceptible to errors. PA spectroscopy was shown to be a viable alternative method of analysis in this work: $CaCO_3$ can be identified by its characteristic band at 2513 $cm^{-1}$ in spectra of binary $CaCO_3/CaO$ mixtures. Norton and McClelland obtained a linear calibration curve for limestone concentrations up to 5% in this way. An impressive detection limit of about 0.1% was thought to be achievable with this method.

The above discussion can be briefly summarized through the statement that PA infrared spectroscopy has been successfully utilized for quantitative analysis by many research groups over the last two decades, and that a wide variety of samples are amenable to this approach. The greatest success has been achieved for analyte concentrations up to a few percent; in favorable cases, much higher concentrations have been examined successfully. On the other hand, saturation limits the capability of PA infrared spectroscopy for quantitative analysis in some circumstances.

## 7.3. QUANTITATIVE ANALYSIS AT HIGHER CONCENTRATIONS

Obviously, it would be desirable to widen the concentration range in which PA infrared spectroscopy might be used for quantitative analysis. A numerical method that addresses this objective was developed by Belton and his co-workers in the 1980s and subsequently implemented in the current author's laboratory. This procedure is summarized in this section.

As mentioned in the previous discussion on PA spectra of food products, Belton and Tanner (1983) used near-infrared PA spectroscopy to determine the water content in starch. Their results showed that the intensity of the characteristic 1.9-μm water band initially increased linearly with concentration, but that the intensification proceeded in a more gradual manner as the amount of water in the samples was further increased. This behavior—which resembles that in several other studies discussed in the present chapter—limited the suitability of PA spectroscopy for quantitative analysis and led the authors to develop the theory described in the following paragraphs.

The general expression for PA intensity $H$, resulting from the absorption of light by a thermally thick sample (Poulet et al., 1980), may be written as

$$H = AI_o 2^{1/2} \mu\beta / [(\mu\beta)^2 + (\mu\beta + 2)^2]^{1/2} \tag{7.1}$$

where $A$ is a composite term that relates the thermal wave and PA signal intensities, $I_o$ is the intensity of the incident light, and $\mu$ is the thermal diffusion length. The absorption coefficient for a chromophore mixed with a diluent is given by

$$\beta = 2.303 \in W_{\mathrm{A}}\rho / M W_{\mathrm{T}} \tag{7.2}$$

where $M$ denotes the molecular weight of the absorbing species, $W_{\mathrm{A}}$ and $W_{\mathrm{T}}$ are the masses of the absorber and the total mixture, respectively, $\in$ is molar absorptivity, and $\rho$ is the density of the mixture.

It can be observed that Eq. (7.1) simplifies to

$$H = AI_{\mathrm{o}}\mu\beta / 2^{1/2} \tag{7.3}$$

for $\mu\beta \ll 1$, and to

$$H = AI_{\mathrm{o}}2^{1/2}\mu\beta / (\mu\beta + 2) \tag{7.4}$$

for $(\mu\beta)^2 \ll (\mu\beta + 2)^2$. The first simplification corresponds to the weak absorption case in which the PA spectrum is proportional to the traditional absorption spectrum. In this situation, PA intensity is expected to vary linearly with concentration. Belton and Tanner plotted Eqs. (7.1), (7.3), and (7.4) against $\mu\beta$ and found the latter relationship to be linear over a wide range of conditions. Moreover, Eq. (7.4) can be rewritten to show that a plot of $1/H$ vs. $1/\beta$ should be linear. This equation also pertains to higher concentrations, where PA intensity does not increase in direct proportion with analyte concentration.

The applicability of Eq. (7.4) was subsequently investigated by Belton et al. (1987, 1988). These authors compared PA, ATR, and diffuse reflectance infrared spectroscopies with regard to the quantitative analysis of protein–starch mixtures and other food samples. They observed that the reciprocal of the sum of the amide I and II band intensities varied linearly with the reciprocal of the amount of protein that was present in the sample mixtures. A similar relation held for the other mixtures. Thus the range of concentrations in which PA spectroscopy could reliably be used for quantitative analysis was considerably extended through the use of this technique.

This method was subsequently used to analyze kaolin–silica and kaolin–KBr mixtures in the author's laboratory (Saffa and Michaelian, 1994a,b). Plots of $1/H$ vs. $W_{\mathrm{T}}/W_{\mathrm{A}}$ (the latter quantity being proportional to $1/\beta$) were shown to be linear over wide concentration ranges for both mixtures. The kaolin bands at 550 and 924 $\mathrm{cm}^{-1}$ were used in this analysis; typical results are summarized in Table 7.1. Even though kaolin concentrations as high as 80% were included in this work, the average error in the results was a comparatively moderate 12%. It can therefore be concluded that this method is generally applicable to the quantitative analysis of highly concentrated solid mixtures, as well as the more dilute samples that are generally considered to be amenable to PA infrared spectroscopy.

**Table 7.1. Quantitation of Kaolin (%) in Binary Mixtures[a]**

| Kaolin/Silica | | Kaolin/KBr | | |
|---|---|---|---|---|
| | PA Result | | PA Result | |
| Actual Value | 550 $cm^{-1}$ | Actual Value | 550 $cm^{-1}$ | 924 $cm^{-1}$ |
| 32.7 | 37.3 | 14.6 | 12.4 | 12.0 |
| 64.1 | 71.3 | 80.0 | 75.4 | 71.5 |
| 76.0 | 78.6 | | | |

[a] The infrared frequencies refer to the kaolin bands used in the calculations.
From Saffa and Michaelian (1994a,b).

## REFERENCES


Adams, M. J., Beadle, B. C., and Kirkbright, G. F. (1978). Optoacoustic spectrometry in the near-infrared region. *Anal. Chem.* **50**: 1371–1374.

Belton, P. S., and Tanner, S. F. (1983). Determination of the moisture content of starch using near infrared photoacoustic spectroscopy. *Analyst* **108**: 591–596.

Belton, P. S., Saffa, A. M., and Wilson, R. H. (1987). Use of Fourier transform infrared spectroscopy for quantitative analysis: A comparative study of different detection methods. *Analyst* **112**: 1117–1120.

Belton, P. S., Saffa, A. M., and Wilson, R. H. (1988). Quantitative analysis by Fourier transform infrared photoacoustic spectroscopy. *SPIE* **917**: 72–77.

Castleden, S. L., Kirkbright, G. F., and Menon, K. R. (1980). Determination of moisture in single-cell protein utilising photoacoustic spectroscopy in the near-infrared region. *Analyst* **105**: 1076–1081.

Castleden, S. L., Kirkbright, G. F., and Long, S. E. (1982). Quantitative assay of propanolol by photoacoustic spectroscopy. *Can. J. Spectrosc.* **27**: 245–248.

Gardella, J. A., Jiang, D.-Z., and Eyring, E. M. (1983). Quantitative determination of catalytic surface adsorption sites by Fourier transform infrared photoacoustic spectroscopy. *Appl. Spectrosc.* **37**: 131–133.

Gordon, S. H., Greene, R. V., Freer, S. N., and James, C. (1990). Measurement of protein biomass by Fourier transform infrared-photoacoustic spectroscopy. *Biotech. Appl. Biochem.* **12**: 1–10.

Highfield, J. G., and Moffat, J. B. (1985). The influence of experimental conditions in quantitative analysis of powdered samples by Fourier transform infrared photoacoustic spectroscopy. *Appl. Spectrosc.* **39**: 550–552.

Jin, Q., Kirkbright, G. F., and Spillane, D. E. M. (1982). The determination of moisture in some solid materials by near infrared photoacoustic spectroscopy. *Appl. Spectrosc.* **36**: 120–124.

Letzelter, N. S., Wilson, R. H., Jones, A. D., and Sinnaeve, G. (1995). Quantitative determination of the composition of individual pea seeds by Fourier transform infrared photoacoustic spectroscopy. *J. Sci. Food Agric.* **67**: 239–245.

McClelland, J. F., Luo, S., Jones, R. W., and Seaverson, L. M. (1991). FTIR photoacoustic spectroscopy applications in qualitative and quantitative analyses of solid samples. *SPIE* **1575**: 226–227.

McClelland, J. F., Luo, S., Jones, R. W., and Seaverson, L. M. (1992). A tutorial on the state-of-the-art of FTIR photoacoustic spectroscopy. In: *Photoacoustic and Photothermal Phenomena III.* D. Bićanić (ed.). Springer, Berlin, pp. 113–124.

Norton, G. A., and McClelland, J. F. (1997). Rapid determination of limestone using photoacoustic spectroscopy. *Miner. Eng.* **10**: 237–240.

Pandurangi, R. S., and Seehra, M. S. (1992). Quantitative analysis of silica in silica-kaolin mixtures by photoacoustic and diffuse reflectance spectroscopies. *Appl. Spectrosc.* **46**: 1719–1723.

Poulet, P., Chambron, J., and Unterreiner, R. (1980). Quantitative photoacoustic spectroscopy applied to thermally thick samples. *J. Appl. Phys.* **51**: 1738–1742.

Riseman, S. M., Massoth, F. E., Dhar, G. M., and Eyring, E. M. (1982). Fourier transform infrared photoacoustic spectroscopy of pyridine adsorbed on silica-alumina and $\gamma$-alumina. *J. Phys. Chem.* **86**: 1760–1763.

Rockley, M. G., Richardson, H. H., and Davis, D. M. (1980). Fourier-transformed infrared photoacoustic spectroscopy, the technique and its applications. *1980 Ultrasonics Symposium Proceedings*, 649–651.

Rockley, M. G., Davis, D. M., and Richardson, H. H. (1981). Quantitative analysis of a binary mixture by Fourier transform infrared photoacoustic spectroscopy. *Appl. Spectrosc.* **35**: 185–186.

Rosenthal, R. J., Carl, R. T., Beauchaine, J. P., and Fuller, M. P. (1988). Quantitative applications of photoacoustic spectroscopy in the infrared. *Mikrochim. Acta* **2**: 149–153.

Saffa, A. M., and Michaelian, K. H. (1994a). Quantitative analysis of clay mixtures by photoacoustic FT-IR spectroscopy. *SPIE* **2089**: 566–567.

Saffa, A. M., and Michaelian, K. H. (1994b). Quantitative analysis of kaolinite/silica and kaolinite/KBr mixtures by photoacoustic FT-IR spectroscopy. *Appl. Spectrosc.* **48**: 871–874.

Yokoyama, Y., Kosugi, M., Kanda, H., Ozasa, M., and Hodouchi, K. (1984). Infrared photoacoustic spectrometry of powdered samples. Calibration curves of sulfate compounds. *Bunseki Kagaku* **33**: E1–E7.

## RECOMMENDED READING

Bajic, S. J., Luo, S., Jones, R. W., and McClelland, J. F. (1995). Analysis of underground storage tank waste simulants by Fourier transform infrared photoacoustic spectroscopy. *Appl. Spectrosc.* **49**: 1000–1005.

Favier, J. P., Bicanic, D., Cozijnsen, J., van Veldhuizen, B., and Helander, P. (1998). $CO_2$ laser infrared optothermal spectroscopy for quantitative adulteration studies in binary mixtures of extra-virgin olive oil. *J. Am. Oil Chem. Soc.* **75**: 359–362.

CHAPTER

8

# SPECIAL TOPICS

This chapter discusses two topics that have not been dealt with so far in this book. The first is PA microspectroscopy—that is, measurement of PA spectra for microscopic samples, or for very small regions within larger samples. This application can be described as a major departure from PA spectroscopy as it is most commonly practiced, with macroscopic (~10–100 mg) samples. The second technique is synchrotron-based PA infrared spectroscopy, a method that has only recently been described in the literature. This experiment utilizes the high brightness of synchrotron radiation to improve the sensitivity of far- and mid-infrared PA spectroscopy. At the time of this writing, both PA microspectroscopy and synchrotron infrared PA spectroscopy can be described as emerging techniques; in light of the initial success of both methods, it is anticipated that they will be utilized by an increasing number of researchers in PA infrared spectroscopy in future work.

## 8.1. PA INFRARED MICROSPECTROSCOPY

PA infrared spectroscopy is routinely implemented as a "bulk" sampling technique in which solid or liquid samples with cross-sectional areas on the order of 10–20 $mm^2$ are examined. A question frequently asked of PA researchers pertains to the suitability of the method for smaller samples; this in turn raises the more fundamental issue of improvement of spatial resolution. Four recently published works that are relevant to this discussion will be briefly summarized in this section.

In what is believed to be the first published report on microsampling in PA infrared spectroscopy, Jiang (1999) presented results obtained with an accessory that is commercially available from MTEC Photoacoustics. In this pioneering work, single-particle and single-fiber microsamplers (Fig. 8.1) were used in conjunction with an MTEC 300 PA cell and a Bio-Rad FTIR spectrometer. PA spectra were obtained for samples that included a coated bead, grease-coated nylon, and human hair. The objectives of the investigation were to demonstrate the use of the microsampling accessory and to carry out depth profiling studies in both rapid-scan and step-scan experi-

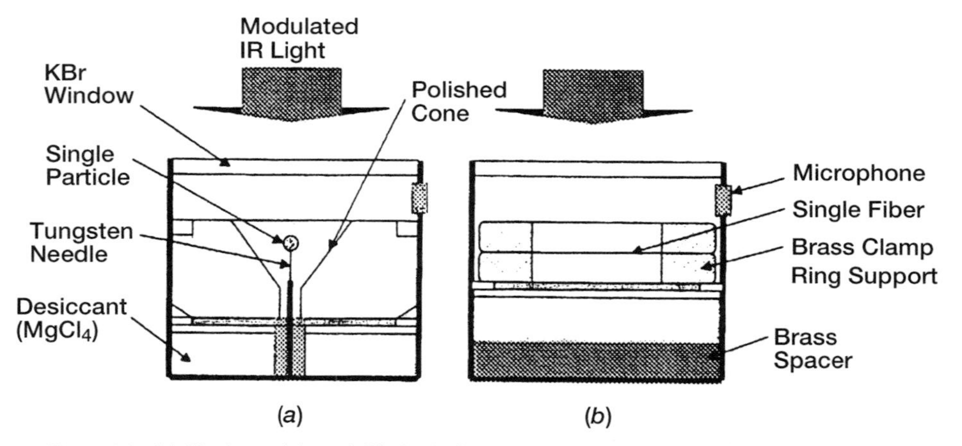


**Figure 8.1.** (*a*) Single-particle and (*b*) single-fiber microsamplers available from MTEC Photoacoustics. (Reproduced from Jiang, E. Y., *Appl. Spectrosc.* **53**: 583–587, by permission of the Society for Applied Spectroscopy; copyright © 1999.)

ments. A sample area as small as $25 \times 25$ μm can be analyzed with the equipment utilized in this study.

Because both microsamplers reduce sample cup volume, effectively concentrate the incident light on the sample, and provide better thermal isolation than ordinary sample cups, the signal-to-noise ratios in PA spectra obtained with this accessory turn out to be much higher than those obtained for larger quantities of a given sample. To illustrate this trend, results reported by Jiang for a 150-μm bead are summarized in Table 8.1. The single-particle accessory yielded a signal-to-noise ratio more than twice that obtained with the smaller (5-mm diameter) of two conventional sample cups, which contains approximately 70 mg when filled with an ordinary solid. A further reduction in signal was observed with the larger (10-mm diameter) sample cup. These important results show that single-particle PA infrared

**Table 8.1. Signal-to-Noise Ratios in PA Infrared Spectra of a 150-μm Bead Observed with Various Sampling Methods**

| Sample Accessory | Signal-to-Noise Ratio |
|---|---|
| Single particle | 6528 |
| Small sample cup | 2980 |
| Large sample cup | 2584 |

From Jiang (1999).

spectroscopy is actually preferable to bulk PA infrared analysis in some circumstances.

Several depth profiling experiments were carried out in this investigation. For example, in rapid-scan PA experiments on the coated bead, bands due to the bead intensified at lower mirror velocities (greater penetration depths) as would be expected. Step-scan experiments with phase modulation and digital signal processor (DSP) detection allowed variation of the sampling depth between 9.5 and 25.1 μm for a silicon-greased wire-cored nylon fiber; these depths were calculated on the basis of the known value of 0.002 $cm^2/s$ for the thermal diffusivity of grease. Similarly, frequency- and phase-resolved step-scan experiments were used to analyze human hair coated with a commercial gel. DSP demodulation separated the various harmonics, with higher-frequency components emphasizing surface species. Surface and bulk absorption were also discriminated by the phase spectrum and phase rotation methods described in Chapter 4. The excellent data presented in this study suggest that microsampling will become common in future applications of PA infrared spectroscopy.

Ordinarily, the best spatial resolution achievable in infrared microspectroscopy corresponds to the diffraction limit, that is, the smallest observable sample dimension is approximately equal to the infrared wavelength of interest. The reader is probably aware that much higher (up to several orders of magnitude) spatial definition is routinely achieved in scanning probe microscopy: If similar resolution could be obtained in the infrared, the diffraction limit would effectively be overcome. This was the motivation for a pioneering investigation by Hammiche et al. (1999), who used a miniature Wollaston wire thermometer to record infrared spectra of polymers by detecting photothermally induced temperature fluctuations at the sample surface. The objective of this work was to obtain infrared spectra with a spatial resolution determined by the size of the contact between probe and sample, which, at a few hundred nanometers, is well below the diffraction limit in the infrared region. The experiment performed by Hammiche et al. is more accurately described as photothermal, rather than photoacoustic, spectroscopy; while it did not involve the use of a gas-microphone cell or piezoelectric detection, it is based on photothermal detection of infrared spectra and is therefore related to several topics discussed elsewhere in this book.

The experimental arrangement is depicted in Figures 8.2 and 8.3. As the sample absorbed infrared radiation, the resulting temperature rise of the probe was detected. This induced changes in its electrical resistance, which were amplified and used to produce an inteferogram. Fourier transformation produced an absorptive, or positive-going, spectrum. Spatial resolution was affected by a number of factors, most having to do with temperature distri-

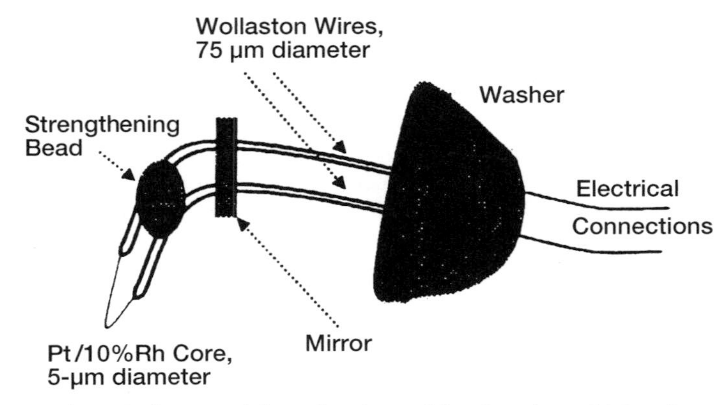


**Figure 8.2.** Schematic diagram of thermal probe used for photothermal infrared spectroscopy. (Reproduced from Hammiche, A. et al., *Appl. Spectrosc.* **53**: 810–815, by permission of the Society for Applied Spectroscopy; copyright © 1999.)

bution and contact area. The detected interferogram was quite weak, to the point where it was affected by harmonics of the line frequency. Photothermal spectra of polymers were compared with the corresponding ATR spectra; in some cases, major differences were observed. The exact reasons why the two types of spectra were so different are not yet known; it can be noted that the bands in the photothermal spectra are rather broad, suggesting that they are partially saturated. The photothermal spectra obviously reflect both optical and thermal characteristics of the sample, and hence this method could in principle be utilized to study both properties.

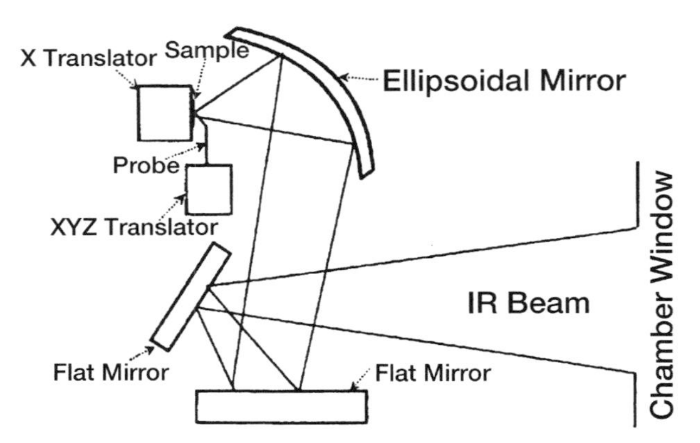


**Figure 8.3.** Probe and sample arrangement for photothermal infrared spectroscopy. (Reproduced from Hammiche, A. et al., *Appl. Spectrosc.* **53**: 810–815, by permission of the Society for Applied Spectroscopy; copyright © 1999.)

This experiment was refined and improved in two subsequent works. Hammiche et al. (2000) described the use of cantilever-type probes for the characterization of polymers, and presented photothermal spectra of single and bilayer polymer samples. The photothermal spectra obtained in this research are of good quality, although the bands again are broadened with respect to those in corresponding ATR spectra. In this and subsequent work (Bozec et al., 2001) a bilayer system (polyisobutylene on polystyrene) was studied. The case where the top layer is both optically transparent and thermally thin was examined in detail. An equation that related the amplitude of the signal from the bottom layer to the thickness of the top layer was obtained. The fit of the experimental data to this expression yielded a thermal diffusion length of 8.9 μm for the top layer, which agrees well with typical values for polymers at the modulation frequencies used in this experiment.

The relative intensities of neighboring bands in the polystyrene spectrum obtained by this technique are undoubtedly affected by saturation. As is often the case in PA spectra obtained with a gas-microphone cell, relative band intensities are significantly different from those in other types of infrared spectra (transmission, diffuse reflectance, etc.). When the polyisobutylene overlayer thickness was increased to the point that it equaled the thermal diffusion length, the perturbation in the intensities of the polystyrene bands was reduced but not entirely eliminated. This improvement can be put down to the fact that only the surface layer of the polystyrene is probed in this experiment.

The results obtained so far by this group are very encouraging. As this technique is optimized, the spatial resolution will be determined by the size of the probe and the temperature distribution within the sample. The spatial resolution of photothermal infrared spectroscopy will be improved and begin to approach that in Raman microscopy. This objective is particularly important, inasmuch as many industrial samples vary on a lateral dimension that is smaller than that usually accessible in infrared microspectroscopy.

## 8.2. SYNCHROTRON PA INFRARED SPECTROSCOPY

Synchrotron radiation (SR) possesses several properties that make it an attractive source for many optical experiments. Among the most important of these are its high brightness and continuous tunability across the entire electromagnetic spectrum, from X-ray wavelengths to the far infrared; the fact that the emission from a synchrotron is both pulsed and polarized enables and facilitates a wide variety of spectroscopic measurements. In the mid-infrared, the most important application of SR so far has certainly been in microspectroscopy, where the high radiance (radiant power per unit area per unit solid angle) of SR makes it superior to an ordinary thermal (globar) source. Because of this advantage, both the SR microspectroscopy method and its applications are currently resulting in an increasing number of publications in the scientific literature.

The justification for the utilization of SR in PA infrared spectroscopy must still be demonstrated. Since PA spectra tend to be significantly weaker than infrared spectra measured with optical detectors, it might appear that the high brightness of SR immediately offers an advantage in PA spectroscopy. However, this assertion is tempered by the fact that the total power emitted by a globar in the infrared is actually greater than that available in an infrared beam from a synchrotron. This makes the choice between the two sources less than obvious. Until very recently, it was not clear whether there is any advantage to the use of SR in PA infrared spectroscopy; hence, this question was recently investigated by the present author and two collaborators. A synopsis of this study is presented in the following paragraphs.

In the initial investigation of synchrotron infrared PA spectroscopy (Jackson et al., 2001; Michaelian et al., 2001), measurements were performed at beamline U10A at the National Synchrotron Light Source, Brookhaven National Laboratory. A Bruker IFS 66V/s FTIR spectrometer and an MTEC 100 PA cell were used to obtain mid- and far-infrared spectra of carbon-filled rubber, glassy carbon, hydrocarbon coke, and clay samples. Both SR and globar sources were utilized in this study. The objectives of the work were to determine the feasibility of synchrotron infrared PA spectroscopy and to ascertain the conditions in which SR might be superior to thermal radiation as a source in PA infrared spectroscopy.

A typical result for the far-infrared region is shown in Figure 8.4. These PA spectra of glassy carbon were obtained with a beam size of 5 mm, which results from the slight focusing of the infrared beam by the standard mirror in the PA cell that directs light onto the sample. (It should, however, be pointed out that the diameter of the SR beam in this experiment was narrower than this size.) The features in the spectra in this figure, all of which arise from absorption by the beamsplitter or by residual water vapor in the PA cell, are not of much interest. It is more important to observe that the SR PA spectrum is more intense than the globar spectrum at frequencies up to approximately 200 $cm^{-1}$, whereas the latter spectrum is stronger at higher frequencies. The ratio of these two spectra is plotted in Figure 8.5. The value of the SR/globar intensity ratio is almost equal to 7 at the low-frequency end of the spectrum, decreasing smoothly as wavenumber increases. This result shows that SR is preferable to a thermal source in far-infrared PA spectroscopy under the conditions employed in this experiment. Hence, affirmative

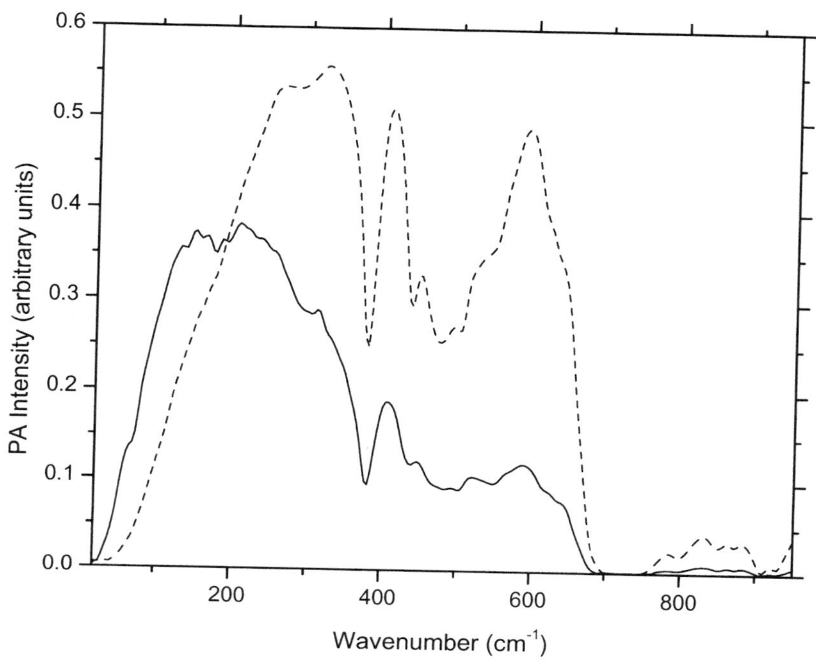


**Figure 8.4.** Far-infrared PA spectra of glassy carbon, obtained using synchrotron radiation (solid line) and globar source (dashed line).

answers were obtained for both of the objectives described above for this work.

The upper limit of the region in which the SR PA spectrum is stronger than the corresponding globar spectrum is termed the "crossing frequency" since it is the point at which the two curves in Figure 8.4 cross. This frequency was determined for the four samples mentioned above, using a series of brass apertures to limit the size of the beam that impinged on the sample in the PA cell. The results are summarized in Figure 8.6. The experimental points were fitted by a straight line corresponding to the equation

$$\log \nu_c = -1.34198 \log x + 3.21178 \tag{8.1}$$

where $\nu_c$ is the crossing frequency in reciprocal centimeters ($cm^{-1}$) and $x$ is the aperture size in millimeters. Data obtained with a 1-mm aperture show that the entire fingerprint region (up to approximately 1800 $cm^{-1}$) is more intense in SR PA spectra than in PA spectra obtained with a globar source.

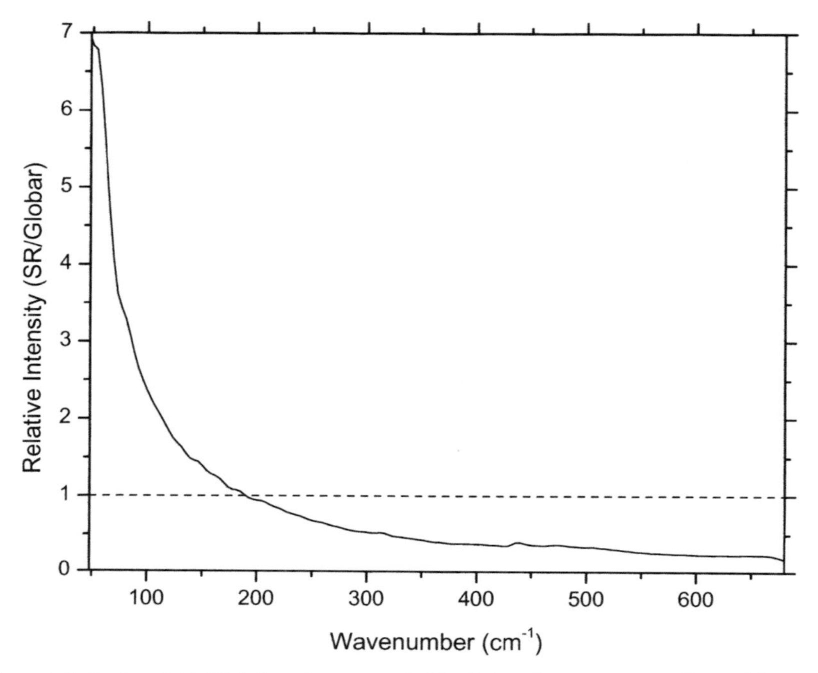


**Figure 8.5.** Ratio of SR PA infrared spectrum in Fig. 8.4 to globar spectrum. Dashed line at 1.0 intersects curve at the "crossing frequency," the wavenumber at which SR PA intensity equals globar PA intensity.

Moreover, this equation leads to an important prediction: SR PA spectra should be stronger than globar spectra throughout the entire mid-infrared region for aperture (or sample) sizes less than 0.5 mm. Hence the choice of SR as a source in PA infrared spectroscopy depends on both aperture size and spectral coverage; with smaller apertures and at longer wavelengths, SR is the preferred source, whereas larger samples and shorter wavelengths suggest the use of a globar source. In the latter case, it is logical that the experiment be carried out with a conventional laboratory-based spectrometer.

This preliminary investigation has clearly established the feasibility of SR PA infrared spectroscopy. Since SR tends to surpass a globar source in PA spectroscopy as the aperture size is reduced, there is obviously no need to utilize conventional sample cups in this experiment. Instead, a much smaller sample holder can be employed, such as the single-particle and single-fiber accessories described in the first part of this chapter. The microsampling accessories should be compatible with the smaller SR beam and produce spectra superior to those obtained in this preliminary study.

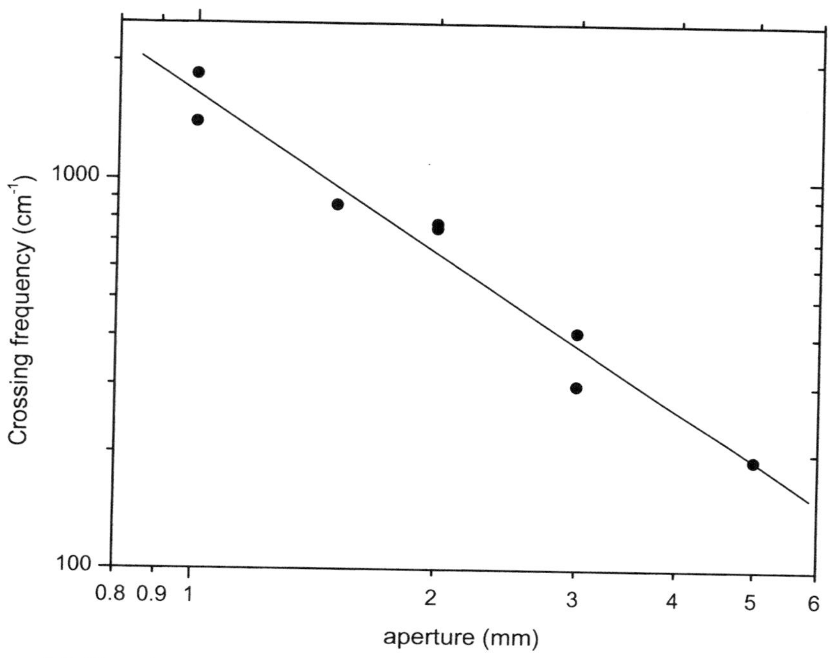


**Figure 8.6.** Dependence of crossing frequency on aperture (sample) size. Straight line is a least-squares fit to data (see text).

## REFERENCES


Bozec, L., Hammiche, A., Pollock, H. M., Conroy, M., Chalmers, J. M., Everall, N. J., and Turin, L. (2001). Localized photothermal infrared spectroscopy using a proximal probe. *J. Appl. Phys.* **90**: 5159–5165.

Hammiche, A., Pollock, H. M., Reading, M., Claybourn, M., Turner, P. H., and Jewkes, K. (1999). Photothermal FT-IR spectroscopy: A step towards FT-IR microscopy at a resolution better than the diffraction limit. *Appl. Spectrosc.* **53**: 810–815.

Hammiche, A., Bozec, L., Conroy, M., Pollock, H. M., Mills, G., Weaver, J. M. R., Price, D. M., Reading, M., Hourston, D. J., and Song, M. (2000). Highly localized thermal, mechanical, and spectroscopic characterization of polymers using miniaturized thermal probes. *J. Vac. Sci. Technol. B* **18**: 1322–1332.

Jackson, R. S., Michaelian, K. H., and Homes, C. C. (2001). Photoacoustic spectroscopy using a synchrotron light source. In: *Fourier Transform Spectroscopy*, OSA Technical Digest. Optical Society of America, Washington, DC, pp. 161–163.

Jiang, E. Y. (1999). Heterogeneity studies of a single particle/fiber by using Fourier transform infrared micro-sampling photoacoustic spectroscopy. *Appl. Spectrosc.* **53**: 583–587.

Michaelian, K. H., Jackson, R. S., and Homes, C. C. (2001). Synchrotron infrared photoacoustic spectroscopy. *Rev. Sci. Instrum.* **72**: 4331–4336.

## RECOMMENDED READING

Anderson, M. S. (2000). Infrared spectroscopy with an atomic force microscope. *Appl. Spectrosc.* **54**: 349–352.

Carr, G. L. (2001). Resolution limits for infrared microspectroscopy explored with synchrotron radiation. *Rev. Sci. Instrum.* **72**: 1613.

Huang, Y., and Xian, D. (1996). Synchrotron radiation X-ray photoacoustic effect. *Prog. Nat. Sci.* **6**(Suppl.): 763–766.

Masujima, T., Yoshida, H., Kawata, H., Amemiya, Y., Katsura, T., Ando, M., Nanba, T., Fukui, K., and Watanabe, M. (1989). Photoacoustic detector for synchrotron-radiation research. *Rev. Sci. Instrum.* **60**: 2318–2320.

Pollock, H. M., and Hammiche, A. (2001). Micro-thermal analysis: Techniques and applications. *J. Phys. D: Appl. Phys.* **34**: R23–R53.

# APPENDIX 1 GLOSSARY

|  |  |
|---|---|
| *Amplitude modulation* | Modulation of the radiation from an interferometer, laser, etc. by a chopper, shutter, or other device capable of completely interrupting the beam |
| *Depth profiling* | Characterization of a (solid) sample at selected distances from its surface |
| *Optically opaque* | Most of the incident light is absorbed within a distance that is small compared with the physical length of the sample |
| *Optically transparent* | Light is absorbed throughout the length of the sample; some light is transmitted through the sample |
| *Optoacoustic* | An alternative term for photoacoustic, implying the conversion of (absorbed) light into sound |
| *Phase modulation* | Modulation of the radiation from an interferometer by oscillating the moving mirror about each equilibrium position in step-scan mode |
| *Photopyroelectric* | Production of a temperature-dependent voltage is caused by the absorption of modulated radiation |
| *Photothermal* | An effect in which acoustic or elastic waves are generated by the absorption of incident radiation |
| *Photothermal radiometry* | Infrared detection of the temperature perturbation in a sample caused by the absorption of light |
| *Piezoelectric* | An effect in which a strain is converted to a voltage |
| *Rapid-scan* | An FTIR scan mode in which the motion of the moving mirror is rapid and continuous; named to contrast with step-scan |

| | |
|---|---|
| *Saturation* | Observed spectrum is independent of the optical absorption coefficient |
| *Step-scan* | An FTIR scan mode in which the position of the moving mirror is varied in a stepwise fashion |
| *Thermal conductivity* | Proportionality constant between the rate of heat flow and a temperature gradient. Symbol: $\kappa$ |
| *Thermal diffusion length* | Active thermal length; the distance over which the amplitude of a thermal wave decays to $1/e$ of its original magnitude. Symbol: $\mu$ |
| *Thermal diffusivity* | Ratio of thermal conductivity ($\kappa$) and thermal capacitance ($\rho C$, where $\rho$ is density and $C$ is heat capacity). Symbol: $\alpha$ |
| *Thermal effusivity* | Measure of thermal impedance or the ability of a sample to exchange heat with the environment; mathematically equal to $\sqrt{(\rho C \kappa)}$. Symbol: $e$ |
| *Thermally thick* | Heat generated only in a region near the surface of the sample can escape; the physical length of the sample is greater than the thermal diffusion length |
| *Thermally thin* | Heat is conducted throughout the length of the sample; the thermal diffusion length is greater than the physical length of the sample |

# APPENDIX 2 LITERATURE GUIDE

The following table lists 543 references in PA infrared spectroscopy that were published in the period beginning in 1978 and ending in early 2002. Bibliographic data consists of authors' names and publication details (year of publication; name of journal, conference proceedings, or book; volume number; and pagination). The majority of the these publications are discussed within the main part of this book, where the titles of the articles are given in the reference lists.

The presentation or discussion of experimental results for a particular sample type is indicated by an "X" in the appropriate column. The correspondence between most of the columns and the appropriate sections in Chapter 6 is obvious. It should be noted that coals are classified together with hydrocarbons and fuels in this appendix. Studies that discuss silica are cited under the catalysts and/or clays and minerals classifications as appropriate. References on carbonyl compounds are entered in the inorganics category, even though some samples contained both organic and inorganic ligands.

Works that describe the PA infrared spectra of gases are not included in this appendix. As discussed in the text, only a subset (consisting of articles published between 1994 and 2001) of the literature on this subject has been included in this book. The entire literature on the PA spectroscopy of gases is well beyond the scope of the present publication; the *Chemical Analysis* volume edited by M. W. Sigrist, or the proceedings of earlier conferences on PA and photothermal phenomena, should be consulted for more information on this subject. Similarly, publications that discuss only theoretical aspects of PA infrared spectroscopy, and those that describe instrumentation but do not present spectra of condensed-phase samples, are also not included. Several government publications, reviews, and articles in less accessible journals have also been omitted.

This listing is extensive but not comprehensive. The interested reader is invited to communicate the publication details of additional references on PA infrared spectroscopy to the author, who promises to gratefully acknowledge all contributions.

| Citation | Carbons | Catalysts | Clays, Minerals | Food Products | Hydro-carbons, Fuels | Inorganics | Medical, Biological | Organics | Polymers | Textiles | Wood, Paper |
|---|---|---|---|---|---|---|---|---|---|---|---|
| Abu-Zeid, M. E., Nofal, E. E., Marafi, M. A., Tahseen, L. A., Abdul-Rasoul, F. A., and Ledwith, A. (1984). *J. Appl. Polym. Sci.* **29**: 2431–2442 | | | | | | | | | X | | |
| Abu-Zeid, M. E., Tahseen, L. A., and Anani, A. A. (1985). *Colloids Surf.* **16**: 301–307 | | | X | | | | | X | | | |
| Abu-Zeid, M. E., Nofal, E. E., Tahseen, L. A., and Abdul-Rasoul, F. A. (1985). *J. Appl. Polym. Sci.* **30**: 3791–3800 | | | | | | | | | X | | |
| Adams, M. J., Beadle, B. C., and Kirkbright, G. F. (1978). *Anal. Chem.* **50**: 1371–1374 | | | X | | X | X | | | | | |
| Adams, M. J. (1982). *Prog. Analyt. Atom. Spectrosc.* **5**: 153–204 | | | | X | X | X | X | | | | |
| Aho, M., Kortelainen, P., Rantanen, J., and Linna, V. (1988). *J. Anal. Appl. Pyrolysis* **15**: 297–306 | | | | | X | | | | | | |
| Amer, N. M. (1984). *U.S. Pat. Appl.* 545338 | | | | | X | | | | | | |
| Anderson, M. S. (2000). *Appl. Spectrosc.* **54**: 349–352 | | | | | | | | | X | | |
| Angle, C. W., Donini, J. C., and Hamza, H. A. (1988). *Colloids Surfaces* **30**: 373–385 | | | | | X | | | | | | |

| Reference | | | | | |
|---|---|---|---|---|---|
| Annyas, J., Bićanić, D., and Schouten, F. (1999). *Appl. Spectrosc.* **53**: 339–343 | | | X | | |
| Ardeleanu, M., Morisset, R., and Bertrand, L. (1992). In: *Photoacoustic and Photothermal Phenomena III*, pp. 81–84. Springer, Berlin | | | X | | |
| Ashizawa, K. (1989). *J. Pharm. Sci.* **78**: 256–260 | | | X | | |
| Bain, C. D., Davies, P. B., and Ong, T. H. (1992). In: *Photoacoustic and Photothermal Phenomena III*, pp. 158–160. Springer, Berlin | | | | X | |
| Bajic, S. J., Luo, S., Jones, R. W., and McClelland, J. F. (1995). *Appl. Spectrosc.* **49**: 1000–1005 | | X | | | |
| Bajic, S. J., McClelland, J. F., and Jones, R. W. (1997). *Mikrochim. Acta* **14**(Suppl.): 611–612 | | X | | | |
| Bajic, S. J., Jones, R. W., McClelland, J. F., Hames, B. R., and Meglen, R. R. (1998). *AIP Conf. Proc.* **430**: 466–469 | | | | | X |
| Bandyopadhyay, S., Massoth, F. E., Pons, S., and Eyring, E. M. (1985). *J. Phys. Chem.* **89**: 2560–2564 | X | | | | |

*(Continued)*

| Citation | Carbons | Catalysts | Clays, Minerals | Food Products | Hydro-carbons, Fuels | Inorganics | Medical, Biological | Organics | Polymers | Textiles | Wood, Paper |
|---|---|---|---|---|---|---|---|---|---|---|---|
| Bauer, D. R., Paputa Peck, M. C., and Carter, R. O. (1987). *J. Coatings Technol.* **59**: 103–109 | | | | | | | | | X | | |
| Belton, P. S., and Tanner, S. F. (1983). *Analyst* **108**: 591–596 | | | | X | | | | | | | |
| Belton, P. S. (1984). In: *Biophysical Methods in Food Research*, pp. 123–135. Blackwell Scientific, Oxford | | | | X | | | | | | | |
| Belton, P. S., Saffa, A. M., and Wilson, R. H. (1987). *Analyst* **112**: 1117–1120 | | | | X | | | | | | | |
| Belton, P. S., Wilson, R. H., and Saffa, A. M. (1987). *Anal. Chem.* **59**: 2378–2382 | | | | X | | | | | | | |
| Belton, P. S., Saffa, A. M., Wilson, R. H., and Ince, A. D. (1988). *Food Chem.* **28**: 53–61 | | | | X | | | | | | | |
| Belton, P. S., Saffa, A. M., and Wilson, R. H. (1988). *SPIE* **917**: 72–77 | X | | | X | | | | | | | |
| Belton, P. S., Saffa, A. M., and Wilson, R. H. (1988). In: *Analytical Applications of Spectroscopy*, pp. 245–250. The Royal Society of Chemistry, Cambridge | | | | | | | X | | | | |

| | | | | | | | |
|---|---|---|---|---|---|---|---|
| Bensebaa, F., Majid, A., and Deslandes, Y. (2001). *Spectrochim. Acta A* **57**: 2695–2702 | | X | | X | | | |
| Benziger, J. B., McGovern, S. J., and Royce, B. S. H. (1985). In: *Catalyst Characterization Science*, pp. 449–463. American Chemical Society, Washington DC | X | X | | | | | |
| Berbenni, V., Marini, A., Bruni, G., Bini, M., Magnone-Grato, A., and Villa, M. (1996). *Appl. Spectrosc.* **50**: 871–879 | | | | | X | | |
| Bertrand, L., Monchalin, J.-P., and Lepoutre, F. (1982). *Appl. Opt.* **21**: 248–252 | | X | | | | | |
| Bertrand, L. (1988). *Appl. Spectrosc.* **42**: 134–138 | | X | | | | | X |
| Bićanić, D., Krüger, S., Torfs, P., Bein, B., and Harren, F. (1989). *Appl. Spectrosc.* **43**: 148–153 | | | | | | X | |
| Bićanić, D., Jalink, H., Chirtoc, M., Sauren, H., Lubbers, M., Quist, J., Gerkema, E., van Asselt, K., Miklós, A., Sólyom, A., Angeli, Gy. Z., Helander, P., and Vargas, H. (1992). In: *Photoacoustic and Photothermal Phenomena III*, pp. 20–27. Springer, Berlin | | | X | | X | | |

(Continued)

| Citation | Carbons | Catalysts | Clays, Minerals | Food Products | Hydro-carbons, Fuels | Inorganics | Medical, Biological | Organics | Polymers | Textiles | Wood, Paper |
|---|---|---|---|---|---|---|---|---|---|---|---|
| Bićanić, D., Chirtoc, M., Chirtoc, I., Favier, J. P., and Helander, P. (1995). *Appl. Spectrosc.* **49**: 1485–1489 | | | | | | | | X | | | |
| Bićanić, D., Fink, T., Franko, M., Močnik, G., van de Bovenkamp, P., van Veldhuizen, B., and Gerkema, E. (1999). *AIP Conf. Proc.* **463**: 637–639 | | | | X | | | | | | | |
| Blank, R. E., and Wakefield, T. (1979). *Anal. Chem.* **51**: 50–54 | | | | | | X | | | | | |
| Bordeleau, A., Rousset, G., Bertrand, L., and Crine, J. P. (1986). *Can. J. Phys.* **64**: 1093–1097 | | | | | | | | | X | | |
| Bordeleau, A., Bertrand, L., and Sacher, E. (1987). *Spectrochim. Acta A* **43**: 1189–1190 | | | | | | X | | | | | |
| Bouzerar, R., Amory, C., Zeinert, A., Benlahsen, M., Racine, B., Durand-Drouhin, O., and Clin, M. (2001). *J. Non-Cryst. Solids* **281**: 171–180 | X | | | | | | | | | | |
| Bowen, J. M., Compton, S. V., and Blanche, M. S. (1989). *Anal. Chem.* **61**: 2047–2050 | | | X | | | | | X | | | |

| Reference | | | | | | |
|---|---|---|---|---|---|---|
| Bozec, L., Hammiche, A., Pollock, H. M., Conroy, M., Chalmers, J. M., Everall, N. J., and Turin, L. (2001). *J. Appl. Phys.* **90**: 5159–5165 | | | | X | X | |
| Burggraf, L. W., and Leyden, D. E. (1981). *Anal. Chem.* **53**: 759–764 | | | | | | |
| Burggraf, L. W., Leyden, D. E., Chin, R. L., and Hercules, D. M. (1982). *J. Catal.* **78**: 360–379 | X | | | | | |
| Burt, J. A., Michaelian, K. H., and Zhang, S. L. (1997). *Mikrochim. Acta* **14**(Suppl.): 173–174 | | X | | | | |
| Butler, I. S., Xu, Z. H., Werbowyj, R. S., and St.-Germain, F. (1987). *Appl. Spectrosc.* **41**: 149–153 | | | X | | | |
| Butler, I. S., Xu, Z. H., Darensbourg, D. J., and Pala, M. (1987). *J. Raman Spectrosc.* **18**: 357–363 | | | X | | | |
| Butler, I. S., Li, H., and Gao, J. P. (1991). *Appl. Spectrosc.* **45**: 223–226 | | | X | | | |
| Butler, I. S., Gilson, D. F. R., and Lafleur, D. (1992). *Appl. Spectrosc.* **46**: 1605–1607 | | | X | | | |
| Cardamone, J. M., Gould, J. M., and Gordon, S. H. (1987). *Textile Res. J.* **57**: 235–239 | | | | | | X |

*(Continued)*

| Citation | Carbons | Catalysts | Clays, Minerals | Food Products | Hydro-carbons, Fuels | Inorganics | Medical, Biological | Organics | Polymers | Textiles | Wood, Paper |
|---|---|---|---|---|---|---|---|---|---|---|---|
| Carter, E. A., Fredericks, P. M., and Church, J. S. (1996). *Textile Res. J.* **66**: 787–794 | | | | | | | | | | X | |
| Carter, R. O., and Bauer, D. R. (1987). *Polym. Mater. Sci. Eng.* **57**: 875–879 | | | | | | | | | X | | |
| Carter, R. O., and Paputa Peck, M. C. (1989). *Appl. Spectrosc.* **43**: 468–473 | X | | | | | | | | X | | |
| Carter, R. O., Paputa Peck, M. C., and Bauer, D. R. (1989). *Polym. Degrad. Stability* **23**: 121–134 | | | | | | | | | X | | |
| Carter, R. O., Paputa Peck, M. C., Samus, M. A., and Killgoar, P. C. (1989). *Appl. Spectrosc.* **43**: 1350–1354 | X | | | | | | | | X | | |
| Carter, R. O., and Wright, S. L. (1991). *Appl. Spectrosc.* **45**: 1101–1103 | X | | | | | | | | X | | |
| Carter, R. O. (1992). *Appl. Spectrosc.* **46**: 219–224 | | | | | | | | | X | | |
| Castleden, S. L., Kirkbright, G. F., and Menon, K. R. (1980). *Analyst* **105**: 1076–1081 | | | | X | | | | | | | |
| Castleden, S. L., Kirkbright, G. F., and Long, S. E. (1982). *Can. J. Spectrosc.* **27**: 244–248 | | | | | | | X | | | | |

| | | | | | |
|---|---|---|---|---|---|
| Cella, N., Vargas, H., Galembeck, E., Galembeck, R., and Miranda, L. C. M. (1989). *J. Polym. Sci. Polym. Lett.* **27**: 313–320 | | | | X | |
| Chalmers, J. M., Stay, B. J., Kirkbright, G. F., Spillane, D. E. M., and Beadle, B. C. (1981). *Analyst* **106**: 1179–1186 | | | | X | |
| Chalmers, J. M., and Wilson, J. (1988). *Mikrochim. Acta* **2**: 1099–1111 | X | | | X | |
| Chalmers, J. M., and Mackenzie, M. W. (1988). In: *Advances in Applied Fourier Transform Infrared Spectroscopy*, pp. 105–188. Wiley, Chichester | X | | X | X | |
| Chatzi, E. G., Urban, M. W., Ishida, H., and Koenig, J. L. (1986). *Polymer* **27**: 1850–1854 | | | | | X |
| Chatzi, E. G., Urban, M. W., Ishida, H., Koenig, J. L., Laschewski, A., and Ringsdorf, H. (1988). *Langmuir* **4**: 846–855 | | | | X | |
| Chien, P.-L., Markuszewski, R., and McClelland, J. F. (1985). *Preprints, ACS Div. Fuel Chem.* **30**: 13–20 | | X | | | |

*(Continued)*

| Citation | Carbons | Catalysts | Clays, Minerals | Food Products | Hydro-carbons, Fuels | Inorganics | Medical, Biological | Organics | Polymers | Textiles | Wood, Paper |
|---|---|---|---|---|---|---|---|---|---|---|---|
| Chien, P.-L., Markuszewski, R., Araghi, H. G., and McClelland, J. F. (1985). *Proc. 1985 Conf. Coal Sci.*, 818–821 | | | | | X | | | | | | |
| Choquet, M., Rousset, G., and Bertrand, L. (1985). *SPIE* **553**: 224–225 | | | X | | | | | | | | |
| Choquet, M., Rousset, G., and Bertrand, L. (1986). *Can. J. Phys.* **64**: 1081–1085 | | | X | | | | | | | | |
| Church, J. S., and Evans, D. J. (1995). *J. Appl. Polym. Sci.* **57**: 1585–1594 | | | | | | | | | | X | |
| Cody, G. D., Larsen, J. W., and Siskin, M. (1989). *Energy Fuels* **3**: 544–551 | | | | | X | | | | | | |
| Cook, L. E., Luo, S. Q., and McClelland, J. F. (1991). *Appl. Spectrosc.* **45**: 124–126 | | | | | | | | | X | | |
| Cooper, A. I., Howdle, S. M., Hughes, C., Jobling, M., Kazarian, S. G., Poliakoff, M., Shepherd, L. A., and Johnston, K. P. (1993). *Analyst* **118**: 1111–1116 | | | | | | X | | | X | | |
| Cooper, E. A., Urban, M. W., and Provder, T. (1988). *Polym. Mater. Sci. Eng.* **59**: 316–320 | | | | | | | | | X | | |

| | | | | | | | |
|---|---|---|---|---|---|---|---|
| Davidson, R. S., and King, D. (1983). *J. Textile Inst.* **74**: 382–384 | | | | | | | X |
| Davidson, R. S., and Fraser, G. V. (1984). *J. Soc. Dyers Colour* **100**: 167–170 | | | | | | | X |
| Débarre, D., Boccara, A. C., and Fournier, D. (1981). *Appl. Opt.* **20**: 4281–4286 | | | X | | | | |
| Débarre, D., Boccara, A. C., and Fournier, D. (1984). *Proc. Int. Conf. Photoacoustic Effect*, 147–153 | | | X | | | | |
| DeBellis, A. D., and Low, M. J. D. (1987). *Infrared Phys.* **27**: 181–191 | | | X | X | X | | |
| DeBellis, A. D., and Low, M. J. D. (1988). *Infrared Phys.* **28**: 225–237 | X | | X | | X | | |
| Delgado, A. H., Paroli, R. M., and Beaudoin, J. J. (1996). *Appl. Spectrosc.* **50**: 970–976 | | X | X | | | | |
| Delprat, P., and Gardette, J.-L. (1993). *Polymer* **34**: 933–937 | | | | | | X | |
| Deng, Z., Spear, J. D., Rudnicki, J. D., McLarnon, F. R., and Cairns, E. J. (1996). *J. Electrochem. Soc.* **143**: 1514–1521 | | | X | | | | |
| De Oliveira, M. G., Pessoa, O., Vargas, H., and Galembeck, F. (1988). *J. Appl. Polym. Sci.* **35**: 1791–1802 | | | | | | X | |

*(Continued)*

| Citation | Carbons | Catalysts | Clays, Minerals | Food Products | Hydro-carbons, Fuels | Inorganics | Medical, Biological | Organics | Polymers | Textiles | Wood, Paper |
|---|---|---|---|---|---|---|---|---|---|---|---|
| Dittmar, R. M., Chao, J. L., and Palmer, R. A. (1991). *Appl. Spectrosc.* **45**: 1104–1110 | | | | | | | | | X | | |
| Dittmar, R. M., Palmer, R. A., and Carter, R. O. (1994). *Appl. Spectrosc. Rev.* **29**: 171–231 | | | | | | | | | X | | |
| Dóka, O., Bićanić, D., Szücs, M., and Lubbers, M. (1998). *Appl. Spectrosc.* **52**: 1526 | | | X | | | | | | | | |
| Donini, J. C., and Michaelian, K. H. (1984). *Infrared Phys.* **24**: 157–163 | | | | | | | | X | X | | |
| Donini, J. C., and Michaelian, K. H. (1985). *SPIE* **553**: 344–345 | | | X | | | | | | | | |
| Donini, J. C., and Michaelian, K. H. (1986). *Infrared Phys.* **26**: 135–140 | | | X | | X | | | | | | |
| Donini, J. C., and Michaelian, K. H. (1988). *Appl. Spectrosc.* **42**: 289–292 | | | X | | | | | | | | |
| Drapcho, D. L., Curbelo, R., Jiang, E. Y., Crocombe, R. A., and McCarthy, W. J. (1997). *Appl. Spectrosc.* **51**: 453–460 | | | | | | | | | X | | |

| | | | | |
|---|---|---|---|---|
| Dubois, M., Enguehard, F., Bertrand, L., Choquet, M., and Monchalin, J.-P. (1994). *J. Phys. IV* **C7**: 377–380 | X | | | X |
| Duerst, R. W., and Mahmoodi, P. (1984). *Polym. Prepr.* **25**: 194–195 | | | | X |
| Duerst, R. W., Mahmoodi, P., and Duerst, M. D. (1987). In: *Fourier Transform Infrared Characterization of Polymers*, pp. 113–122. Plenum, New York | | | | X |
| Eckhardt, H., and Chance, R. R. (1983). *J. Chem. Phys.* **79**: 5698–5704 | | | | X |
| Einsiedel, H., Kreiter, M., Leclerc, M., and Mittler-Neher, S. (1998). *Opt. Mater.* **10**: 61–68 | | | | X |
| Esumi, K., Nichina, S., Sakurada, S., Meguro, K., and Honda, H. (1987). *Carbon* **25**: 821–825 | X | | | |
| Eyring, E. M., Riseman, S. M., and Massoth, F. E. (1984). *ACS Symp. Ser.* **248**: 399–410 | | X | | |
| Factor, A., Tilley, M. G., and Codella, P. J. (1991). *Appl. Spectrosc.* **45**: 135–138 | | | | X |
| Fairbrother, J. E. (1983). *Pharm. J.* **230**: 326–329 | | | X | |

*(Continued)*

| Citation | Carbons | Catalysts | Clays, Minerals | Food Products | Hydro-carbons, Fuels | Inorganics | Medical, Biological | Organics | Polymers | Textiles | Wood, Paper |
|---|---|---|---|---|---|---|---|---|---|---|---|
| Fangxin, L., Jinlong, Y., and Tianpeng, Z. (1997). *Phys. Rev. B* **55**: 8847–8851 | | | | | | X | | | | | |
| Fathallah, M., Rezig, B., Zouaghi, M., Amer, N. M., Roger, J. P., Boccara, A. C., and Fournier, D. (1988). In: *Photoacoustic and Photothermal Phenomena*, pp. 260–262. Springer, Berlin | | | | | | X | | | | | |
| Favier, J. P., Bićanić, D., van de Bovenkamp, P., Chirtoc, M., and Helander, P. (1996). *Anal. Chem.* **68**: 729–733 | | | | X | | | | | | | |
| Favier, J. P., Bićanić, D., Cozijnsen, J., van Veldhuizen, B., and Helander, P. (1998). *J. Am. Oil Chem. Soc.* **75**: 359–362 | | | | X | | | | | | | |
| Forsskåhl, I., Kenttä, E., Kyyrönen, P., and Sundström, O. (1995). *Appl. Spectrosc.* **49**: 163–170 | | | | | | | | | | | X |
| Fournier, D., Boccara, A. C., and Badoz, J. (1982). *Appl. Opt.* **21**: 74–76 | | | | | | X | | | | | |
| Fowkes, F. M., Huang, Y. C., Shah, B. A., Kulp, M. J., and Lloyd, T. B. (1988). *Colloids. Surf.* **29**: 243–261 | | | | | | X | | | | | |

| Reference | Col. 1 | Col. 2 | Col. 3 | Col. 4 |
| --- | --- | --- | --- | --- |
| Friesen, W. I., and Michaelian, K. H. (1986). *Infrared Phys.* **26**: 235–242 | X | X | | |
| Friesen, W. I., and Michaelian, K. H. (1991). *Appl. Spectrosc.* **45**: 50–56 | | X | | |
| Fukuyama, A., Akashi, Y., Suemitsu, M., and Ikari, T. (2000). *J. Cryst. Growth* **210**: 255–259 | | | X | |
| Gaboury, S. R., and Urban, M. W. (1988). *Polym. Prepr.* **29**: 356–357 | | | | X |
| Gaboury, S. R., and Urban, M. W. (1989). *Polym. Mater. Sci. Eng.* **60**: 875–879 | | | | X |
| Gagarin, S. G., Gladun, T. G., Friesen, W. I., and Michaelian, K. H. (1993). *Coke Chem.* 9–15 | | X | | |
| Gagarin, S. G., Friesen, W. I., Michaelian, K. H., and Gladun, T. G. (1994). *Solid Fuel Chem.* **28**: 35–42 | | X | | |
| Gagarin, S. G., Friesen, W. I., and Michaelian, K. H. (1995). *Coke Chem.* 6–16 | | X | | |
| Gagarin, S. G., Friesen, W. I., and Michaelian, K. H. (1995). *Coke Chem.* 23–28 | | X | | |
| Garbassi, F., and Occhiello, E. (1987). *Anal. Chim. Acta* **197**: 1–42 | | | | X |

(*Continued*)

| Citation | Carbons | Catalysts | Clays, Minerals | Food Products | Hydro-carbons, Fuels | Inorganics | Medical, Biological | Organics | Polymers | Textiles | Wood, Paper |
|---|---|---|---|---|---|---|---|---|---|---|---|
| Garcia, J. A., Mandelis, A., Marinova, M., Michaelian, K. H., and Afrashtehfar, S. (1998). *Appl. Spectrosc.* **52**: 1222–1229 | | | | | | | | | | | X |
| Garcia, J. A., Mandelis, A., Marinova, M., Michaelian, K. H., and Afrashtehfar, S. (1999). *AIP Conf. Proc.* **463**: 395–397 | | | | | | | | | | | X |
| Gardella, J. A., Eyring, E. M., Klein, J. C., and Carvalho, M. B. (1982). *Appl. Spectrosc.* **36**: 570–573 | | | X | | | | | | | | |
| Gardella, J. A., Jiang, D.-Z., McKenna, W. P., and Eyring, E. M. (1983). *Appl. Surf. Sci.* **15**: 36–49 | | X | | | | X | | | | | |
| Gardella, J. A., Jiang, D.-Z., and Eyring, E. M. (1983). *Appl. Spectrosc.* **37**: 131–133 | | X | | | | | | | | | |
| Gardella, J. A., Grobe, G. L., Hopson, W. L., and Eyring, E. M. (1984). *Anal. Chem.* **56**: 1169–1177 | | | | | | | | | X | | |
| Gerson, D. J. (1984). *Appl. Spectrosc.* **38**: 436–437 | | | | | | | | | X | | |

| | | | | | | |
|---|---|---|---|---|---|---|
| Gerson, D. J., McClelland, J. F., Veysey, S., and Markuszewski, R. (1984). *Appl. Spectrosc.* **38**: 902–904 | | | X | | | |
| Gillis-D'Hamers, I., Vrancken, K. C., Vansant, E. F., and De Roy, G. (1992). *J. Chem. Soc. Faraday Trans.* **88**: 2047–2050 | X | | | | | |
| Gonon, L., Vasseur, O. J., and Gardette, J.-L. (1999). *Appl. Spectrosc.* **53**: 157–163 | | | | | X | |
| Gonon, L., Mallegol, J., Commereuc, S., and Verney, V. (2001). *Vib. Spectrosc.* **26**: 43–49 | | | | | X | |
| Goodarzi, F., and McFarlane, R. A. (1991). *Int. J. Coal Geol.* **19**: 283–301 | | | X | | | |
| Gordon, S. H. (1987). *Appl. Spectrosc.* **41**: 195–199 | | | | | | X |
| Gordon, S. H., Greene, R. V., Freer, S. N., and James, C. (1990). *Biotech. Appl. Biochem.* **12**: 1–10 | | | | X | | |
| Gordon, S. H., Schudy, R. B., Wheeler, B. C., Wicklow, D. T., and Greene, R. V. (1997). *Int. J. Food Microbiol.* **35**: 179–186 | | X | | X | | |
| Gordon, S. H., Wheeler, B. C., Schudy, R. B., Wicklow, D. T., and Greene, R. V. (1998). *J. Food Prot.* **61**: 221–230 | | X | | X | | |

*(Continued)*

| Citation | Carbons | Catalysts | Clays, Minerals | Food Products | Hydro-carbons, Fuels | Inorganics | Medical, Biological | Organics | Polymers | Textiles | Wood, Paper |
|---|---|---|---|---|---|---|---|---|---|---|---|
| Gordon, S. H., Jones, R. W., McClelland, J. F., Wicklow, D. T., and Greene, R. V. (1999). *J. Agric. Food Chem.* **47**: 5267–5272 | | | | X | | | X | | | | |
| Gosselin, F., DiRenzo, M., Ellis, T. H., and Lubell, W. D. (1996). *J. Org. Chem.* **61**: 7980–7981 | | | | | | | | X | | | |
| Graf, R. T., Koenig, J. L., and Ishida, H. (1987). *Polym. Sci. Technol.* **36**: 1–32 | | | | | | | | | X | | |
| Graham, J. A., Grim, W. M., and Fateley, W. G. (1985). In: *Fourier Transform Infrared Spectroscopy*, Vol. 4, pp. 345–392 | | | | | | | | | | | |
| Graves, D. J., and Luo, S. (1994). *Biochem. Biophys. Res. Comm.* **205**: 618–624 | | | | | | | X | | | | |
| Greene, R. V., Freer, S. N., and Gordon, S. H. (1988). *FEMS Microbiol. Lett.* **52**: 73–78 | | | | | | | X | | | | |
| Greene, R. V., Gordon, S. H., Jackson, M. A., Bennett, G. A., McClelland, J. F., and Jones, R. W. (1992). *J. Agric. Food Chem.* **40**: 1144–1149 | | | | X | | | X | | | | |

|  |  |  |  |  |  |  |  |  |  |  |  |
|---|---|---|---|---|---|---|---|---|---|---|---|
| Gregoriou, V. G., Daun, M., Schauer, M. W., Chao, J. L., and Palmer, R. A. (1993). *Appl. Spectrosc.* **47**: 1311–1316 |  |  |  |  |  |  |  |  | X |  |  |
| Gregoriou, V. G., and Hapanowicz, R. (1996). *Prog. Nat. Sci.* **6**(Suppl.): S10–S13 |  |  |  |  |  |  |  |  | X |  |  |
| Gregoriou, V. G., and Hapanowicz, R. (1997). *Macromol. Symp.* **119**: 101–111 |  |  |  |  |  |  |  |  | X |  |  |
| Griffiths, P. R., and de Haseth, J. A. (1986). *Fourier Tranform Infrared Spectrom.* 312–337 | X | X |  |  |  | X |  | X | X |  |  |
| Grobe, G. L., Gardella, J. A., Hopson, W. L., and McKenna, W. P. (1987). *J. Biomed. Mater. Res.* **21**: 211–229 |  |  |  |  |  |  | X |  |  |  |  |
| Gurnagul, N., St-Germain, F. G. T., and Gray, D. G. (1986). *J. Pulp Paper Sci.* **12**: J156–J159 |  |  |  |  |  |  |  |  |  |  | X |
| Haas, U., and Seiler, H. (1984). *Z. Naturforsch.* **39a**: 1242–1249 |  |  |  |  |  |  |  | X |  |  |  |
| Halttunen, M., Tenhunen, J., Saarinen, T., and Stenius, P. (1999). *Vibr. Spectrosc.* **19**: 261–269 |  |  |  |  |  |  |  |  |  |  | X |
| Hammiche, A., Pollock, H. M., Reading, M., Claybourn, M., Turner, P. H., and Jewkes, K. (1999). *Appl. Spectrosc.* **53**: 810–815 |  |  |  |  |  |  |  |  | X |  |  |

(*Continued*)

| Citation | Carbons | Catalysts | Clays, Minerals | Food Products | Hydro-carbons, Fuels | Inorganics | Medical, Biological | Organics | Polymers | Textiles | Wood, Paper |
|---|---|---|---|---|---|---|---|---|---|---|---|
| Hammiche, A., Bozec, L., Conroy, M., Pollock, H. M., Mills, G., Weaver, J. M. R., Price, D. M., Reading, M., Hourston, D. J., and Song, M. (2000). *J. Vac. Sci. Technol. B* **18**: 1322–1332 | | | | | | | | | X | | |
| Hamza, H. A., Michaelian, K. H., and Andersen, N. E. (1983). *Proc. 1983 Conf. Coal Sci.* 248–251 | | | | | X | | | | | | |
| Harbour, J. R., Hopper, M. A., Marchessault, R. H., Dobbin, C. J., and Anczurowski, E. (1985). *J. Pulp Paper Sci.* **11**: J42–J47 | | | | | | | | | | | X |
| Harris, M., Pearson, G. N., Willetts, D. V., Ridley, K., Tapster, P. R., and Perrett, B. (2000). *Appl. Opt.* **39**: 1032–1041 | | | | | | | | X | | | |
| Hauser, M., and Oelichmann, J. (1987). *Mikrochim. Acta* **1**: 39–43 | | | | | | | | | X | | |
| Helander, P. (1993). *Meas. Sci. Technol.* **4**: 178–185 | | | | | | | | | | | |
| Herres, W., and Zachmann, G. (1984). *LaborPraxis* 632–638 | | | | | X | | | | X | | |

| Reference | | | |
|---|---|---|---|
| Highfield, J. G., and Moffat, J. B. (1984). *J. Catal.* **89**: 185–195 | X | | |
| Highfield, J. G., and Moffat, J. B. (1985). *Appl. Spectrosc.* **39**: 550–552 | X | | |
| Highfield, J. G., and Moffat, J. B. (1985). *J. Catal.* **95**: 108–119 | X | | |
| Highfield, J. G., and Moffat, J. B. (1986). *J. Catal.* **98**: 245–258 | X | | |
| Hocking, M. B., Syme, D. T., Axelson, D. E., and Michaelian, K. H. (1990). *J. Polym. Sci.: Part A: Polym. Chem.* **28**: 2949–2968 | | | X |
| Hocking, M. B., Syme, D. T., Axelson, D. E., and Michaelian, K. H. (1990). *J. Polym. Sci.: Part A: Polym. Chem.* **28**: 2969–2982 | | | X |
| Hocking, M. B., Syme, D. T., Axelson, D. E., and Michaelian, K. H. (1990). *J. Polym. Sci.: Part A: Polym. Chem.* **28**: 2983–2996 | | | X |
| Honda, F., Imada, Y., and Nakajima, K. (1988). *Hyomen Kagaku* **9**: 356–361 | | X | |

*(Continued)*

| Citation | Carbons | Catalysts | Clays, Minerals | Food Products | Hydro-carbons, Fuels | Inorganics | Medical, Biological | Organics | Polymers | Textiles | Wood, Paper |
|---|---|---|---|---|---|---|---|---|---|---|---|
| Hou, R., Wu, J., Soloway, R. D., Guo, H., Zhang, Y., Du, Y., Liu, F., and Xu, G. (1988). *Mikrochim. Acta* **2**: 133–136 | | | | | | | X | | | | |
| Huvenne, J. P., and Lacroix, B. (1988). *Spectrochim. Acta A* **44**: 109–113 | | | | | | | X | | | | |
| Imhof, R. E., McKendrick, A. D., and Xiao, P. (1995). *Rev. Sci. Instrum.* **66**: 5203–5213 | | | | | | | | | X | | |
| Irudayaraj, J., and Yang, H. (2000). *Appl. Spectrosc.* **54**: 595–600 | | | | X | | | | | | | |
| Irudayaraj, J., Sivakesava, S., Kamath, S., and Yang, H. (2001). *J. Food Sci.* **66**: 1416–1421 | | | | X | | | | | | | |
| Irudayaraj, J., Yang, H., and Sivakesava, S. (2002). *J. Mol. Struct.* **606**: 181–188 | | | | X | | | | | | | |
| Jackson, R. S., Michaelian, K. H., and Homes, C. C. (2001). *OSA Technical Digest*, 161–163 | X | | X | | | | | | | | |

| Reference | | | | | | |
|---|---|---|---|---|---|---|
| Jasse, B. (1989). *J. Macromol. Sci.-Chem.* **A26**: 43–67 | X | | | | X | |
| Jiang, E. Y., Palmer, R. A., and Chao, J. L. (1995). *J. Appl. Phys.* **78**: 460–469 | | | | | X | |
| Jiang, E. Y., and Palmer, R. A. (1997). *Anal. Chem.* **69**: 1931–1935 | | | | | X | |
| Jiang, E. Y., Palmer, R. A., Barr, N. E., and Morosoff, N. (1997). *Appl. Spectrosc.* **51**: 1238–1244 | | | | | X | |
| Jiang, E. Y., McCarthy, W. J., Drapcho, D. L., and Crocombe, R. A. (1997). *Appl. Spectrosc.* **51**: 1736–1740 | | | | | X | |
| Jiang, E. Y., Drapcho, D. L., McCarthy, W. J., and Crocombe, R. A. (1998). *AIP Conf. Proc.* **430**: 381–384 | | | | | X | |
| Jiang, E. Y. (1999). *Appl. Spectrosc.* **53**: 583–587 | | | | X | | |
| Jin, Q., Kirkbright, G. F., and Spillane, D. E. M. (1982). *Appl. Spectrosc.* **36**: 120–124 | | X | | | | X |
| Johgo, A., Ozawa, E., Ishida, H., and Shoda, K. (1987). *J. Mater. Sci. Lett.* **6**: 429–430 | | | X | | | |
| Jones, R. W., and McClelland, J. F. (1996). *Appl. Spectrosc.* **50**: 1258–1263 | | | | | X | |

*(Continued)*

| Citation | Carbons | Catalysts | Clays, Minerals | Food Products | Hydro-carbons, Fuels | Inorganics | Medical, Biological | Organics | Polymers | Textiles | Wood, Paper |
|---|---|---|---|---|---|---|---|---|---|---|---|
| Jones, R. W., and McClelland, J. F. (2001). *Appl. Spectrosc.* **55**: 1360–1367 | X | | | | | | | | X | | |
| Jones, R. W., and McClelland, J. F. (2002). *Appl. Spectrosc.* **56**: 409–418 | | | | | | | | | X | | |
| Jurdana, L. E., Ghiggino, K. P., Leaver, I. H., Barraclough, C. G., and Cole-Clarke, P. (1994). *Appl. Spectrosc.* **48**: 44–49 | | | | | | | X | | | X | |
| Jurdana, L. E., Ghiggino, K. P., Leaver, I. H., and Cole-Clarke, P. (1995). *Appl. Spectrosc.* **49**: 361–366 | | | | | | | X | | | X | |
| Kanstad, S. O., and Nordal, P.-E. (1977). *Int. J. Quantum Chem.* **12**(Suppl. 2): 123–130 | | | | | | X | | | | | |
| Kanstad, S. O., and Nordal, P.-E. (1978). *Opt. Comm.* **26**: 367–371 | | | | | | X | | X | | | |
| Kanstad, S. O., and Nordal, P.-E. (1979). *Infrared Phys.* **19**: 413–422 | | | | | | X | | | | | |
| Kanstad, S. O., and Nordal, P.-E. (1980). *Phys. Technol.* **11**: 142–147 | | | | | | X | | | | | |

| | | | | | | | | |
|---|---|---|---|---|---|---|---|---|
| Kanstad, S. O., and Nordal, P.-E. (1980). *Appl. Surf. Sci.* **5**: 286–295 | | | | X | | | | |
| Kanstad, S. O., Nordal, P.-E., Hellgren, L., and Vincent, L. (1981). *Naturwiss.* **68**: 47–48 | | | | | X | | | |
| Kaplanová, M., and Katuščáková, G. (1992). In: *Photoacoustic and Photothermal Phenomena III*, pp. 180–182. Springer, Berlin | | | | | | | | |
| Kendall, D. S., Leyden, D. E., Burggraf, L. W., and Pern, F. J. (1982). *Appl. Spectrosc.* **36**: 436–440 | | X | | | | X | | |
| Kinney, J. B., Staley, R. H., Reichel, C. L., and Wrighton, M. S. (1981). *J. Am. Chem. Soc.* **103**: 4273–4275 | | X | | X | | X | | |
| Kinney, J. B., and Staley, R. H. (1983). *Anal. Chem.* **55**: 343–348 | X | X | | | | X | | |
| Kinney, J. B., and Staley, R. H. (1983). *J. Phys. Chem.* **87**: 3735–3740 | X | | | | | | | |
| Kirkbright, G. F. (1978). *Optica Pura y Aplicada* **11**: 125–136 | | X | | X | X | X | | X |
| Koenig, J. L. (1985). *Pure Appl. Chem.* **57**: 971–976 | | X | | | | | X | |
| Krishnan, K. (1981). *Appl. Spectrosc.* **35**: 549–557 | | X | X | | X | | X | X |

(*Continued*)

| Citation | Carbons | Catalysts | Clays, Minerals | Food Products | Hydro-carbons, Fuels | Inorganics | Medical, Biological | Organics | Polymers | Textiles | Wood, Paper |
|---|---|---|---|---|---|---|---|---|---|---|---|
| Krishnan, K., Hill, S. L., Witek, H., and Knecht, J. (1981). *SPIE* **289**: 96–98 | | | | | | | | | X | | |
| Krishnan, K., Hill, S., Hobbs, J. P., and Sung, C. S. P. (1982). *Appl. Spectrosc.* **36**: 257–259 | | | | | | | | | X | | |
| Kuo, M.-L., McClelland, J. F., Luo, S., Chien, P.-L., Walker, R. D., and Hse, C.-Y. (1988). *Wood Fiber Sci.* **20**: 132–145 | | | | | | | | | | | X |
| Kuwahata, H., Muto, N., and Uehara, F. (2000). *Jpn. J. Appl. Phys.* **39**: 3169–3171 | | | | | | X | | | | | |
| Lai, E. P., Chan, B. L., and Hadjmohammadi, M. (1985). *Appl. Spectrosc. Rev.* **21**: 179–210 | | | X | | | | | | X | | |
| Larsen, J. W. (1988). *Preprints, ACS Div. Fuel Chem.* **33**: 400–406 | | | | | X | | | | | | |
| Laufer, G., Huneke, J. T., Royce, B. S. H., and Teng, Y. C. (1980). *Appl. Phys. Lett.* **37**: 517–519 | | | | | | X | | | | | |
| Lerner, B., Perkins, J. H., Pariente, G. L., and Griffiths, P. R. (1989). *SPIE* **1145**: 476–477 | | | | | | | | | X | | |

| | | | | | | |
|---|---|---|---|---|---|---|
| Letzelter, N. S., Wilson, R. H., Jones, A. D., and Sinnaeve, G. (1995). *J. Sci. Food Agric.* **67**: 239–245 | | | X | | | |
| Lewis, L. N. (1982). *J. Organometall. Chem.* **234**: 355–365 | | | | X | | |
| Li, G., Burggraf, L. W., and Baker, W. P. (2000). *Appl. Phys. Lett.* **76**: 1122–1124 | | | | | X | |
| Li, H., and Butler, I. S. (1992). *Appl. Spectrosc.* **46**: 1785–1789 | | | | X | | |
| Li, H., and Butler, I. S. (1993). *Appl. Spectrosc.* **47**: 218–221 | | | | X | | |
| Lin, J. W. P., and Dudek, L. P. (1985). *J. Polymer Sci., Polym. Chem. Ed.* **23**: 1589–1597 | | | | | | X |
| Linton, R. W., Miller, M. L., Maciel, G. E., and Hawkins, B. L. (1985). *Surf. Interf. Anal.* **7**: 196–203 | | X | | | | |
| Liu, D. K., Wrighton, M. S., McKay, D. R., and Maciel, G. E. (1984). *Inorg. Chem.* **23**: 212–220 | | | | X | | |
| Lloyd, L. B., Yeates, R. C., and Eyring, E. M. (1982). *Anal. Chem.* **54**: 549–552 | | | | | X | X |
| Lochmüller, C. H., and Wilder, D. R. (1980). *Anal. Chim Acta* **116**: 19–24 | X | X | | | | |

(Continued)

| Citation | Carbons | Catalysts | Clays, Minerals | Food Products | Hydro-carbons, Fuels | Inorganics | Medical, Biological | Organics | Polymers | Textiles | Wood, Paper |
|---|---|---|---|---|---|---|---|---|---|---|---|
| Lochmüller, C. H., and Wilder, D. R. (1980). *Anal. Chim Acta* **118**: 101–108 | | X | X | | | | | | | | |
| Lochmüller, C. H., Thompson, M. M., and Kersey, M. T. (1987). *Anal. Chem.* **59**: 2637–2638 | | X | X | | | | | | | | |
| Lotta, T. I., Tulkki, A. P., Virtanen, J. A., and Kinnunen, P. K. J. (1990). *Chem. Phys. Lipids* **52**: 11–27 | | | | | | | X | | | | |
| Low, M. J. D., and Parodi, G. A. (1978). *Spectrosc. Lett.* **11**: 581–588 | X | X | X | | | | | | | | |
| Low, M. J. D., and Parodi, G. A. (1980). *Infrared Phys.* **20**: 333–340 | X | | | | | X | X | | X | | |
| Low, M. J. D., and Parodi, G. A. (1980). *Spectrosc. Lett.* **13**: 151–158 | | | | | | | X | | | | |
| Low, M. J. D., and Parodi, G. A. (1980). *Spectrosc. Lett.* **13**: 663–669 | X | | | | | | | | | | |
| Low, M. J. D., and Parodi, G. A. (1980). *Appl. Spectrosc.* **34**: 76–80 | | | X | | X | | | | | | |
| Low, M. J. D., and Parodi, G. A. (1980). *J. Mol. Struct.* **61**: 119–124 | | | X | | | | | | | | |

| | | | | | | | |
|---|---|---|---|---|---|---|---|
| Low, M. J. D., and Lacroix, M. (1982). *Infrared Phys.* **22**: 139–147 | X | | | | X | X | |
| Low, M. J. D., Lacroix, M., and Morterra, C. (1982). *Appl. Spectrosc.* **36**: 582–584 | | | X | | X | | X |
| Low, M. J. D., Lacroix, M., Morterra, C., and Severdia, A. G. (1982). *Am. Lab.*, 16–27 | | | X | | | X | |
| Low, M. J. D., and Severdia, A. G. (1982). *J. Mol. Struct.* **80**: 209–212 | | | X | | | | |
| Low, M. J. D., Lacroix, M., and Morterra, C. (1982). *Spectrosc. Lett.* **15**: 57–64 | | | | | | X | |
| Low, M. J. D., Morterra, C., and Lacroix, M. (1982). *Spectrosc. Lett.* **15**: 159–164 | | | | | | X | |
| Low, M. J. D., Morterra, C., and Severdia, A. G. (1982). *Spectrosc. Lett.* **15**: 415–421 | X | X | | | | | |
| Low, M. J. D., and Parodi, G. A. (1982). *J. Photoacoustics* **1**: 131–144 | X | | | X | X | | X |
| Low, M. J. D. (1983). *Spectrosc. Lett.* **16**: 913–922 | X | | | | | | |
| Low, M. J. D., and Morterra, C. (1983). *Carbon* **21**: 275–281 | X | | | | | | |
| Low, M. J. D., Arnold, T. H., and Severdia, A. G. (1983). *Infrared Phys.* **23**: 199–206 | | | | | X | | |
| Low, M. J. D. (1984). *Spectrosc. Lett.* **17**: 455–461 | X | | | | | | |

*(Continued)*

| Citation | Carbons | Catalysts | Clays, Minerals | Food Products | Hydro-carbons, Fuels | Inorganics | Medical, Biological | Organics | Polymers | Textiles | Wood, Paper |
|---|---|---|---|---|---|---|---|---|---|---|---|
| Low, M. J. D., Morterra, C., and Severdia, A. G. (1984). *Mater. Chem. Phys.* **10**: 519–528 | | | | | | | X | X | X | X | X |
| Low, M. J. D. (1985). *Spectrosc. Lett.* **18**: 619–625 | X | | | | | | | | | | |
| Low, M. J. D., and Morterra, C. (1985). *Carbon* **23**: 311–316 | X | | | | | | | | | | |
| Low, M. J. D., and Tascon, J. M. D. (1985). *Phys. Chem. Minerals* **12**: 19–22 | | | X | | | | | | | | |
| Low, M. J. D. (1986). *Appl. Spectrosc.* **40**: 1011–1019 | X | | | | X | | | | | | |
| Low, M. J. D., Morterra, C., and Khosrofian, J. M. (1986). *IEEE Trans. UFFC* **33**: 573–584 | X | | X | | | | | | | | |
| Low, M. J. D., and Morterra, C. (1986). *IEEE Trans. UFFC* **33**: 585–589 | X | | | | | | | | | | |
| Low, M. J. D., and Morterra, C. (1987). *Appl. Spectrosc.* **41**: 280–287 | X | | | | | X | | X | | | X |
| Low, M. J. D., and DeBellis, A. D. (1987). *Spectrosc. Lett.* **20**: 213–219 | | | | | | | | | X | | |
| Low, M. J. D., and Glass, A. S. (1989). *Spectrosc. Lett.* **22**: 417–429 | X | | | | | | | | | | |

| | | | | | | |
|---|---|---|---|---|---|---|
| Low, M. J. D., and Morterra, C. (1989). In: *Structure and Reactivity of Surfaces*, pp. 601–609. Elsevier, Amsterdam | X | | | | | |
| Low, M. J. D., Politou, A. S., Varlashkin, P. G., and Wang, N. (1990). *Spectrosc. Lett.* **23**: 527–531 | X | | | | | |
| Low, M. J. D., and Wang, N. (1990). *Spectrosc. Lett.* **23**: 983–990 | X | | | | | |
| Low, M. J. D. (1993). *Spectrosc. Lett.* **26**: 453–459 | | | X | | | |
| Lowry, S. R., Mead, D. G., and Vidrine, D. W. (1981). *SPIE* **289**: 102–104 | | X | | | | X |
| Lowry, S. R., Mead, D. G., and Vidrine, D. W. (1982). *Anal. Chem.* **54**: 546–548 | | X | | | | X |
| Luo, S., Liao, C. X., McClelland, J. F., and Graves, D. J. (1987). *Int. J. Pept. Protein Res.* **29**: 728–733 | | | | | X | |
| Luo, S., Huang, C.-Y. F., McClelland, J. F., and Graves, D. J. (1994). *Anal. Biochem.* **216**: 67–76 | | | | | X | |
| Lynch, B. M., MacEachern, A. M., MacPhee, J. A., Nandi, B. N., Hamza, H., and Michaelian, K. H. (1983). *Proc. 1983 Conf. Coal Sci.* 653–654 | | | | X | | |

*(Continued)*

| Citation | Carbons | Catalysts | Clays, Minerals | Food Products | Hydro-carbons, Fuels | Inorganics | Medical, Biological | Organics | Polymers | Textiles | Wood, Paper |
|---|---|---|---|---|---|---|---|---|---|---|---|
| Lynch, B. M., Lancaster, L.-I., and Fahey, J. T. (1986). *Preprints, ACS Div. Fuel Chem.* **31**: 43–48 | | | | | | X | | | | | |
| Lynch, B. M., Lancaster, L.-I., and MacPhee, J. A. (1987). *Fuel* **66**: 979–983 | | | | | | X | | | | | |
| Lynch, B. M., Lancaster, L.-I., and MacPhee, J. A. (1987). *Preprints, ACS Div. Fuel Chem.* **32**: 138–145 | | | | | | X | | | | | |
| Lynch, B. M., MacPhee, J. A., and Martin, R. R. (1987). *Proc. Int. Conf. Coal. Sci.* 19–22 | | | | | X | | | | | | |
| Lynch, B. M., Lancaster, L.-I., and MacPhee, J. A. (1988). *Energy Fuels* **2**: 13–17 | | | | | | X | | | | | |
| Lynch, B. M., and MacPhee, J. A. (1989). In: *Chemistry of Coal Weathering*, pp. 83–106. Elsevier, Amsterdam | | | | | | X | | | | | |
| Mackenzie, M. W., and Sellors, J. (1988). *Polym. Degrad. Stab.* **22**: 303–312 | | | | | | | | | X | | |
| Mahmoodi, P., Duerst, R. W., and Meiklejohn, R. A. (1984). *Appl. Spectrosc.* **38**: 437–438 | | | | | | | | | X | | |

| Reference | | | | | |
|---|---|---|---|---|---|
| Manning, C. J., Palmer, R. A., and Chao, J. L. (1991). *Rev. Sci. Instrum.* **62**: 1219–1229 | | | | | X |
| Manning, C. J., Dittmar, R. M., Palmer, R. A., and Chao, J. L. (1992). *Infrared Phys.* **33**: 53–62 | | | | | X |
| Manning, C. J., Palmer, R. A., Chao, J. L., and Charbonnier, F. (1992). *J. Appl. Phys.* **71**: 2433–2440 | | | | X | |
| Manning, C. J., Charbonnier, F., Chao, J. L., and Palmer, R. A. (1992). In: *Photoacoustic and Photothermal Phenomena III*, pp. 161–164. Springer, Berlin | | | | X | |
| Manzanares, C., Blunt, V. M., and Peng, J. (1993). *Spectrochim. Acta A* **49**: 1139–1152 | | | | X | |
| Marchand, H., Cournoyer, A., Enguehard, F., and Bertrand, L. (1997). *Opt. Eng.* **36**: 312–320 | | | | | X |
| Martin, M. A., Childers, J. W., and Palmer, R. A. (1987). *Appl. Spectrosc.* **41**: 120–126 | | X | X | | |
| Masujima, T., Yoshida, H., Kawata, H., Amemiya, Y., Katsura, T., Ando, M., Nanba, T., Fukui, K., and Watanabe, M. (1989). *Rev. Sci. Instrum.* **60**: 2318–2320 | X | | | | |

(Continued)

| Citation | Carbons | Catalysts | Clays, Minerals | Food Products | Hydro-carbons, Fuels | Inorganics | Medical, Biological | Organics | Polymers | Textiles | Wood, Paper |
|---|---|---|---|---|---|---|---|---|---|---|---|
| McAskill, N. A. (1987). *Appl. Spectrosc.* **41**: 313–317 | | | | | X | | | | | | |
| McClelland, J. F. (1983). *Anal. Chem.* **55**: 89A–105A | | | | | X | | | | X | | |
| McClelland, J. F., Luo, S., Jones, R. W., and Seaverson, L. M. (1991). *SPIE* **1575**: 226–227 | | | | | | | | | X | | |
| McClelland, J. F., Luo, S., Jones, R. W., and Seaverson, L. M. (1992). In: *Photo-acoustic and Photothermal Phenomena III*, pp. 113–124. Springer, Berlin | | | X | | | | | | X | | |
| McClelland, J. F., Jones, R. W., Luo, S., and Seaverson, L. M. (1993). In: *Practical Sampling Techniques for Infrared Analysis*, pp. 107–144. CRC Press, Bcca Raton, FL | | | | | | | | | | | |
| McClelland, J. F., Jones, R. W., and Ochiai, S. (1994). *SPIE* **2089**: 302–303 | | | | | | | | | X | | |
| McClelland, J. F., Jones, R. W., Bajic, S. J., and Power, J. F. (1997). *Mikrochim. Acta* **14**(Suppl.): 613–614 | | | | | | | | | X | | |

| Reference | | | | | |
|---|---|---|---|---|---|
| McClelland, J. F., Bajic, S. J., Jones, R. W., and Seaverson, L. M. (1998). In: *Modern Techniques in Applied Molecular Spectroscopy*, pp. 221–265. Wiley, New York | | | | | |
| McClelland, J. F., Jones, R. W., and Bajic, S. J. (2001). In: *Handbook of Vibrational Spectroscopy*, Vol. 2, pp. 1231–1251. Wiley, Chichester | | | | | |
| McDonald, W. F., Goettler, H., and Urban, M. W. (1989). *Appl. Spectrosc.* **43**: 1387–1393 | | | | | X |
| McDonald, W. F., and Urban, M. W. (1989). *Polym. Mater. Sci. Eng.* **60**: 739–743 | | | | | |
| McFarlane, R. A., Gentzis, T., Goodarzi, F., Hanna, J. V., and Vassallo, A. M. (1993). *Int. J. Coal Geol.* **22**: 119–147 | | | | X | |
| McGovern, S. J., Royce, B. S. H., and Benziger, J. B. (1984). *Appl. Surf. Sci.* **18**: 401–413 | | X | X | | |
| McGovern, S. J., Royce, B. S. H., and Benziger, J. B. (1985). *J. Appl. Phys.* **57**: 1710–1718 | | | X | | |
| McGovern, S. J., Royce, B. S. H., and Benziger, J. B. (1985). *Appl. Opt.* **24**: 1512–1514 | X | | X | | |

*(Continued)*

| Citation | Carbons | Catalysts | Clays, Minerals | Food Products | Hydro-carbons, Fuels | Inorganics | Medical, Biological | Organics | Polymers | Textiles | Wood, Paper |
|---|---|---|---|---|---|---|---|---|---|---|---|
| McKenna, W. P., Bandyopadhyay, S., and Eyring, E. M. (1984). *Appl. Spectrosc.* **38**: 834–837 | | X | | | | | | | | | |
| McKenna, W. P., Gale, D. J., Rivett, D. E., and Eyring, E. M. (1985). *Spectrosc. Lett.* **14**: 687–694 | | | | | | | | | | | |
| McKenna, W. P., Gale, D. J., Rivett, D. E., and Eyring, E. M. (1985). *Spectrosc. Lett.* **18**: 115–122 | | | | | | | | | | X | |
| McKenna, W. P., and Eyring, E. M. (1985). *J. Mol. Catal.* **29**: 363–369 | | X | | | | | | | | | |
| McKenna, W. P., Higgins, B. E., and Eyring, E. M. (1985). *J. Mol. Catal.* **31**: 199–206 | | X | | | | | | | | | |
| McQueen, D. H., Wilson, R., and Kinnunen, A. (1995). *Trends Anal. Chem.* **14**: 482–492 | | | | X | | | | | | | |
| McQueen, D. H., Wilson, R., Kinnunen, A., and Jensen, E. P. (1995). *Talanta* **42**: 2007–2015 | | | | X | | | | | | | |

| Reference | | | | | | |
|---|---|---|---|---|---|---|
| Mead, D. G., Lowry, S. R., Vidrine, D. W., and Mattson, D. R. (1979). *Fourth Int. Conf. Infrared Millimeter Waves Appl.* 231. IEEE, New York | | | | X | | |
| Mead, D. G., Lowry, S. R., and Anderson, C. R. (1981). *Int. J. Infrared Millimeter Waves* **2**: 23–34 | | | | | X | |
| Mehicic, M., Kollar, R. G., and Grasselli, J. G. (1981). *SPIE* **289**: 99–101 | | X | | X | | X |
| Michaelian, K. H., and Friesen, W. I. (1985). *SPIE* **553**: 260–261 | | | X | | | |
| Michaelian, K. H. (1987). *Infrared Phys.* **27**: 287–296 | | | | X | | |
| Michaelian, K. H., Bukka, K., and Permann, D. N. S. (1987). *Can. J. Chem.* **65**: 1420–1423 | | | X | | | |
| Michaelian, K. H. (1989). *Appl. Spectrosc.* **43**: 185–190 | X | | X | X | | X |
| Michaelian, K. H. (1989). *Infrared Phys.* **29**: 87–100 | X | | X | X | | |
| Michaelian, K. H. (1990). *Infrared Phys.* **30**: 181–186 | | | X | | | |
| Michaelian, K. H. (1990). In: *Vibrational Spectra and Structure*, Vol. 18, pp. 81–126. Elsevier, Amsterdam | | | X | X | | |
| Michaelian, K. H., and Friesen, W. I. (1990). *Fuel* **69**: 1271–1275 | | | | X | | |

*(Continued)*

| Citation | Carbons | Catalysts | Clays, Minerals | Food Products | Hydro-carbons, Fuels | Inorganics | Medical, Biological | Organics | Polymers | Textiles | Wood, Paper |
|---|---|---|---|---|---|---|---|---|---|---|---|
| Michaelian, K. H. (1991). *Appl. Spectrosc.* **45**: 302–304 | | | | | X | | | | | | |
| Michaelian, K. H., Yariv, S., and Nasser, A. (1991). *Can. J. Chem.* **69**: 749–754 | | | X | | | | | | | | |
| Michaelian, K. H., and Birch, J. R. (1991). *Infrared Phys.* **31**: 527–537 | | | | | X | | | | | | |
| Michaelian, K. H., Ogunsola, O. I., and Bartholomew, R. J. (1995). *Can. J. Appl. Spectrosc.* **40**: 94–99 | | | | | X | | | | | | |
| Michaelian, K. H., Friesen, W. I., Zhang, S. L., Gentzis, T., Crelling, J. C., and Gagarin, S. G. (1995). In: *Coal Science*, pp. 255–258. Elsevier, Amsterdam | | | | | X | | | | | | |
| Michaelian, K. H., Akers, K. L., Zhang, S. L., Yariv, S., and Lapides, I. (1997). *Mikrochim. Acta* **14**(Suppl.): 211–212 | | | X | | | | | | | | |
| Michaelian, K. H., Lapides, I., Lahav, N., Yariv, S., and Brodsky, I. (1998). *J. Coll. Interf. Sci.* **204**: 389–393 | | | X | | | | | | | | |
| Michaelian, K. H., Zhang, S. L., Hall, R. H., and Bulmer, J. T. (2001). *Can. J. Anal. Sci. Spectrosc.* **46**: 10–22 | | | | | X | | | | | | |

| Reference | | | | | | |
|---|---|---|---|---|---|---|
| Michaelian, K. H. (2001). In: *Frontiers in Science and Technology*. Stefan University Press, La Jolla, CA | | | X | X | | |
| Michaelian, K. H., Jackson, R. S., and Homes, C. C. (2001). *Rev. Sci. Instrum.* **72**: 4331–4336 | X | | X | | | X |
| Michaelian, K. H., Hall, R. H., and Bulmer, J. T. (2002). *J. Therm. Anal. Calorim.* **69**: 135–147 | X | | | X | | |
| Michaelian, K. H., Hall, R. H., and Bulmer, J. T. (2003). *Spectrochim. Acta A*, in press | | | | X | | |
| Mikula, R. J., Axelson, D. E., and Michaelian, K. H. (1985). In: *Proc. 1985 Int. Conf. Coal Sci.*, pp. 495–498. Pergamon, Sydney | | | | X | | |
| Minato, H., and Ishido, Y. (2001). *Rev. Sci. Instrum.* **72**: 2889–2892 | | | | | X | |
| Moffat, J. B., and Highfield, J. G. (1984). *Stud. Surf. Sci. Catal.* **19**: 77–84 | | X | | | | |
| Moffat, J. B., and Highfield, J. G. (1984). *Preprints, ACS Div. Fuel Chem.* **29**: 254–260 | | X | | | | |
| Moffat, J. B. (1989). *J. Mol. Catal.* **52**: 169–191 | | X | | | | |
| Mohamed, M. M. (1995). *Spectrochim. Acta A* **51**: 1–9 | | X | | | | |

*(Continued)*

| Citation | Carbons | Catalysts | Clays, Minerals | Food Products | Hydro-carbons, Fuels | Inorganics | Medical, Biological | Organics | Polymers | Textiles | Wood, Paper |
|---|---|---|---|---|---|---|---|---|---|---|---|
| Monchalin, J.-P., Gagné, J.-M., Parpal, J.-L., and Bertrand, L. (1979). *Appl. Phys. Lett.* **35**: 360–363 | | | X | | | | | | | | |
| Monchalin, J.-P., Bertrand, L., Rousset, G., and Lepoutre, F. (1984). *J. Appl. Phys.* **56**: 190–210 | | | X | | | | | | | | |
| Mongeau, B., Rousset, G., and Bertrand, L. (1986). *Can. J. Phys.* **64**: 1056–1058 | | | | | | X | | | | | |
| Moore, P., Poslusny, M., Daugherty, K. E., Venables, B. J., and Okuda, T. (1990). *Appl. Spectrosc.* **44**: 326–328 | | | | | | | | | | | X |
| Morterra, C., and Low, M. J. D. (1982). *Spectrosc. Lett.* **15**: 689–697 | X | | | | | | | | | | |
| Morterra, C., Low, M. J. D., and Severdia, A. G. (1982). *Infrared Phys.* **22**: 221–227 | X | | X | | | X | | | | | |
| Morterra, C., and Low, M. J. D. (1983). *Carbon* **21**: 283–288 | X | | | | | | | | | | |
| Morterra, C., Low, M. J. D., and Severdia, A. G. (1984). *Carbon* **22**: 5–12 | X | | | | | | | | | | |
| Morterra, C., and Low, M. J. D. (1985). *Mater. Chem. Phys.* **12**: 207–233 | X | | | | X | | | | | | |

| | | | | | |
|---|---|---|---|---|---|
| Morterra, C., and Low, M. J. D. (1985). *Carbon* **23**: 301–310 | X | | | | |
| Morterra, C., and Low, M. J. D. (1985). *Carbon* **23**: 335–341 | X | | | | |
| Morterra, C., and Low, M. J. D. (1985). *Carbon* **23**: 525–530 | X | | | | |
| Morterra, C., and Low, M. J. D. (1985). *Langmuir* **1**: 320–326 | X | | | | |
| Morterra, C., and Low, M. J. D. (1985). *Mater. Chem. Phys.* **12**: 207–233 | | | | | |
| Morterra, C., O'Shea, M. L., and Low, M. J. D. (1988). *Mater. Chem. Phys.* **20**: 123–144 | X | | | | |
| Moyer, D. J. D., and Wightman, J. P. (1989). *Surf. Interf. Anal.* **14**: 496–504 | X | | | | X |
| Muraishi, S. (1984). *Bunko Kenkyu* **33**: 269–270 | | | | | X |
| Nadler, M. P., Nissan, R. A., and Hollins, R. A. (1988). *Appl. Spectrosc.* **42**: 634–642 | | X | | X | |
| Nakanaga, T., Matsumoto, M., Kawabata, Y., Takeo, H., and Matsumura, C. (1989). *Chem. Phys. Lett.* **160**: 129–133 | | | | X | |
| Natale, M., and Lewis, L. N. (1982). *Appl. Spectrosc.* **36**: 410–413 | | | X | | |

(Continued)

| Citation | Carbons | Catalysts | Clays, Minerals | Food Products | Hydro-carbons, Fuels | Inorganics | Medical, Biological | Organics | Polymers | Textiles | Wood, Paper |
|---|---|---|---|---|---|---|---|---|---|---|---|
| Nelson, J. H., MacDougall, J. J., Baglin, F. G., Freeman, D. W., Nadler, M., and Hendrix, J. L. (1982). *Appl. Spectrosc.* **36**: 574–576 | X | | | X | | | | | | | |
| Neubert, R., Collin, B., and Wartewig, S. (1997). *Pharm. Res.* **14**: 946–948 | | | | | | | X | | | | |
| Nguyen, T. T. (1989). *J. Appl. Polym. Sci.* **38**: 765–768 | | | | | | | | | X | | |
| Nishikawa, Y., Kimura, K., Matsuda, A., and Kenpo, T. (1992). *Appl. Spectrosc.* **46**: 1695–1698 | | | | | | X | | X | X | | |
| Nishio, E., Abe, I., Ikuta, N., Koga, J., Okabayashi, H., and Nishikida, K. (1991). *Appl. Spectrosc.* **45**: 496–497 | | | X | | | | X | | | | |
| Noda, I., Story, G. M., Dowrey, A. E., Reeder, R. C., and Marcott, C. (1997). *Macromol. Symp.* **119**: 1–13 | | | | | | | | | X | | |
| Nordal, P.-E., and Kanstad, S. O. (1977). *Opt. Comm.* **22**: 185–189 | | | | X | | X | | | | | |
| Nordal, P.-E., and Kanstad, S. O. (1977). *Int. J. Quantum Chem.* **12**(Suppl. 2): 115–121 | | | | | | X | | X | | | |

| Reference | | | | | | |
|---|---|---|---|---|---|---|
| Nordal, P.-E., and Kanstad, S. O. (1978). *Opt. Comm.* **24**: 95–99 | | | | X | | |
| Norton, G. A., and McClelland, J. F. (1997). *Miner. Eng.* **10**: 237–240 | | X | | | | |
| Ochiai, S. (1985). *Toso Kogaku* **20**: 192–195 | | | | | | X |
| Olson, E. S., Diehl, J. W., and Froehlich, M. L. (1988). *Fuel* **67**: 1053–1061 | | | X | | | |
| O'Shea, M. L., Low, M. J. D., and Morterra, C. (1989). *Mater. Chem. Phys.* **23**: 499–516 | X | | | | | |
| O'Shea, M. L., Morterra, C., and Low, M. J. D. (1990). *Mater. Chem. Phys.* **25**: 501–521 | X | | | | | |
| O'Shea, M. L., Morterra, C., and Low, M. J. D. (1990). *Mater. Chem. Phys.* **26**: 193–209 | X | | | | | |
| O'Shea, M. L., Morterra, C., and Low, M. J. D. (1991). *Mater. Chem. Phys.* **27**: 155–179 | X | | | | | |
| O'Shea, M. L., Morterra, C., and Low, M. J. D. (1991). *Mater. Chem. Phys.* **28**: 9–31 | X | | | | | |
| Palmer, R. A., and Smith, M. J. (1986). *Can. J. Phys.* **64**: 1086–1092 | | | | X | X | |

*(Continued)*

| Citation | Carbons | Catalysts | Clays, Minerals | Food Products | Hydro-carbons, Fuels | Inorganics | Medical, Biological | Organics | Polymers | Textiles | Wood, Paper |
|---|---|---|---|---|---|---|---|---|---|---|---|
| Palmer, R. A., Smith, M. J., Manning, C. J., Chao, J. L., Boccara, A. C., and Fournier, D. (1988). In: *Photoacoustic and Photothermal Phenomena*, pp. 50–52. Springer, Berlin | X | | | | | X | | | | | |
| Palmer, R. A., and Dittmar, R. M. (1993). *Thin Solid Films* **223**: 31–38 | | | | | | | | | X | | |
| Palmer, R. A., Chao, J. L., Dittmar, R. M., Gregoriou, V. G., and Plunkett, S. E. (1993). *Appl. Spectrosc.* **47**: 1297–1310 | | | | | | | | | X | | |
| Palmer, R. A., Jiang, E. Y., and Chao, J. L. (1994). *SPIE* **2089**: 250–251 | | | | | | | | | X | | |
| Palmer, R. A., Boccara, A. C., and Fournier, D. (1996). *Prog. Nat. Sci.* **6**(Suppl.): S3–S9 | | | | | | X | | | | | |
| Palmer, R. A., Jiang, E. Y., and Chao, J. L. (1997). *Mikrochim. Acta* **14**(Suppl.): 591–594 | | | | | | | | | X | | |
| Pandurangi, R. S., and Seehra, M. S. (1990). *Anal. Chem.* **62**: 1943–1947 | | | X | | | | | | | | |

| Reference | | | | |
|---|---|---|---|---|
| Pandurangi, R. S., and Seehra, M. S. (1991). *Appl. Spectrosc.* **45**: 673–676 | | X | | |
| Pandurangi, R. S., and Seehra, M. S. (1992). *Appl. Spectrosc.* **46**: 1719–1723 | | X | | |
| Papendorf, U., and Riepe, W. (1989). In: *Proc. 1989 Int. Conf. Coal Science* **2**: 1111–1113 | X | | | |
| Peoples, M. E., Smith, M. J., and Palmer, R. A. (1987). *Appl. Spectrosc.* **41**: 1257–1259 | X | | | X |
| Perkins, J. A., and Griffiths, P. R. (1989). *SPIE* **1145**: 360–361 | | | | X |
| Pesce-Rodriguez, R. A., and Fifer, R. A. (1991). *Appl. Spectrosc.* **45**: 417–419 | | | X | |
| Philippaerts, J., Vansant, E. F., Peeters, G., and Vanderheyden, E. (1987). *Anal. Chim. Acta* **195**: 237–246 | | X | | |
| Philippaerts, J., Vanderheyden, E., and Vansant, E. F. (1988). *Mikrochim. Acta* **2**: 145–148 | | | | X |
| Philippaerts, J., Vanderheyden, E., and Vansant, E. F. (1988). In: *Photoacoustic and Photothermal Phenomena*, pp. 33–34. Springer, Berlin | | | | X |

*(Continued)*

| Citation | Carbons | Catalysts | Clays, Minerals | Food Products | Hydro-carbons, Fuels | Inorganics | Medical, Biological | Organics | Polymers | Textiles | Wood, Paper |
|---|---|---|---|---|---|---|---|---|---|---|---|
| Politou, A. S., Morterra, C., and Low, M. J. D. (1990). *Carbon* **28**: 529–538 | X | | | | | | | | | | |
| Politou, A. S., Morterra, C., and Low, M. J. D. (1990). *Carbon* **28**: 855–865 | X | | | | | | | | | | |
| Politou, A. S., Morterra, C., and Low, M. J. D. (1991). *Polym. Degrad. Stab.* **32**: 331–356 | X | | | | | | | | | | |
| Pollock, H. M., and Hammiche, A. (2001). *J. Phys. D: Appl. Phys.* **34**: R23–R53 | | | | | | | | | X | | |
| Porter, M. D., Karweik, D. H., Kuwana, T., Theis, W. B., Norris, G. B., and Tiernan, T. D. (1984). *Appl. Spectrosc.* **38**: 11–16 | | | | | | X | | | | | |
| Poslusny, M., and Daugherty, K. E. (1988). *Appl. Spectrosc.* **42**: 1466–1469 | | | | | | | | | | | X |
| Poulet, P., Chambron, J., and Unterreiner, R. (1980). *J. Appl. Phys.* **51**: 1738–1742 | | | | | | | | | | | |
| Raveh, A., Martinu, L., Domingue, A., Wertheimer, M. R., and Bertrand, L. (1992). In: *Photoacoustic and Photothermal Phenomena III*, pp. 151–154. Springer, Berlin | X | | | | | | | | | | |

| Reference | | | | | |
|---|---|---|---|---|---|
| Reddy, K. T. R., Chalapathy, R. B. V., Slifkin, M. A., Weiss, A. W., and Miles, R. W. (2001). *Thin Solid Films* **387**: 205–207 | | | X | | |
| Renugopalakrishnan, V., and Bhatnagar, R. S. (1984). *J. Am. Chem. Soc.* **106**: 2217–2219 | | | | X | |
| Renugopalakrishnan, V., Horowitz, P. M., and Glimcher, M. J. (1985). *J. Biol. Chem.* **260**: 11406–11413 | | | | X | |
| Renugopalakrishnan, V., Chandrakasan, G., Moore, S., Hutson, T. B., Berney, C. V., and Bhatnager, R. S. (1989). *Macromolecules* **22**: 4121–4124 | | | | X | |
| Riseman, S. M., and Eyring, E. M. (1981). *Spectrosc. Lett.* **14**: 163–185 | X | | | | |
| Riseman, S. M., Yaniger, S. I., Eyring, E. M., MacInnes, D., MacDiarmid, G., and Heeger, A. J. (1981). *Appl. Spectrosc.* **35**: 557–559 | | | | | X |
| Riseman, S. M., Massoth, F. E., Dhar, G. M., and Eyring, E. M. (1982). *J. Phys. Chem.* **86**: 1760–1763 | | X | | | |

*(Continued)*

| Citation | Carbons | Catalysts | Clays, Minerals | Food Products | Hydro-carbons, Fuels | Inorganics | Medical, Biological | Organics | Polymers | Textiles | Wood, Paper |
|---|---|---|---|---|---|---|---|---|---|---|---|
| Riseman, S. M., Bandyopadhyay, S., Massoth, F. E., and Eyring, E. M. (1985). *Appl. Catal.* **16**: 29–37 | | X | | | | | | | | | |
| Rockley, M. G. (1979). *Chem. Phys. Lett.* **68**: 455–456 | | | | | | | | | X | | |
| Rockley, M. G. (1980). *Appl. Spectrosc.* **34**: 405–406 | X | | | | | | X | | X | | |
| Rockley, M. G., and Devlin, J. P. (1980). *Appl. Spectrosc.* **34**: 407–408 | | | | | X | | | | | | |
| Rockley, M. G., Richardson, H. H., and Davis, D. M. (1980). *Ultrasonics Symp. Proc.* **2**: 649–651 | | | X | | | | | | | | |
| Rockley, M. G. (1980). *Chem. Phys. Lett.* **75**: 370–372 | | | | | | X | | | | | |
| Rockley, M. G., Davis, D. M., and Richardson, H. H. (1981). *Appl. Spectrosc.* **35**: 185–186 | | | | | | X | | | | | |
| Rockley, M. G., Richardson, H. H., and Davis, D. M. (1982). *J. Photoacoustics* **1**: 145–149 | | | X | | | | | | | | |
| Rockley, M. G., Woodard, M., Richardson, H. H., Davis, D. M., Purdie, N., and Bowen, J. M. (1983). *Anal. Chem.* **55**: 32–34 | | | | X | | | | X | | | |

| | | | | | | | | |
|---|---|---|---|---|---|---|---|---|
| Rockley, M. G., Ratcliffe, A. E., Davis, D. M., and Woodard, M. K. (1984). *Appl. Spectrosc.* **38**: 553–556 | X | | | | | | | |
| Rockley, N. L., Woodard, M. K., and Rockley, M. G. (1984). *Appl. Spectrosc.* **38**: 329–334 | | | | | | | X | X |
| Rockley, N. L., and Rockley, M. G. (1987). *Appl. Spectrosc.* **41**: 471–475 | | | X | | | | | |
| Röhl, R., Childers, J. W., and Palmer, R. A. (1982). *Anal. Chem.* **54**: 1234–1236 | | | | | X | | | |
| Rosenthal, R. J., and Lowry, S. R. (1987). *Mikrochim. Acta* **2**: 291–302 | | | | | | | | X |
| Rosenthal, R. J., Carl, R. T., Beauchaine, J. P., and Fuller, M. P. (1988). *Mikrochim. Acta* **2**: 149–153 | | | | | | X | | |
| Roush, P. B., and Oelichmann, J. (1987). *Mikrochim. Acta* **1**: 49–52 | | | | | | | | |
| Royce, B. S. H., Teng, Y. C., and Enns, J. (1980). *Ultrason. Symp. Proc.*, 652–657 | | X | | X | X | | | X |
| Royce, B. S. H., Teng, Y. C., and Ors, J. A. (1981). *Ultrason. Symp. Proc.*, 784–787 | | | | | | | | X |
| Royce, B. S. H., and Benziger, J. B. (1986). *IEEE Trans. UFFC* **33**: 561–572 | X | X | | | | | | |

*(Continued)*

| Citation | Carbons | Catalysts | Clays, Minerals | Food Products | Hydro-carbons, Fuels | Inorganics | Medical, Biological | Organics | Polymers | Textiles | Wood, Paper |
|---|---|---|---|---|---|---|---|---|---|---|---|
| Royce, B. S. H., and Alexander, J. (1988). In: *Photoacoustic and Photothermal Phenomena*, pp. 9–18. Springer, Berlin | | | X | | | | | | | | |
| Ryczkowski, J. (1994). *SPIE* **2089**: 182–183 | | X | | | | | | X | | | |
| Sadler, A. J., Horsch, J. G., Lawson, E. Q., Harmatz, D., Brandau, D. T., and Middaugh, C. R. (1984). *Anal. Biochem.* **138**: 44–51 | | | | | | | X | | | | |
| Saffa, A. M., and Michaelian, K. H. (1994). *SPIE* **2089**: 566–567 | | | X | | | | | | | | |
| Saffa, A. M., and Michaelian, K. H. (1994). *Appl. Spectrosc.* **48**: 871–874 | | | X | | | | | | | | |
| Salazar-Rojas, E. M., and Urban, M. W. (1987). *Preprints, ACS Div. Polym. Chem.* **28**: 1–2 | | | | | | | | | | | |
| Salazar-Rojas, E. M., and Urban, M. W. (1989). *Prog. Org. Coatings* **16**: 371–386 | | | | | | | | | X | | |
| Salnick, A. O., and Faubel, W. (1995). *Appl. Spectrosc.* **49**: 1516–1524 | | | X | | | | | | | | |

| | | | | | |
|---|---|---|---|---|---|
| Salnick, A., and Faubel, W. (1996). *Prog. Nat. Sci.* **6**(Suppl.): S14–S17 | X | | | | |
| Sarma, T. V. K., Sastry, C. V. R., and Santhamma, C. (1987). *Spectrochim. Acta A* **43**: 1059–1065 | | | | X | |
| Saucy, D. A., Simko, S. J., and Linton, R. W. (1985). *Anal. Chem.* **57**: 871–875 | | | | | X |
| Saucy, D. A., Cabaniss, G. E., and Linton, R. W. (1985). *Anal. Chem.* **57**: 876–879 | | | | X | |
| Schendzielorz, A., Hanh, B. D., Neubert, R. H. H., and Wartewig, S. (1999). *Pharm. Res.* **16**: 42–45 | | | X | | |
| Schüle, G., Schmitz, B., and Steiner, R. (1999). *AIP Conf. Proc.* **463**: 615–617 | | | X | | |
| Seaverson, L. M., McClelland, J. F., Burnet, G., Anderegg, J. W., and Iles, M. K. (1985). *Appl. Spectrosc.* **39**: 38–45 | X | X | | | |
| Seehra, M. S., and Pandurangi, R. (1989). *J. Phys.: Condens. Matter* **1**: 5301–5304 | X | | | | |
| Shoval, S., Yariv, S., Michaelian, K. H., Boudeuille, M., and Panczer, G. (1999). *Clay Miner.* **34**: 551–563 | X | | | | |

*(Continued)*

| Citation | Carbons | Catalysts | Clays, Minerals | Food Products | Hydro-carbons, Fuels | Inorganics | Medical, Biological | Organics | Polymers | Textiles | Wood, Paper |
|---|---|---|---|---|---|---|---|---|---|---|---|
| Shoval, S., Yariv, S., Michaelian, K. H., Lapides, I., Boudeuille, M., and Panczer, G. (1999). *J. Coll. Interf. Sci.* **212**: 523–529 | | | X | | | | | | | | |
| Shoval, S., Yariv, S., Michaelian, K. H., Boudeuille, M., and Panczer, G. (2001). *Clays Clay Miner.* **49**: 347–354 | | | X | | | | | | | | |
| Shoval, S., Yariv, S., Michaelian, K. H., Boudeuille, M., and Panczer, G. (2002). *Clays Clay Miner.* **50**: 56–62 | | | X | | | | | | | | |
| Shoval, S., Michaelian, K. H., Boudeuille, M., Panczer, G., Lapides, I., and Yariv, S. (2002). *J. Thermal Anal. Calorim.* **69**: 205–225 | | | X | | | | | | | | |
| Siew, D. C. W., Heilmann, C., Easteal, A. J., and Cooney, R. P. (1999). *J. Agric. Food Chem.* **47**: 3432–3440 | | | | X | | | | | | | |
| Small, R. D., and Ors, J. A. (1984). *Org. Coat. Appl. Polym. Sci. Proc.* **48**: 678–686 | | | | | | | | | X | | |
| Smith, M. J., and Palmer, R. A. (1987). *Appl. Spectrosc.* **41**: 1106–1113 | | | | | | | | X | | | |

| | | | | | | |
|---|---|---|---|---|---|---|
| Smith, M. J., Manning, C. J., Palmer, R. A., and Chao, J. L. (1988). *Appl. Spectrosc.* **42**: 546–555 | | | X | | | |
| Smith, M. J., and Palmer, R. A. (1988). In: *Photoacoustic and Photothermal Phenomena*, pp. 211–213. Springer, Berlin | | | | | | |
| Solomon, P. R., and Carangelo, R. M. (1982). *Fuel* **61**: 663–669 | | X | | | | |
| Sowa, M. G., and Mantsch, H. H. (1993). *SPIE* **2089**: 128–129 | | | | X | | |
| Sowa, M. G., and Mantsch, H. H. (1994). *Appl. Spectrosc.* **48**: 316–319 | | | | X | | |
| Sowa, M. G., Wang, J., Schultz, C. P., Ahmed, M. K., and Mantsch, H. H. (1995). *Vib. Spectrosc.* **10**: 49–56 | | | | X | | |
| Sowa, M. G., Fischer, D., Eysel, H. H., and Mantsch, H. H. (1996). *J. Mol. Struct.* **379**: 77–85 | | | | | X | |
| Spencer, N. D. (1986). *CHEMTEC* **16**: 378–384 | X | | | | | |
| St.-Germain, F. G. T., and Gray, D. G. (1987). *J. Wood Chem. Technol.* **7**: 33–50 | | | | | | X |
| Story, G. M., and Marcott, C. (1998). *AIP Conf. Proc.* **430**: 513–515 | | | | X | X | |

*(Continued)*

| Citation | Carbons | Catalysts | Clays, Minerals | Food Products | Hydro-carbons, Fuels | Inorganics | Medical, Biological | Organics | Polymers | Textiles | Wood, Paper |
|---|---|---|---|---|---|---|---|---|---|---|---|
| Story, G. M., Marcott, C., and Noda, I. (1994). *SPIE* **2089**: 242–243 | | | | | | | | | X | | |
| Stout, P. J., and Crocombe, R. A. (1994). *SPIE* **2089**: 300–301 | | | | | | | | | X | | |
| Szurkowski, J., and Wartewig, S. (1999). *Instrument. Sci. Technol.* **27**: 311–317 | | | | X | | | | | | | |
| Szurkowski, J., and Wartewig, S. (1999). *AIP Conf. Proc.* **463**: 618–620 | | | | X | | | | | | | |
| Szurkowski, J., Pawelska, I., Wartewig, S., and Pogorzelski, S. (2000). *Acta Phys. Polon. A* **97**: 1073–1082 | | | | X | | | | | | | |
| Tanaka, K., Gotoh, T., Yoshida, N., and Nonomura, S. (2002). *J. Appl. Phys.* **91**: 125–128 | | | | | | X | | | | | |
| Tanaka, S., and Teramae, N. (1984). *Preprints, ACS Div. Polym. Chem.* **25**: 190–191 | | | | | | | | | X | | |
| Teng, Y. C., and Royce, B. S. H. (1982). *Appl. Opt.* **21**: 77–80 | | | | | | | | | X | | |
| Teramae, N., and Tanaka, S. (1981). *Spectrosc. Lett.* **14**: 687–694 | | | | | | | | | | X | |

| Reference | | | |
|---|---|---|---|
| Teramae, N., Hiroguchi, M., and Tanaka, S. (1982). *Bull. Chem. Soc. Jpn.* **55**: 2097–2100 | | X | |
| Teramae, N., Yamamoto, T., Hiroguchi, M., Matsui, T., and Tanaka, S. (1982). *Chem. Lett.* 37–40 | | X | |
| Teramae, N., and Tanaka, S. (1984). *Bunseki Kagaku* **33**: E397–E400 | | X | |
| Teramae, N., and Tanaka, S. (1985). *Appl. Spectrosc.* **39**: 797–799 | | X | |
| Teramae, N., and Tanaka, S. (1985). *Anal. Chem.* **57**: 95–99 | | X | |
| Teramae, N., and Tanaka, S. (1987). In: *Fourier Transform Infrared Characterization of Polymers*, pp. 315–340. Plenum, New York | | X | |
| Teramae, N., and Tanaka, S. (1988). *Mikrochim. Acta* **2**: 159–162 | X | | |
| Thompson, M. M., and Palmer, R. A. (1988). *Appl. Spectrosc.* **42**: 945–951 | | | |
| Tiefenthaler, A. M., and Urban, M. W. (1988). *Polym. Mater. Sci. Eng.* **59**: 311–315 | | | X |
| Tiefenthaler, A. M., and Urban, M. W. (1989). *Composites* **20**: 145–150 | | X | |

*(Continued)*

| Citation | Carbons | Catalysts | Clays, Minerals | Food Products | Hydro-carbons, Fuels | Inorganics | Medical, Biological | Organics | Polymers | Textiles | Wood, Paper |
|---|---|---|---|---|---|---|---|---|---|---|---|
| Tsuge, A., Uwamino, Y., and Ishizuka, T. (1988). *Appl. Spectrosc.* **42**: 168–169 | | | X | | | X | | | | | |
| Urban, M. W., and Koenig, J. L. (1986). *Appl. Spectrosc.* **40**: 513–519 | | | X | | | | | X | | | |
| Urban, M. W., and Koenig, J. L. (1986). *Appl. Spectrosc.* **40**: 851–856 | | | X | | | | | X | | | |
| Urban, M. W., and Koenig, J. L. (1986). *Appl. Spectrosc.* **40**: 994–998 | | | | | | | | | X | | |
| Urban, M. W., Chatzi, E. G., Perry, B. C., and Koenig, J. L. (1986). *Appl. Spectrosc.* **40**: 1103–1107 | | | | | | | | | X | | |
| Urban, M. W. (1987). *J. Coatings Technol.* **59**: 29–34 | | | | | | | | | X | | |
| Urban, M. W., and Koenig, J. L. (1988). *Anal. Chem.* **60**: 2408–2412 | | | X | | | | | | | | |
| Urban, M. W. (1989). *Prog. Org. Coatings* **16**: 321–353 | | | | | | | | | | | |
| Urban, M. W., and Gaboury, S. R. (1989). *Macromolecules* **22**: 1486–1487 | | | | | | | | | X | | |
| Urban, M. W. (1997). *Prog. Org. Coatings* **32**: 215–229 | | | | | | | | | X | | |

| Reference | | | | | | | | | |
|---|---|---|---|---|---|---|---|---|---|
| Urban, M. W., Allison, C. L., Johnson, G. L., and Di Stefano, F. (1999). *Appl. Spectrosc.* **53**: 1520–1527 | | | | | | | | X | |
| van Dalen, G. (2000). *Appl. Spectrosc.* **54**: 1350–1356 | | | | | | X | | | X |
| Van Der Voort, P., Swerts, J., Vrancken, K. C., Vansant, E. F., Geladi, P., and Grobet, P. (1993). *J. Chem. Soc. Faraday Trans.* **89**: 63–68 | | | X | | | | | | |
| Vargas, H., and Miranda, L. C. M. (1988). *Phys. Rep.* **161**: 43–101 | | | | X | | | X | | |
| Varlashkin, P. G., Low, M. J. D., Parodi, G. A., and Morterra, C. (1986). *Appl. Spectrosc.* **40**: 636–641 | X | | X | | | X | | | |
| Varlashkin, P. G., and Low, M. J. D. (1986). *Appl. Spectrosc.* **40**: 1170–1176 | X | | | | X | X | | | |
| Vidrine, D. W. (1980). *Appl. Spectrosc.* **34**: 314–319 | | | | X | | X | | X | |
| Vidrine, D. W. (1981). *SPIE* **289**: 355–360 | | X | | | | | | X | |
| Vidrine, D. W. (1982). In: *Fourier Transform Infrared Spectroscopy* **3**: 125–148 | | | | | | | | | |
| Vidrine, D. W., and Lowry, S. R. (1983). *Adv. Chem. Ser. (Polym. Charact.)* **203**: 595–613 | | | | | | | | X | |

*(Continued)*

| Citation | Carbons | Catalysts | Clays, Minerals | Food Products | Hydro-carbons, Fuels | Inorganics | Medical, Biological | Organics | Polymers | Textiles | Wood, Paper |
|---|---|---|---|---|---|---|---|---|---|---|---|
| Vidrine, D. W. (1984). *Polym. Prepr.* **25**: 147–148 | | | | | | | | | X | | |
| Wahls, M. W. C., Toutenhoofd, J. P., Leyte-Zuiderweg, L. H., de Bleijser, J., and Leyte, J. C. (1997). *Appl. Spectrosc.* **51**: 552–557 | X | | | | | | | | X | | |
| Wahls, M. W. C., and Leyte, J. C. (1998). *J. Appl. Phys.* **83**: 504–509 | | | | | | | | | X | | |
| Wahls, M. W. C., Weisman, J. L., Jesse, W. J., and Leyte, J. C. (1998). *AIP Conf. Proc.* **430**: 392–394 | X | | | | | | | | X | | |
| Wahls, M. W. C., and Leyte, J. C. (1998). *Appl. Spectrosc.* **52**: 123–127 | | | | | | | | | X | | |
| Wahls, M. W. C., Kenttä, E., and Leyte, J. C. (2000). *Appl. Spectrosc.* **54**: 214–220 | | | | | | | | | | | X |
| Wahls, M. W. C., and Leyte, J. C. (2001). In: *Frontiers in Science and Technology*. Stefan University Press, La Jolla, CA | | | | | | | | | X | | X |
| Wang, H. P., Eyring, E. M., and Huai, H. (1991). *Appl. Spectrosc.* **45**: 883–885 | | X | | | | | | | | | |

| Reference | | | | |
|---|---|---|---|---|
| Wang, J., Ahmed, M. K., Sowa, M. G., and Mantsch, H. H. (1994). *SPIE* **2089**: 492–493 | | | X | |
| Wang, N., and Low, M. J. D. (1989). *DOE/PC/79920-10, DE90 006239*, 1–14 | X | | | |
| Wang, N., and Low, M. J. D. (1990). *Mater. Chem. Phys.* **26**: 117–130 | X | | | |
| Wang, N., and Low, M. J. D. (1990). *Mater. Chem. Phys.* **26**: 465–481 | X | | | |
| Wang, N., and Low, M. J. D. (1991). *Mater. Chem. Phys.* **27**: 359–374 | X | | | |
| Wentrup-Byrne, E., Rintoul, L., Smith, J. L., and Fredericks, P. M. (1995). *Appl. Spectrosc.* **49**: 1028–1036 | | | X | |
| Wentrup-Byrne, E., Rintoul, L., Gentner, J. M., Smith, J. L., and Fredericks, P. M. (1997). *Mikrochim. Acta* **14**(Suppl.): 615–616 | | | X | |
| Wetzel, D. L., and Carter, R. O. (1998). *AIP Conf. Proc.* **430**: 567–570 | | | | X |
| White, R. L. (1985). *Anal. Chem.* **57**: 1819–1822 | | X | | |
| Will, F. G., McDonald, R. S., Gleim, R. D., and Winkle, M. R. (1983). *J. Chem. Phys.* **78**: 5847–5852 | | | | X |
| Will, F. G. (1985). *J. Electrochem. Soc.* **132**: 518–519 | | | | X |

*(Continued)*

| Citation | Carbons | Catalysts | Clays, Minerals | Food Products | Hydro-carbons, Fuels | Inorganics | Medical, Biological | Organics | Polymers | Textiles | Wood, Paper |
|---|---|---|---|---|---|---|---|---|---|---|---|
| Woo, S. I., and Hill, C. G. (1985). *J. Mol. Catal.* **29**: 209–229 | | X | | | | | | | | | |
| Xu, Z. H., Butler, I. S., and St.-Germain, F. G. T. (1986). *Appl. Spectrosc.* **40**: 1004–1009 | | | | | | X | | | | | |
| Xu, Z., Butler, I. S., Wu, J., and Xu, G. (1988). *Mikrochim. Acta* **2**: 171–174 | | | | | | X | | | | | |
| Yamada, O., Yasuda, H., Soneda, Y., Kobayashi, M., Makino, M., and Kaiho, M. (1996). *Preprints, ACS Div. Fuel Chem.* **41**: 93–97 | | | | | X | | | | | | |
| Yang, C. Q., Ellis, T. J., Bresee, R. R., and Fateley, W. G. (1985). *Polym. Mater. Sci. Eng.* **53**: 169–175 | | | | | | | | | X | X | |
| Yang, C. Q., and Fateley, W. G. (1986). *J. Mol. Struct.* **141**: 279–281 | | | X | | | | | | | | |
| Yang, C. Q., and Fateley, W. G. (1986). *J. Mol. Struct.* **146**: 25–39 | | | X | | | | | | | | |
| Yang, C. Q., and Fateley, W. G. (1986). *Polym. Mater. Sci. Eng.* **54**: 404–410 | | | | | | | | | X | | |
| Yang, C. Q., and Bresee, R. R. (1987). *J. Coated Fabrics* **17**: 110–128 | | | | | | | | | | X | |

| Reference | | | | |
| --- | --- | --- | --- | --- |
| Yang, C. Q., Bresee, R. R., and Fateley, W. G. (1987). *Appl. Spectrosc.* **41**: 889–896 | | | X | X |
| Yang, C. Q., and Fateley, W. G. (1987). *Anal. Chim. Acta* **194**: 303–309 | | | X | X |
| Yang, C. Q., Bresee, R. R., and Fateley, W. G. (1987). *ACS Symp. Ser.* **340**: 214–232 | | | | X |
| Yang, C. Q., Perenich, T. A., and Fateley, W. G. (1989). *Textile Res. J.* **59**: 562–568 | | | | X |
| Yang, C. Q., Bresee, R. R., and Fateley, W. G. (1990). *Appl. Spectrosc.* **44**: 1035–1039 | | | X | |
| Yang, C. Q. (1991). *Appl. Spectrosc.* **45**: 102–108 | | | X | X |
| Yang, C. Q., and Freeman, J. M. (1991). *Appl. Spectrosc.* **45**: 1695–1698 | | | | X |
| Yang, C. Q., and Simms, J. R. (1993). *Carbon* **31**: 451–459 | X | | | |
| Yang, C. Q., and Simms, J. R. (1995). *Fuel* **74**: 543–548 | X | | | |
| Yang, H., and Irudayaraj, J. (2000). *J. Am. Oil Chem. Soc.* **77**: 291–295 | | X | | |
| Yang, H., and Irudayaraj, J. (2000). *Trans. Am. Soc. Agric. Eng.* **43**: 953–961 | | X | | |
| Yang, H., Irudayaraj, J., and Sakhamuri, S. (2001). *Appl. Spectrosc.* **55**: 571–583 | | X | | |

*(Continued)*

| Citation | Carbons | Catalysts | Clays, Minerals | Food Products | Hydro-carbons, Fuels | Inorganics | Medical, Biological | Organics | Polymers | Textiles | Wood, Paper |
|---|---|---|---|---|---|---|---|---|---|---|---|
| Yang, H., and Irudayaraj, J. (2001). *J. Am. Oil Chem. Soc.* **78**: 889–895 | | | | X | | | | | | | |
| Yang, H., and Irudayaraj, J. (2001). *Lebensm.-Wiss. u.-Technol.* **34**: 402–409 | | | | X | | | | | | | |
| Yaniger, S. I., Riseman, S. M., Frigo, T., and Eyring, E. M. (1982). *J. Chem. Phys.* **76**: 4298–4299 | | | | | | | | | X | | |
| Yaniger, S. I., Rose, D. J., McKenna, W. P., and Eyring, E. M. (1984). *Macromolecules* **17**: 2579–2583 | | | | | | | | | X | | |
| Yaniger, S. I., Rose, D. J., McKenna, W. P., and Eyring, E. M. (1984). *Appl. Spectrosc.* **38**: 7–11 | | | | | | | | | X | | |
| Yariv, S., Nasser, A., Michaelian, K. H., Lapides, I., Deutsch, Y., and Lahav, N. (1994). *Thermochim. Acta* **234**: 275–285 | | | X | | | | | | | | |
| Yeboah, S. A., Griffiths, P. R., Krishnan, K., and Kuehl, D. (1981). *SPIE* **289**: 105–107 | | | | | | | | X | X | | |
| Ying, J. Y., Benziger, J. B., and Gleiter, H. (1993). *Phys. Rev. B* **48**: 1830–1836 | | | | | | X | | | | | |

| | | | | | | | |
|---|---|---|---|---|---|---|---|
| Yokoyama, Y., Kosugi, M., Kanda, H., Ozasa, M., and Hodouchi, K. (1984). *Bunseki Kagaku* **33**: E1–E7 | | | | | X | | |
| Yoshino, K., Fukuyama, A., Yokoyama, H., Meada, K., Fons, P. J., Yamada, A., Niki, S., and Ikari, T. (1999). *Thin Solid Films* **343–344**: 591–593 | | | | | X | | |
| Zachmann, G. (1984). *J. Mol. Struct.* **115**: 465–468 | | | | | | | X |
| Zerlia, T. (1985). *Fuel* **64**: 1310–1312 | | | | X | | | |
| Zerlia, T. (1986). *Appl. Spectrosc.* **40**: 214–217 | | | | X | | | |
| Zerlia, T., and Girelli, A. (1988). *Mikrochim. Acta* **2**: 175–178 | | X | | X | | | |
| Zerlia, T., Carimati, A., Marengo, S., Martinengo, S., and Zanderighi, L. (1988). *Struct. React. Surf.* **48**: 943–953 | | X | | | | | |
| Zhang, G., Wang, Q., Yu, X., Su, D., Li, Z., and Zhang, G. (1991). *Spectrochim. Acta A* **47**: 737–741 | | | | | | X | |
| Zhang, S. L., Michaelian, K. H., and Burt, J. A. (1997). *Opt. Eng.* **36**: 321–325 | | | X | | | | |
| Zhou, W., Xie, S., Qian, S., Wang, G., Qian, L., Sun, L., Tang, D., and Liu, Z. (2000). *J. Phys. Chem. Solids* **61**: 1165–1169 | X | | | | | | |

# AUTHOR INDEX

(Page numbers in bold contain references)

# SUBJECT INDEX